SELECTED PAPERS ON EPISTEMOLOGY AND PHYSICS

VIENNA CIRCLE COLLECTION

VOLUME 7

VOLUME-EDITOR: HENK L. MULDER

BÉLA VON JUHOS

BÉLA *von* JUHOS

SELECTED PAPERS
ON EPISTEMOLOGY AND
PHYSICS

Edited and with an Introductory Essay by

GERHARD FREY

D. REIDEL PUBLISHING COMPANY

DORDRECHT-HOLLAND/BOSTON-U.S.A.

Library of Congress Cataloging in Publication Data

Juhos, Béla von, 1901–1971.
 Selected papers.

  (Vienna circle collection ; v. 7)
  Includes bibliographical references and indexes.
    1. Physics—Philosophy—Addresses, essays,
lectures.   2.   Science—Philosophy—Addresses, essays,
lectures.   3.   Science—Methodology—Addresses,
essays, lectures.   4.   Knowledge, Theory of—Addresses,
essays, lectures.   5.   Probabilities—Addresses, essays, lectures.
I.   Series.
QC6.2.J84   1976   501   76–17019
ISBN 90–277–0686–7
ISBN 90–277–0687–5 pbk.

Published by D. Reidel Publishing Company,
P.O. Box 17, Dordrecht, Holland

Sold and distributed in the U.S.A., Canada, and Mexico
by D. Reidel Publishing Company, Inc.
Lincoln Building, 160 Old Derby Street, Hingham,
Mass. 02043, U.S.A.

Printed in the Netherlands

TABLE OF CONTENTS

PREFACE

It was as a result of having known Juhos personally over many years that I became familiar with his thought. I met him and Viktor Kraft in Vienna soon after the War and through their acquaintance I first came into contact with the tradition of the Vienna Circle. To their conversation too I owe much as regards the clarification of my own views, even if in the end these took quite a different turn in many essentials.

At this point my gratitude goes first of all to Mrs. Lia Juhos for the generous help she has given me and the editors of the Vienna Circle collection in selecting the contents of this volume. Next, we owe a special debt to Dr. Paul Foulkes for his splendid translation of the text. Finally, I wish to thank Dr. Veit Pittioni for his constant assistance. As Juhos' last student, he was thoroughly familiar with his supervisor's mode of thought and has significantly furthered the assembly and execution of this book.

Innsbruck, May 1976   G. FREY

TRANSLATOR'S NOTE

Beyond the normal problems, the original text presents one terminological difficulty that deserves special mention. Juhos' central theme in the theory of knowledge is logical criticism. The term he uses to describe his method might be rendered as 'cognito-logical' or 'epistemo-logical', both of them unusual. The second without hyphen is our ordinary adjective for 'pertaining to theory of knowledge'. We here draw attention to the author's critical aims, so that in translation we may refrain from burdening the text with a constantly repeated neologism. We therefore use the ordinary term 'epistemological'. With the above reminder there will be no danger of misconstruing the position.

London, June 1976   PAUL FOULKES

INTRODUCTION

Béla von Juhos (22 November 1901 – 27 May 1971) belonged to a Hungarian family but was born in Vienna. He passed his childhood in Budapest and in 1909 returned to Vienna, where he spent the rest of his life.

At the University of Vienna he studied mathematics, physics and philosophy, which soon brought him into contact with Moritz Schlick and gave him access to the Vienna Circle forming round that philosopher. When Schlick was assassinated in 1936, the Circle began to dissolve, and almost all its members emigrated. The German occupation of 1938, with the attendant absorption of Austria into the German Empire, made it in any case impossible for the Circle to continue in Vienna, because its critical and rational philosophy was disliked by National Socialism, as indeed by all other ideologies. Of the original members, only Viktor Kraft and Juhos remained behind during and after the war years. This, in part, is probably why Juhos became a lecturer only in 1948 and then taught in Vienna until the end. Of independent means all his life, he was able to do without official academic appointment, so that he had the time to accomplish his extensive scientific oeuvre. The list of his publications comprises eight books and some 110 articles and dissertations.

Juhos' central interest lay in the logical analysis of knowledge, with which all his writings are concerned and which he applied particularly to physics. Most of the papers collected in the present volume deal with general epistemology, save the long dissertation on 'Fictitious Predicates', which also includes some detailed discussion of physical problems. The choice reflects our view that his main significance lies in the field of general epistemology. Even some of his epistemological work in physics remains undoubtedly controversial.

In line with his central interest, Juhos called his own philosophical method 'epistemo-logical' or 'epistemo-analytic'. He distinguishes formal from material analysis. The former concerns the question 'What linguistic logical stipulations characterize the expressions (propositions, concepts) of a science, and to what language system do they belong?', which he sees as closely related to Carnap's attempt at a 'logical syntax of language'. As to the analysis of content, he views it semantically: 'What are the objects denoted by expressions, in particular by the propositions of a science?'. This analytic method he brings to bear especially on the natural sciences. His analysis of the scientific concept of law leads him to the distinction between first and second-order laws. An empirical hypothetical statement 'all instances (events) of the kind A are always followed, if the same circumstances recur, by instances (events) of the kind B' he calls a first-order law, from which we can only ever derive predictions that sequences of events stated in the law will be repeated under the same or similar circumstances. Accordingly, these are called first-order predictions. As an example of such a proposition having the form of a first-order law we may cite the expansion of nitrogen by one part in 273 of its volume for a $1\,^{\circ}C$ rise in its temperature.

Empirical hypothetical propositions that record the comparison of events, usually as a functional connection, Juhos calls second-order laws. A classical example is Newton's law of gravitation, which asserts the continuous change of gravitational attraction with given changes of the masses and their mutual distance. From second-order laws we can derive first-order laws, but not in general the reverse.

From second-order laws we can make second-order predictions, relating to events that are not explicitly mentioned in the laws, nor were used for establishing them. If a domain of events can be described only by first-order laws, then Juhos says that it is subject to a first-order determinism. If the account requires second-order laws as well, then we have a second-order determinism. His analysis of physics leads him to the observation that classical physics tries to describe all connections between phenomena by means of second-order laws, which turn out to be laws of proximate.action. Whether Juhos' identification of second-order laws with proximate action stands up to close examination is, however, doubtful, even if we, as he does, include the concept of continuity in that of second-order law. We should have to introduce a wider concept of law that did not use continuous functional connections. Such non-continuous second-order laws are indeed used in quantum mechanics.

Starting from classical physics, Juhos conducted detailed enquiries on probability statements and probable inference, trying to reduce probability to classical two-valued logic, which he always regarded as alone valid. In other words, in all his epistemological enquiries he relies on a logic roughly corresponding to that of Russell's *Principia Mathematica*. One condition for laws of proximate action is that all physical states can be sufficiently marked by unambiguous and exact values of measurement, a characterization expressed by conjunctive connection of individual statements. All inference based on proximate action (second-order laws) is then to be regarded as drawn from conjunctive classes of propositions. In contrast, probabilities, such as average values in thermo-dynamics, are marked by disjunctive classes of propositions, that is disjunctive connections of value assignments. Given the characterizing descriptions E_1, E_2,... E_n, a 'probable inference', according to Juhos, has the form $E_1 \vee E_2 \vee ... \vee E_n \rightarrow E_1$. This is not meant as a way of establishing the probability calculus, for he holds that a probability metric must in any case be presupposed, either of the *a priori* or of the empirical-statistical kind. Juhos gave particular attention to relativity theory, analysing the concepts 'rest', 'motion' and 'space'. Since in the theory these concepts have only relative meaning, if follows that there can be no 'absolute space' or 'time' of the kind still presupposed by Newton. Henry More and Newton had theological reasons for assuming space as absolute. Accordingly Juhos sees the transition from Newtonian to Einsteinian mechanics as a step forward because it dissolved absolute concepts which he regarded as metaphysical. In classical physics all such absolute concepts are based on the notion of actions at a distance spreading at infinite speed, as Newton assumed. Proximate action, as stated above, is based on the assumption of continuity. Juhos observes that the general theory of relativity presupposes 'absolute physical continuity', which he recognizes as yet another metaphysical presupposition that must be relativized in its turn. Against his view that by thus relativizing absolute notions such metaphysical concepts can be banished, it may be said that in our empirical theories there always seem to arise new metaphysical concepts and propositions. We must seriously envisage the possibility that metaphysical presuppositions can never be completely excluded from empirical theories and that perhaps exclusion of some of these (absolute concepts) necessarily brings in others of the same kind – a surmise which of course cannot be discussed in detail here.

As criterion for distinguishing scientific from metaphysical propositions Juhos adopted testability. In his earlier phase he largely shared the well-known position of Schlick and the Vienna Circle. In later years he moved somewhat closer to K.R. Popper's views on this matter. At the same time, he offered telling criticism of Popper's claim that verification and falsification were mutually asymmetrical, showing that this was not applicable to scientific theories (cf. p. 134). He is in any case not satisfied by such merely formal examinations of scientific and metaphysical propositions, even if he allowed that in principle they were correct. Metaphysics doubtless has a certain function in human life (Carnap, too, had maintained this view). We need an epistemological analysis of that function to show what is its non-formal character or content. An analysis of the content of the language in each case leads Juhos to distinguish between a 'logical function' of language in science and a 'metaphysical function' in metaphysics.

The 'logical function' of language relates to natural events. It concerns empirical-hypothetical propositions that may be true or false and must be testable: insofar as they are general statements, the most important of these propositions are second-order laws. Metaphysical propositions, as mentioned above, cannot be tested or checked and therefore, scientifically speaking, are neither true nor false. Like the propositions of the humanities, they relate to valuative events. "Every metaphysic seeks the meaning it attaches to the essence of being, and similar expressions, in a special experience. In metaphysics, this 'ultimate' experience is moreover usually to count as the 'all-decisive' highest value and thus to serve as a yardstick for valuing all individual being" (1956, *Das Wertgeschehen und seine Erfassung*, p. 25). The metaphysical function of language therefore exhibits two characteristics, "first, the non-logical connection of words for the purpose of triggering off certain value experiences with which we are to become acquainted, and second, the claim that the verbal connections thus established are to be regarded as absolutely true knowledge, in order thereby to achieve a reinforced mental inclination towards the valuation that the metaphysician seeks to induce" (ibid., p. 30). Metaphysical propositions are marked in any case by the fact that they receive a double meaning: alongside the ordinary meaning based on the normal linguistic rules, they acquire a metaphysical meaning precisely by breaking those rules. The metaphysician seeks to exert an influence on mind and spirit, "to induce a certain active attitude towards and conduct within the world" (ibid., p. 41). The sham logic called 'dialectic', which aims at persuading people to a certain form of conduct, thus arises from rhetoric. This same sham logic with the same persuasive function is also evident in metaphysical language as represented and sketched by Juhos. Comparison with poetry, he asserts, shows that only the metaphysical-dialectic method provides the adequate medium for exerting influence on mind and spirit, a 'metaphysical art of persuasion'. "The method of using sham propositions to trigger off certain experiences, produce certain mental attitudes (valuations) and arouse in people thus influenced the belief that their conduct is now the result of true knowledge – this is the only way to achieve an influence on feeling and will, as well as on intellect (and therefore a lasting influence)" (ibid, p. 68). We may question whether Juhos is right in saying that only sham propositions lend themselves to this purpose, since most metaphysical systems are put together from empirical propositions, rational logical arguments and sham propositions in his sense. On some important points, his conception of metaphysics will thus need revision.

The range of what is expressed in metaphysics is to include valuations. By contrast, empirical enquiry into what happens in the world is the task of the so-called humanities. Juhos evidently did not succeed in his analysis of this method, for it fails to do justice to the true concerns of the humanities, or so it seems to me. He regards human knowledge as a kind of link between natural scientific knowledge and the 'empathic-psychological method'. Now it is easily shown that the psychological processes that have been called 'empathy' are subjective and therefore cannot serve as basis for a science to be conveyed between people. Against Juhos' view, interpretations turn out to be testable hypotheses, at least in a limited number of cases. However, what seems of some interest, and certainly worth checking more closely, is his view that natural science uses both first- and second-order laws, while the humanities use only the former. For example, he would regard history as involving only first-order predictions and determinism. Likewise, he is surely wrong in equating the difference between natural science and the humanities with that between nomothetic and idiographic science, as has already been pointed out by V. Kraft (obituary notice, p. 173; in *Zeitschr. f. Allg. Wissenschafts-Theorie* 2 (1971), 163–173).

This brief note is merely to sketch some basic features of Juhos' thought, revealing the main tendencies exhibited throughout his work. The selection here assembled provides a survey of his analysis of knowledge, although largely excluding his enquiries on ethics.

G. FREY

CHAPTER I

ORDERS OF CAUSALITY*

What separates recent from classical physics arose from the transformation of a series of basic physical concepts. Above all, the concepts of space and time on one hand and causality on the other were subjected to a critique that led to their being redefined. While, however, the reconstruction of spatio-temporal concepts, following the preparatory work in mathematics and in the philosophy of Mach and Poincaré, succeeded in physics at a stroke as it were, the case there stands otherwise as regards the concept of causality. Physics did indeed at length succeed in indicating very precisely what are the limits beyond which we can no longer speak of causality in a non-tautological sense, but in spite of this very important result for physics, people continue to differ as to what is to be understood by causality and the causal principle.

Let us note in advance that there is a conception of causality that has been highly popular at all times, because it is unconditionally correct and that fact is easily seen, for this view is tautological. The best known example of a tautological view of causality runs as follows: everything happens according to strict laws and even where we cannot ascertain any regularities, laws of nature still hold, only our limited understanding is unable to recognize them. In this form the principle of causality holds under any circumstances, whether events run this way or that, but just because of this it says nothing about reality. The tautological formulation therefore need interest us no longer.

Of the other causal conceptions those of Kant and Bacon are the ones that have been most severely dealt with by recent enquiry. Two questions were mainly discussed: first, in what sense the principle of causality holds, whether we should regard it as an empirical probable proposition or as an *a priori* certain one; secondly, of special interest to us here, what are we to understand by causality, what is the meaning of the concept of causality.

As is well known, Kant answered the first question by saying that the principle of causality is *a priori* synthetic. 'Synthetic' is synonymous with 'non-tautological'. As being *a priori,* the principle of causality is absolutely

---

* Translated from the German original 'Stufen der Kausalität', first published in *Jahresber. der philosoph. Ges. zu Wien* (1931/1932) 1–19.

certain. On this point recent enquiry has shown Kant wrong. Today there remains no compelling reason for cleaving to the unshakeable certainty of the principle of causality; indeed, recent physics has found it expedient to deny any form of causality for certain domains. What epistemological status causality might retain under these circumstances is something about which people do not agree either. In any case, however, it is now plain that the validity of the causal principle cannot be decided independently of experience. Some regard the law of causality as an empirical judgement about whose truth or falsity experience decides, while others think it is not a statement at all but an instruction for forming laws of nature of a certain kind, experience again deciding whether the instruction is expedient.

Let us now turn to the second question: what are we to understand by causality? According to Kant the function of causality consists in determining the objective temporal relations of events; what we immediately experience, our sensations and sense data also have a temporal order, although a subjective one varying from person to person. The function of causality then consists in re-ordering our experiences in such a way that their order is the same for all people, namely objective. Kant's example is well known: if we look at a house in the direction from top to bottom, then its parts are given to us in temporal succession, yet we nevertheless do not suppose that these parts exist in succession, but ascribe simultaneous existence to them. For Kant it is through applying the category of causality that we reach this last, objective order of experiences, by looking for the causal connections of experiences.

It is important that Kant assumes two different orders for experiences, the subjective and the objective. Of course he thinks that both orders are in the same time. That we might here have two different kinds of time is something he did not consider. Just as he knows only of one space, namely the Euclidean type, he takes it for granted that there is only one universal time.

Meanwhile, recent philosophy has examined the role of time in physics. It was found, by Poincaré especially, that in physics time means something quite other than the intuitive time experience that we refer to when we speak of past, present and future. Here we merely remark that physical time is a conceptual order that has nothing to do with the intuitive course of time. Today one therefore distinguishes the abstract conceptual order of physical time from the intuitive course of phenomenal time.

Kant's view of causality might now be formulated thus: the function of causality consists in ranging experiences given in the course of phenomenal time into an ordering schema that we call physical time. Here the question immediately arises, what specific properties an ordering schema must have in order that we may call it an 'order of physical time'. For if any arbitrary order could take on the role of physical time, then in whatever way events actually run, we could always construct an ordering schema into which the series of events do fit. Kant's definition would thus become tautological and useless. The theoretical or formal conception of causality discussed therefore must answer the question what property the temporal (more precisely: the spatio-temporal) order of events must have for us to be able to speak of causality. More simply: what form of laws of nature by which we describe events is to be regarded as the causal criterion?

Before answering this question, we must discuss a second conception of causality that is older and simpler than the theoretical one and was expressed already by the philosophers and enquirers into nature of the renaissance, but expecially by the English empiricists Bacon and Hume, in a form that can be taken as quite up to date even today. The common aim of these thinkers was to find a scientific method that was to enable them to achieve quite specific practical purposes. This aim was the control of nature. Natural science by its laws is to let us see beforehand, what will happen under determined circumstances. A law of nature is to allow us to predict which event B will follow an event A; or, as the phrase goes, what effect will follow a given cause. Here Bacon held the most radical view: if a law of nature allows us to indicate what event will follow a given other event, then, however fantastic or improbable the law concerned may sound, we are entitled to assert the relation of cause and effect between the events. This is a thoroughly practical view, on which the sole function of causality consists in enabling us to predict, and the sole causal criterion is that what has been predicted should come true, that laws of nature should prove themselves in practice.

These two conceptions of causality, the theoretical and the practical, stand in a certain contrast to each other, as is readily seen. The theoretical view seeks a criterion in a definite formal property of the laws of nature, while the practical view sees the causal conditions in the fact that the laws of nature prove themselves in practice whatever their form. Because of this contrast we must be careful. We had better regard both of these causal conceptions provisionally as merely pointing to two different functions of causality. A final definition may perhaps have to take both functions into account, which is further suggested by the fact that if either conception alone is regarded as the only valid definition, the attendant difficulties suggest that these one-sided definitions do not capture what physics understands by causality.

The lie of the land becomes clearer if we briefly look at the difference between phenomenal and physical time. By phenomenal time we are to understand the intuitive experience of time, the flow of time, an experience marked by a privileged point, namely the 'immediate now' or present, that divides the course of time into two qualitatively absolutely different parts of past and future. It is important that we are not free to choose an arbitrary moment as the present, but that the 'now' is imposed and forced on us in experience. If this were not so, we could arbitrarily choose a point in time as 'present' and it would be merely a matter of different ways of speaking whether we call an event 'future' or 'past', and there would be no absolute qualitative difference between past and future.

Here the difference between phenomenal and physical time stands out clearly: for the latter, which is an abstract ordering schema, lacks a privileged point. Just as analytic geometry describes spatial relations of geometrical structures by means of co-ordinates, so time for physics is no more than a co-ordinate axis added to the spatial ones. We can see at once that the points of a co-ordinate axis are all equivalent and none of them can be privileged. If in physics we wish to speak of a present, then we are free to choose an arbitrary point on the time-axis. It follows that in physics there is no absolute difference between past and future, and if we denote an event as past on one occasion and as future on another, these are merely different ways of speaking. It is therefore correct to say that physics knows neither past nor future.

We return to our main problem: the theoretical view wishes to understand by causality a special spatio-temporal order of events. By temporal order we here certainly mean the order of physical time. Suppose now we knew what special property P must belong to an order of events in order that we may speak of causality. What, however, is then the position in the case where a law of nature lacks P but still allows us to predict what event B will follow as effect upon the cause A? Such a case is certainly conceivable. On the theoretical view we should not here be entitled to speak of causality, merely because our law of nature does not satisfy the formal criterion P. Nevertheless on the basis of such a law we could, in ordinary terms, indicate the appropriate effect for any cause. The theoretical view of causality must at all events account for this difficulty.

The practical view of causality becomes enmeshed in a different kind of difficulty: this view wants to speak of causality wherever our predictions come true, the form of the law from which we predict being irrelevant. However, prediction always touches the future, but physics, as we saw, knows no future. If the coming true of predictions is the only causal criterion, the concept of causality seems to have no place in physics. What contradicts this is that by causality we do understand a physical regularity of some kind and that the problem of causality is at the focus of current physical interest.

We find these difficulties discussed in a recent paper of M. Schlick: 'Die Kausalität in der gegenwärtigen Physik' (*Die Naturwissenschaften,* 1931). Schlick first considers the theoretical view and specially criticizes a definition that is particularly important in it, namely Maxwell's definition of causality. According to Maxwell, in physics we speak of causality whereever the functions representing laws of nature do not explicitly contain spatio-temporal co-ordinates. Schlick thinks this definition is defective since one can conceive of cases where the course of events can be described only by functions with explicit spatio-temporal co-ordinates, while such a law of nature still enables us to determine the effect to every cause. If for example (not Schlick's) the laws of free fall wich hold everywhere else happened not to hold in Vienna alone, so that there bodies fell with an acceleration deviating from that obtaining elsewhere and no other explanation could be found for this fact than the spatial position of Vienna, then its spatial co-ordinates would have to figure explicitly in a general law of free fall. Even using everyday language, Vienna would have to be explicitly mentioned in this law; for example: bodies fall according to law A except in Vienna, where they fall according to law B. According to Maxwell we must not here speak of causality, and yet on the basis of this law we can predict how bodies will fall everywhere, Vienna included. As soon as we can predict what effects are to be expected we are according to Schlick entitled to assume that the events are causally conditioned. That is why he regards all attempts at definition of the causal criterion in terms of formal properties of laws of nature as vain: we can always think of cases where a law of nature lacks the property concerned and yet allows us to determine (predict) the effects that will follow any causes. That these predictions come true, that the laws of nature prove themselves, is for him the only causal criterion. He thus holds the practical conception of causality but while with Bacon the relation of the practical concept of cause to physics remains unclarified, Schlick is precisely intent on clarifying the role of causality in physics, and this leads him beyond Bacon's definition in an important respect.

In daily life we understand by prediction always something touching the future. In physics, Schlick thinks, it is otherwise, there we often speak of prediction and causality where future (or past) events are not involved at all. If for example on the basis of three events a, b, c, we have set up a law of nature and from it we can derive an event d not used in the setting up, then physics says that d is predictable from this law whether the derived event d has as yet been observed or not, whether it is past or future.

Laws of nature thus may allow two kinds of predictions to be made; the first predict only such events as have been used for the setting up of the laws (predictions of the first kind), while the second are derivations of events not used in obtaining the law (predictions of the second kind). Predictions of the first kind are quite trivial cases that we meet all the time in daily life. They are also called inductive inferences and can always be stated in the form: if something has always happened in such and such ways in the past, then under similar conditions it will recur likewise in the future. Here we therefore predict only the repetition of frequently observed and noted regularities. Now there are laws that allow only such predictions of the first kind, namely laws that merely note the observed sequence of events. On the basis of such laws we can only ever predict those events already noted; others, not involved in the setting up of the laws, cannot be derived from them either. Predictions of the second kind are therefore impossible from such laws.

If I have rightly understood him, Schlick thinks that although there is a certain difference between the two kinds of prediction, this is not a question of principle. What matters is merely that a law of nature should enable us to make predictions. If these come true, whatever their kind, then we are always entitled to speak of causality: the only causal criterion is that the laws of nature should prove themselves. Still, he adds that what is meant is not that this should happen in future but absolutely, and that occurs when we can derive from the law of nature such events as were not used in setting it up. Yet he thinks that the possibility of such predictions of the second kind is independent of the formal properties of the laws of nature.

Now I think that this pushing aside of all formal properites of laws of nature is incompatible with the causal conception of physics.

In daily life we doubtless speak of causality wherever our predictions come true, and our predictions in practical life are without exception inductive inferences. They only ever predict the repetition of frequently repeated and well-tried regularities. Our practical rules for everyday life and instructions for using instruments, apparatus, medicaments and so on, are laws that note often observed series of events. From these protocols we then gather what will happen under determined circumstances. Indeed not only we, in daily life: there is also a group of experts who work only with this causal criterion. They are, however, not physicists but technologists. The engineer asks only: what will happen if I do such and such? The special form of functional relation between his actions and the consequent effects does not interest him. He too merely registers observed regularities and from these tables (protocols) he reads off his predictions. To the physicist, on the other hand, these mere protocols of observed regularities are not enough; on the contrary, and here he differs in the first instance from the technologist, the physicist examines the relations between the noted regularities and only when these relations have quite determined formal properties does he speak of causality. At all events we can now give a first correct partial answer to the question what is to be understood by causality: in daily life and in technology we understand by it the practical causality of Bacon in the strictest sense.

As a last step, let us add to this practical definition of causality the features that are needed to obtain a causal concept serviceable in physics.

For this we must first take a closer look at the form of laws of nature in physics. Having observed that an event a_1 is always followed by another b_1, we can record this regularity in the form $a_1 \rightarrow b_1$ (from a_1 follows b_1). For example, let $a_1 =$ cooling of an iron rod from $10°C$ to $9°C$, and $b_1 =$ contraction of its length by a certain amount. Continuing systematically with $a_2 =$ cooling of the rod from $9°C$ to $8°C$ and $b_2 =$ corresponding contraction and so on. Writing down the regularities thus obtained: $a_1 \rightarrow b_1$, $a_2 \rightarrow b_2$, ... $a_n \rightarrow b_n$, ..., we see that to the order of causes a_1, a_2, ... a_n that we have chosen and set up ('cooling of the iron rod by $1°$ at a time') there corresponds, let us assume, a very nice and simple order of effects b_1, b_2, ... b_n (for example 'contraction of the rod by $1/n$ of its length each time'). The fact that to the order of causes there corresponds a certain order of effects can in its turn be stated as a law, perhaps thus: on cooling by one degree an iron rod contracts in such and such a way. This law, however, differs in form from the $a_n \rightarrow b_n$, which we shall call first-order laws: they merely record that according to observation an event is regularly followed by another determined event, while the law that assigns to the order of causes (a_1, a_2... a_n) a certain order of effects (b_1, b_2 ... b_n) states a relation between first-order laws. It is thus of higher order than the $a_n \rightarrow b_n$ and we shall call it a second-order law. Thus second-order laws presuppose first-order laws. Obviously, whenever we have first-order laws there holds between them some relation that can be stated by a second-order law. However, not every such second-order law can be used in physics, but only one that satisfies certain quite definite formal criteria. To clarify the difference between first- and second-order laws further, consider cases where such laws do not hold.

The case where first-order laws do not exist arises where we cannot find out by observation that a given cause a_1 is followed by a determined effect b_1, so that on repeated observation the event a_1 is followed randomly by quite different events. If, for example, we wish to ascertain the precision of a gun and the scatter was so great that for some shells the place of impact could not be ascertained, then there would be no way of predicting where the shells will fall for given bearing and cartridge strength. Where no first order regularities can be ascertained, no predictions of any kind are possible, so that not even the criteria of practical causality are fulfilled and we can in no sense speak of causality. (*A fortiori,* second-order laws are impossible, since they presuppose first-order laws.) That there really are such conditions in certain fields is a finding of quantum theory, whose great extra-scientific effect ultimately rests not least on its having made us aware that our everyday practical concept of causality is not applicable in some real cases.

Consider now the interesting case where first-order regularities have indeed been ascertained but second-order laws cannot be set up in a form serviceable for physics. Instead of an iron rod, let us cool a volume of water from $10\,^{\circ}$C to $9\,^{\circ}$C and record the attendant reduction in volume, and next likewise from $9\,^{\circ}$C to $8\,^{\circ}$C and then to $7\,^{\circ}$C. These observations are enough and we see again that to the order of causes there corresponds a simple order of effects and we do not hesitate to state this as a second-order law: On cooling by one degree, water contracts in such and such measure. Now take a further important step and ask what is the size of the volume of water at temperatures not used in the setting up of our second-order law, for example at $0°C$. From our second-order law we can easily calculate this result and since we know that with diminishing volume the specific gravity of a body rises and water at $0°C$ turns into the solid state, we infer that ice must sink in water. We test this by throwing a chunk of ice into water and find that ice floats. Thus our second-order law does not hold. By experiment we now ascertain step by step that, on cooling, water contracts until $+4°C$, but expands again below that, and let us assume that this change in behaviour of the water occurs suddenly or discontinuously. Thus we have found the point at which our second-order law fails. However, we can by no means assert that there are not first-order regularities. If we observe and record the regular behaviour of water on cooling to or below $4°C$, we can predict exactly how big the volume of water will be at any of these temperatures. The criterion for practical causality is thus satisfied, the change of behaviour of water at $+4°C$ gives the engineer no occasion to ask for a cause. What then does the physicist do? He will at all events seek to explain the discontinuous change in the order of events. Suppose he fails, then if he wishes to describe the behaviour of water by a second-order law he can only do it by explicitly mentioning in it a quantity that specially marks the state of water at $4°C$; that is, in this case a first-order law is explicitly mentioned within a second-order law, which latter now simply records the observed behaviour of water at $4°C$. However, we saw that recording is the task of first-order laws, so that there is no difference of sense between these and second-order laws explicitly containing first-order ones: it is pointless to set up such second-order laws, for they achieve no more than first-order ones, namely the prediction of only those events that have been recorded. The physicist, however, demands more than this from his laws, namely that we should be able to derive from them events not used in setting them up; which is just what second-order laws lose when first-order laws, or quantities marking states, explicitly occur in them.

The quantities that physicists assign to states are spatio-temporal. Now we can grasp why Maxwell and other physicists absolutely reject laws of nature in which spatio-temporal co-ordinates figure explicitly: second-order laws thereby lose the property that is for physicists the most valuable, namely that of allowing derivations (predictions) of events not used in their setting up. If a second-order law fails for a particular event that can be described only by having the quantities (spatio-temporal co-ordinates) assigned to it built explicitly into the law, physics calls such an event an anomaly. If a physicist encounters an anomaly he frames a question that the engineer can no longer ask, namely as to the cause of the anomaly. We saw that even in the case of an anomaly, there are first-order laws, and although it is only from a protocol that the engineer knows how events succeed each other, he can read off his predictions, his causal criteria are fulfilled and he can simply not ask for further causes. In contrast, the physicist works with laws of higher order and if at some point he notices a discontinuous change of such a regularity (an anomaly), then he asks for the cause of this change. If he can find none, and cannot make the anomaly vanish, then he calls this discontinuous deviation from a second-order law uncaused or fortuitous.

Thus we see that the physicist speaks of causality only where his laws of nature satisfy definite formal criteria: they must not explicitly contain the co-ordinates assigned to events. Moreover, we see that the possibility of second-order predictions (that is, the derivation of events not used in obtaining the law) is tied to this formal property of laws of nature. In asserting this I am opposed to Schlick: only of laws of nature of this kind can we say they do or do not prove themselves. Denoting practical and physical causality as 'first-order' and 'second-order' respectively, we can summarize the relation between them as follows: first-order causality exists where we can ascertain first order regularities; second-order causality exists where first order-laws are so related that the second-order law stating this relation does not explicitly contain any of the first-order laws.

One of the motive principles of enquiry in theoretical physics is the setting up of only such second-order laws as satisfy the physical criteria of causality. We can call it the 'principle of physical or theoretical causality'. Where this principle is violated by an anomaly, physicists will make every effort to clarify the anomaly so that the principle should triumph once more. Explaining an anomaly always counts as progress in our knowledge of nature and therefore often has sensational effects even outside science. The immense impact of Leverrier's explanation of the anomaly of Uranus is well known: that planet's orbit seemed to show deviations from Newton's law of gravity, which Leverrier attributed to the influence of an as yet unknown planet acting on Uranus according to that law. Observations confirmed his calculations and the new planet was discovered, so that the orbit of Uranus did not signify a deviation from Newton's law but rather confirmed it: the anomaly had been removed. No less sensational was the effect when relativity theory explained the anomaly of Mercury, which likewise showed deviations from Newton's law, but here it proved impossible to regard the disturbance as due to a new planet. One might then simply have treated the particular deviation as an explicit addition to Newton's law; that, significantly, is not what physicists did do: rather, it was decided to give up the well-tried law of Newton, and Einstein put in its place a new law of gravitation from which all planetary orbits can be derived while it does not explicitly contain any co-ordinates. On this new law, the Newtonian law does not hold exactly for the other planets either, but the deviation does not become noticeable until we reach Mercury.

A new and unfamiliar explanation of an anomaly was that of light, which has very recently become world-famous. It consisted in the fact that the classical mechanical laws for addition of velocities could not be confirmed for the velocity of light. However, the laws of relativity theory that explain this anomaly explicitly contain the speed of light c. As we saw, this points to an anomaly and relativity theory seems merely to shift this to a different point, which is H. Dingler's objection to the theory. Now it emerges from the formulae that the obvious anomaly that can be read from them could occur only at velocities greater than that of light, yet on the theory of relativity such velocities cannot be measured: they are unreal. Hence the anomaly of light is shifted into the region of unreality and cannot occur within the domain of what is measurable, which alone interests physics. Therefore the anomaly of light can be taken as eliminated.

It remains to discuss the philosophic significance of the principle of physical causality. As we saw, already Maxwell held that we speak of causality where the functions describing laws of nature do not explicitly contain spatio-temporal co-ordinates. Maxwell read off this criterion from the laws of physics without giving reasons why he requires this condition. We saw that to this form of laws of nature is tied the property that is the most valuable for physics, namely that second-order predictions are possible only where laws do not explicitly contain the quantities assigned to states, that is, where no first-order laws are explicitly mentioned within second-order laws. With this formulation we have gone beyond Maxwell, for if we were to describe events not by spatio-temporal co-ordinates but by some other ordering schema, we should still require that these latter do not explicitly occur in our laws. With this general formulation of the principle of physical causality we have gone beyond the merely physical sphere and we must therefore not confine ourselves to the advantages that physics gains by following the principle but also indicate other reasons that speak in its favour. There is a further reason why this is necessary, namely that in daily life, in technology, in vast areas of experimental physics and of course in all non-physical empirical sciences we do not always keep to the principle of physical causality, but are often content to record observed regularities and consciously give explicit mention to first-order laws within our second-order ones. Just because of this lack of unity in the method of other sciences physics looks down on them with some contempt, asserting that its own method is the only one leading to an explanation of processes in the genuine and legitimate sense. We must examine whether this claim is justified. Why should explanation exist only where we describe by laws that satisfy the criterion of physical causality, and none be obtainable in terms of laws not formed on this priciple? This can be clarified by means of simply constructed examples.

Suppose somebody intended to arrange his library. He can classify his books from various points of view, for example as to subject, or alphabetically according to authors' names or titles and so on. In any case he will assume *a priori* that his library, however arranged, will always contain the same books. He therefore might decide to classify by subject, and go on to arrange and catalogue his books. After some time he finds the classification inexpedient and re-arranges the books into an alphabetic order according to authors, preparing a new list. On comparing the two catalogues, he is to notice something odd: the new one contains entries lacking in the old. Since he has not bought more books he will assume that he overlooked these entries on the previous occasion. Next, the new arrangement does not suit him either, and he returns to the old one, again preparing a catalogue. On comparing catalogues he notices again that the books that had previously appeared all of a sudden have vanished. Now he will prefer to adopt the most unlikely hypotheses: he will for example rather assume that some demon has played tricks on him than decide to ascribe the waxing and waning of his library to the different arrangements. If nevertheless he opts for the latter, then in his catalogue he must explicitly mention the arrangement of the books alongside each title, thus: arranged by subject, my library contains such and such books; arranged alphabetically, it contains such and such others as well. This, however, means that the arrangement of books is no longer a merely rather convenient way of talking about the contents of the library but is itself a part of that content. In general this means that in such a case language itself belongs to the objects about which we speak. For to speak about objects is to arrange them in some way or other. If the arrangement itself belongs to the objects being arranged, this means that we can no longer clearly distinguish language from what is being talked about. In critical cases this may lead to serious nonsense. If, for example, the word 'blackboard' written in chalk could not be distinguished from the object that it denotes, then the proposition 'the blackboard is black' would be true and false at once, which is always a sign of nonsense. Applying this to physics we obtain space and time as the ordering schema for events, the language of physics. If in order to describe events we had to mention spatio-temporal co-ordinates explicitly in our functions, space and time would no longer simply amount to an order of events but would themselves belong amongst events. For such events explanation would indeed be impossible, for to explain means to indicate an order of events, as we saw. If now the events cannot be ordered according to the indicated schema, one can always keep to it while explicitly recording the deviations where they occur, which simply means that the events cannot be explained by it. So long as events can be described only by explicit co-ordinates, there can be no explanation of them. Attempts are of course occasionally made to explain even events such as these, but the explanations are always sham and illegitimate, consisting in the indication of mystical and untestable forces.

Thus it is indeed the case that genuine explanation of events exists only where our laws do not explicitly contain the co-ordinates. The physicists are thus justified in claiming that their method is preferable. Only where laws are formed according to the principle of physical causality can we speak of explanation of processes; if we offend against the principle, we may at best be able to record the observed regularities. From the fact that a description by means of explicit co-ordinates always involves a lack of distinction between form and content of language, we further see that it is a deep logical instinct that leads physicists to require the principle of physical causality as a principle of inquiry.

If we import the distinction of first- and second-order causality to non-physical problems as well, new problems arise and perhaps new possibilities for solving them. For example, the vitalist assertion that in order to describe living processes we must explicitly add certain other quantities to our physical formulae, means simply that there is no second-order causality in biology, so that here we can have only records of observed regularities of first-order laws. Accordingly, biological processes will always deviate from laws formed on the principle of physical causality: they will always remain anomalies with regard to such laws. Nevertheless, living processes occur according to strict regularities. If we have observed the course of such a process and recorded it, we can predict a repetition of it under similar circumstances; but we cannot derive the process from other events that are independent of these observations. Thus biological processes would be inexplicable. As far as this we can test the vitalist assertion: the existence or otherwise of second-order causality is an empirical question. If beyond this, however, the vitalists seek to provide an explanation for living processes, these accounts must be sham and illegitimate, as indeed has been shown from various sides.

The case stands similarly as regards the split of recent psychology into behaviourism and intuitionism. Behaviourism requires a strictly physical formation of laws even in psychology, which means that the principle of physical causality is to hold. Mental events are then physical events and like these last derivable from other events. Intuitionism denies that this is possible: a mental event is said never to be entirely derivable from other events, or, as the phrase goes, a mental phenomenon is never just the sum of its parts. In explaining a mental event by other events we must therefore always explicitly indicate a factor independent of these events, in order to complete the account. Thus, second-order causality does not apply in psychology, although first-order regularities can indeed be ascertained. On this view, mental events will always amount to anomalies with regard to second-order laws formed on the principle of physical causality. Connected with this are the analogous problems of all those sciences that use psychology as an auxiliary discipline.

Finally, our distinction of first- and second-order causality throws new light on one other notorious philosophic problem, namely that of indeterminism. If the indeterminist wishes to speak of free will only where there is no causality, we must now ask him which causality he means. If he is content with the exclusion of physical causality, then indeed phenomena of the will can never be completely described by laws formed on the principle of physical causality, but will always exhibit deviations from such laws. Nevertheless decisions and modes of character reactions would occur in strict regularity, for first-order causality is still to hold. If therefore we have observed how a person of given character behaves under certain circumstances, we can predict how he will react under similar circumstances. However, from events that are independent of these observations we cannot compute the behaviour. This is the point of view adopted by those indeterminists who speak of free will if somebody not under compulsion acts according to his character, where character is denoted as something mystical, inexplicable, that is, as an anomaly to the laws correctly formed on the principle of physical causality. In this form, indeterminism is entirely discussable. In contrast, radical indeterminism which wishes to speak of free will only where no causality exists at all, whether first- or second-order, is not discussable. In this case freedom simply coincides with chance and formulating any question becomes senseless.

CHAPTER II

CRITICAL COMMENTS ON THE PHYSICALIST THEORY OF SCIENCE*

1. VERIFICATION AND ELUCIDATION

From within the Vienna Circle, some of its own epistemological theses have lately attracted critical objections that require a radical transformation of the foundations of neo-positivist philosophy. What provoked these criticisms was the verification theory of logical positivism. According to this view the sense of a proposition consists in its verification. How, then, are we to carry out the 'verification of a proposition' and what does this expression mean? According to the view hitherto held by the Vienna Circle, verification is effected by reducing a proposition to such other propositions as are statements about the immediately observable, or 'given'. By the given they here understand "that which is simplest and cannot be further questioned".[1] Propositions about the given cannot be further reduced to other propositions, but are "verified in virtue of describing the given".[2] Whether such propositions are true or false is decided by comparing them with the given.

Since the verification of all other propositions consists in reducing them to such propositions about the given, what finally counts in any verification is a confrontation of propositions with the given. A proposition is true if it has the same structure as the state of affairs whose existence it asserts.

Only those propositions have sense which either can be reduced to statements about the immediately given ('observable') or are themselves statements that are reducible no further. Reduction of a proposition to other propositions may be effected in two ways: either by transformation according to defining rules, or by deduction according to rules of inference. The truth of reducible propositions always depends on the truth or falsehood of some other propositions; while irreducible proposition are true or false no matter what the truth values of other propositions, that is they are not truth functions of other propositions.

Thus one has understood a proposition if one knows to what statements

* Translated from the German original 'Kritische Bemerkungen zur Wissenschatstheorie des Physikalismus', first published in *Erkenntnis* **4** (1934) 397–418.

about the given it is reducible; but when has one understood a statement about the immediately given which is not further reducible? On the traditional view of the Vienna Circle such a proposition can no longer be made more intelligible by means of other propositions, but only by exhibiting the state of affairs which is stated in that proposition. The act of exhibiting cannot be replaced by description or explanation involving statements. There is indeed a type of linguistic expedient for achieving or at least furthering intelligibility by means of exhibiting, namely by means of elucidations concerning the basic concepts and presuppositions of science. However, elucidations are not statements, even if externally they have the form of statements. Their function consists in drawing the attention to certain aspects of states of affairs (structures) and indirectly presenting these aspects by means of examples, comparisons and analogies. Precisely because elucidations are not propositions having sense but yet have the form of propositions, they are related to the sham propositions of metaphysics. However, it is not only the foundations of individual sciences that require elucidations, they are likewise needed for understanding the forms of the language of science. On the traditional view of the Vienna Circle, statements about the forms of language are impossible, for statements that assert the existence of linguistic forms must themselves possess this form. Thus they are necessarily of the form 'a is a'. Such tautological propositions however state nothing, just as somebody who answers the question 'what is red?' by 'red is red' says nothing. In order to attain a grasp of the forms of language, in case one is not clear about them from the start, one can proceed only by means of exhibiting them in examples of perspicuous complexes of symbols, or, where that too fails, by means of elucidations. Except that here the position is in one respect more complicated. We have logical forms only in so far as they belong to given propositions and propositional functions. Yet one can make quite a few empirically verifiable statements about complexes of signs that represent propositions or concepts. Thus it seems that we can make statements about linguistic forms too. As against this the Vienna Circle has hitherto held that logical forms show themselves in complexes of signs but are not identical with them. So-called 'statements about the forms of language' are thus at best statements about objects of physical reality, but never about the logical forms themselves, which show themselves in the empirically real signs. Therefore logical forms too can only be exhibited. The best tool for this is an appropriately

differentiated symbolism, but alongside it elucidations are indispensable: they direct the attention mainly to what we are at when we
formulate propositions, or more precisely when we operate with logical
signs. Yet it would be a mistake to think that by describing this activity the
forms of language are after all encompassed in statements. The activity
has the form; the form may show itself especially clearly in the activity,
but the two are not identical.

To this theory of verification and elucidation Carnap[3] and Neurath[4],
themselves members of the Vienna Circle, object that in many respects it
rests on metaphysical presuppositions, using sham propositions devoid of
sense in order to establish them: if in the end verification amounts to confrontation of propositions with the given, this is metaphysics, for what are
we to understand by 'the given'? This sounds as though something existed
independently of the cognizing consciousness to which it is given, which is
pure metaphysical realism. Besides, the best proof that metaphysics is indispensable to the Vienna Circle's traditional view, with its Wittgensteinian
bearings, is that for an understanding of the forms of language and the presuppositions of science it requires 'elucidations' admitted to be word sequences empty of sense and thus of the same kind as the sham propositions
of metaphysics which are unverifiable in principle. According to Carnap
and Neurath these metaphysical ingredients are incompatible with the anti-
metaphysical attitude of positivism and spring, so they think, partly from a
misformulation of the problem and partly from imprecise modes of speech
and to that extent eliminate themselves automatically if modes of speech
are correct.

Neurath (l.c.) sketches the general outline for a programme of positivism
free from metaphysics, which for reasons to be discussed later he calls
'physicalism'. Carnap (l.c.) examines the problems and shows how they
are connected, indicating the epistemological method for their solution.
Let us now discuss the corrections thereby made necessary in the theses of
logical positivism.

2. ELIMINATION OF METAPHYSICAL FORMULATIONS
BY MEANS OF THE PHYSICAL LANGUAGE

On the Vienna Circle's view to date we should not be able to make do with a
unified science that uses only the language of physics.[5] We use the language

of physics if physical concepts alone occur in our statements. The concepts of present day physics are exclusively spatio-temporal. It is indifferent whether physics will always use spatio-temporal concepts in its formulations, the question is whether any arbitrarily chosen scientific proposition can in principle be formulated in such a way that it contains only concepts that are of the same kind as the basic concepts of physics at the particular time. As we remarked, on the Vienna Circle's view to date we cannot say yes to this question. Accordingly there are concepts that cannot be constructed from the concepts of physics alone. The statements in which these concepts occur thus describe states of affairs that cannot be described by the language of physics. For example, the concept 'understanding a statement (or a concept)' cannot so be constructed. For if it could, if by describing spatio-temporal processes one could indicate when we understand a proposition p_1, true or false, which is ranged within the system of science, this would lead to circularity, as can be shown. For then the following proposition p_2 holds: 'A understands p_1' is equivalent to 'A behaves in a certain way' (the behaviour of A being taken as a spatio-temporal process). p_2 is surely not equivalent to p_1. On the one hand, in order to understand p_1 one must know that p_2 is true, on the other, in order to decide whether p_2 is true one must already understand p_1. This clearly involves circularity, and the concept 'to understand' can thus not be constructed merely from physical concepts (this argument is the author's own).

As against this point of view, Neurath and Carnap now assert that there is only one scientific language, namely that of physics. The statements of all empirical sciences must be without exception transformable into propositions containing only physical concepts. Where this is impossible, the 'proposition' in question has no testable sense, that is we have a verbal sequence devoid of sense. From the uniformity of language it follows that there is in principle only one science. The language of this 'unified science' is that of physics. Thus all fields of science are parts of unified science, that is of physics.[6] The name that Carnap and Neurath give to their view, that the language of physic is the basic language of science, is 'physicalism'.

To meet the difficulties to which a consistent working out of 'physicalism' leads – as above in the case of 'understanding' – the physicalists use the following consideration. In so far as such concepts occur in scientific statements testable inter-subjectively by observation, they can always be replaced by spatio-temporal concepts. Where this is not possible, we can in principle not ascertain whether a statement is true or false, so that we have then no proposition at all. Thus all statements of psychology must be transformable into behaviourist propositions, otherwise they are devoid of sense. In particular, propositions that speak of 'understanding' (or 'meaning') of statements or concepts, must be equivalent either to statements about the behaviour of the person who understands, in the sense of behaviourism, or they refer to a pre-scientific or pre-linguistic state and for that reason alone cannot a priori be testable significant propositions (cf. Neurath l.cl, p. 396). In science, only inter-subjectively testable statements occur. That is why circularity of the kind mentioned above cannot arise as long as we stay within science; that is, so long as we decide to speak in universally testable propositions. Such reservations thus do not constitute an obstacle to the feasibility of the physicalist programme, according to Carnap and Neurath.

Already at this point one suspects that this will eliminate, as devoid of sense, problems that arise as soon as one tries to understand the foundations of science. This suspicion is reinforced when in addition one meets similar problems at other points at which in the physicalists' opinion the hidden metaphysics of the Vienna Circle stands forth with particular clarity; that is why physicalist reform operates more extensively at these points.

As mentioned above, verification of propositions on the Vienna Circle's view to date is in the end effected by confronting propositions with the given. Under certain circumstances one might test the corrections of a statement p_1 by indicating a proposition p_2 (or a system of propositions) to which p_1 can be so related according to the rules of logic (for example, inference rules) that p_1 is true when p_2 is true (or false). However, there must in principle be propositions that are true independently of whether any other propositions are true or false; in other words, there must be propositions that are not truth functions of other propositions. These propositions are not further reducible to other propositions: their truth cannot be tested by indicating premises from which they can be inferred. The truth or falsehood of these 'atomic propositions' is determined in that their structure shows itself to be the same as (or different from) that of the given state of affairs.

Now Carnap and Neurath consider that talking of 'the given' amounts to metaphysics, a residue of metaphysical realism. The very name 'given' points to something to which it is given, to which it stands in some relation or other, and this generates a host of well-known sham questions – like that of the nature of the 'given' in itself, and so on – all of which are in principle unanswerable. Now it is certain that all the other members of the Vienna Circle would equally reject such a metaphysical basis for the criticized theory of verification.[7] However, according to Carnap and Neurath these consequences are unavoidable as long as the Vienna Circle's formulations to date are maintained. Nevertheless, they can be eliminated if one reflects how we actually go about verifying propositions of science, and then proceeds to alter the formulations of the theory of verification accordingly. There is a further circumstance that urges the physicalists towards taking this step. In contrast with the previous view of the Vienna Circle, Carnap and Neurath consider that it is possible to make significant statements about the forms of language (syntax, grammar). Verification of such statements by confronting propositions with the given is of course not possible, for in that case the logical forms would have to be counted as given in the same sense as the objects of physics. Therefore this aspect too makes it necessary to reform past views of the Vienna Circle.

Neurath considers that the verification of any scientific propositions consists in comparing it with other propositions.[8] The assertion that in order to test the truth of a proposition we compare it with the so-called 'given' is devoid of sense in virtue of the mere fact that anybody can compare the propositions in question only with what is given to him. In consequence inter-subjective verification would be impossible. Rather, what we have, according to the physicalist view, is a 'mass of statements' (Neurath, l.c., p. 397) which we increase by new statements, verifying these by comparing them with the 'traditional' ones. If there is contradiction, then we must either eliminate one or the other statement as false, or we must otherwise modify statements until we reach a consistent system. In this we never fall back on 'the given'. Naturally, all statements without exception must be formulable in the language of physics.

However, the view that verification consists exclusively in the comparison as to mutual agreement involves every proposition of science being a truth function of other such proposition. In what follows we shall discuss the consequences that follow from this view.

3. THE 'PRIVILEGED POSITION' OF PROTOCOL PROPOSITIONS
AND THE PHYSICALIST CRITERION OF PROBATION

The physicalist conception, that we increase the existing mass of scientific statements by new statements, eliminating or modifying propositions until we get a consistent system, leads to the question how we obtain and verify those propositions whose function it is, on the Vienna Circle's view to date, to establish the truth of other propositions and to decide whether our hypotheses and predictions prove adequate. This function belongs to the so-called protocol propositions.

How these are obtained and verified has been thoroughly examined in two recent articles by Carnap and Neurath[9] where their way of regarding these problems is formulated more precisely.

Neurath's account is directed primarily against earlier assertion of Carnap[10] actually intended more as hints and modified since, that in contrast with propositions of a system, protocol propositions are obtained 'directly' and stand in no need of probation.

This view of Carnap still contains a residue of metaphysics easily linked with the scholastic belief in 'immediate experience', 'ultimate elements', in short 'the given' (Neurath, l.c., p. 210). This undermines the anti-metaphysical stance of physicalism. Neurath rightly remarks that propositions that require no probation cannot be modified or cancelled as false in the course of scientific procedure either. Carnap's conception, and of course even more any view that sees in propositions about the 'given' the ultimate foundation of verification, thus results in a privileged position for protocol propositions as opposed to the other propositions of science.

If a protocol proposition contradicts the existing system of propositions, then, according to the view that Neurath opposes, protocol propositions must never be cancelled or modified for the sake of mending the contradiction; transformation must always fall on hypotheses, laws of nature or rules laid down in the system. Neurath protests against this 'inviolability' of protocol propositions (l.c., p. 209). The process of transformation in science consists precisely in the dropping, or replacement by others, of propositions that have been used earlier. "Every law and every physical proposition of unified science or of one of its real branches can undergo such modification, and every protocol proposition too" (Neurath, l.c., p. 208). So-called finally guaranteed and clean protocol propositions as starting points for science do not exist; to suppose there was an ideal language built out of clean atomic propositions is a metaphysical fiction (Neurath, l.c., p. 204, 213). So-called original protocol propositions, such as Carnap assumes, are, according to Neurath, real propositions like any other real propositions and of the same linguistic form: they are formulated by connecting "personal names or names of groups of persons" with other terms that also occur in other scientific propositions, according to certain (syntactic) rules (cf. Neurath, l.c., pp. 207 ff, 213). Since they are the same in kind as other real propositions, protocol propositions have no privileged position over the others but like the rest must stand proof and prove themselves and may at times be modified or cancelled with the same justification. From this equality of status of all real propositions it follows at once that for all propositions without exception (including, that is, protocol propositions) the criterion for whether they have withstood probations can consist only in their mutual compatibility (freedom from contradiction) (cf. Neurath, l.c. pp. 208 ff). However, the detailed reasons that Neurath gives for this way of taking the matter are in my opinion not without risk for his own point of view.

Thus he remarks (l.c. p. 209) that only 'propositions' can 'themselves withstand probation'. 'Propositions' are 'sequences of signs that can be used within the framework of testing a reaction and can be systematically replaced by other signs'. This is followed at once by a sentence that must arouse misgivings with radical physicalists. "'Same propositions' are to be defined as stimuli which in certain tests of reactions provoke the same reactions" (l.c., p. 209). Sameness of propositions is thus here defined by means of sameness fo reactions. Yet what are 'same reactions'? Surely physicalists may speak only of propositions. At most they might be allowed to define sameness of reactions by means of propositions, but not the other way round. In this respect Neurath's definition makes it appear as though reactions were something 'given' and 'observed' before language, on the basis of which sameness and agreement of propositions is ascertained. This would be a relapse into metaphysics. Of course, this may be an imprecise formulation. However, Neurath makes further remarks that lead to interesting consequences as regards the physicalist point of view.

The fate of being cancelled as false can, according to Neurath, overtake protocol propositions as much as any other real propositions, the criterion for proving themselves being mutual compatibility of the propositions. Neurath is of the opinion that there are cases where at least one protocol proposition must be declared false and be cancelled. On p. 209 he says: "two protocol propositions that contradict each other cannot be used in the system of unified science. Although we may not be able to say which of the two is to be eliminated, or wheter both are, it is certain that not both can 'prove themselves', that is are insertable into the system". Even on Neurath's approach the last assertion does not hold without special conventions, since according to this radical view we are entitled to reformulate or modify statements, hypotheses and conventions in order to bring the propositions of the system into agreement. Here we must emphasize that auxiliary propositions thus newly formulated – possibly including quite new protocol propositions formulated by connecting certain terms according to prescribed rules – are on entirely the same footing with the accepted real propositions of the system, for in not contradicting these last they have already proved themselves in the sense of physicalism. However, by formulating especially suitable hypotheses, it is always in principle possible consistently to accomodate within the system any protocol propositions whatever, even if they stand in so-called 'contradiction' to each other. In a formal logical sense, protocol propositions simply cannot contradict each other.

This comes out clearly in Carnap's reply[11] to Neurath's above account, where Carnap tries to prove that his own view, given appropriate formulation, agrees with Neurath's. He there says that what is recorded in protocols (that is protocol propositions obtained by reactions) may become formally contradictory propositions only if, by means of translation rules, they are translated into propositions in language. If now a contradiction appears, the fault must be sought in the translation rules that we have set up and they must be altered accordingly (Carnap l.c., pp. 219, 220). Thus protocol propositions can contradict each other only after they have been translated into the system language, but these contradictions can always be corrected by modifying either the translation rules or other propositions of the system, so that "it is always possible ... to utilize without contradiction the protocol propositions that happen to be to hand" (Carnap, l.c., p. 220).

Yet what follows from this? If probation consists in the accomodation of propositions within the system without contradiction (cf. Neurath, l.c., p. 209) and we can always in principle accomodate any protocol proposition whatsoever, then clearly that criterion of probation can be satisfied in all cases, that is its content is tautological. However, that makes it unusable. Yet it is such a tautological criterion that we inevitably reach it, with Neurath, we deny protocol propositions their privileged position over other real propositions and regard consistency as the sole criterion of propositions proving themselves. For since one can extend a system by formulating new propositions of all kinds (including protocols), it is always in principle possible to obtain consistency within the system whatever the original protocols.

One can escape this difficulty only by acknowledging that protocols do have the privileged position of verifying other propositions while not themselves requiring to be verified, that is their truth value is independent of other propositions. On that view, understanding such protocols is already seeing them to be true. That is why it is no use trying to escape the tautological criterion of propositions proving themselves, by introducing an arbitrary formal restriction, for example demanding that two 'contradictory' protocol propositions must not be used in the same system (cf. Neurath, l.c., pp. 209, 210). Even then, difficulties of principle remain. For now we obtain indefinitely many scientific systems ('unified sciences') that are indeed mutually exclusive – albeit by reason of an arbitrary convention –, but nevertheless all of them equally valid; for each is based on properly formed protocol propositions that agree with the rest of the system. Which of these systems, then, is *the* unified science? What marks this system off from the others? More about this later.[12]

Disregarding the difference in terminology, we have here a situation similar to Dingler's conventionalism. Dingler thinks that the results of our 'observations' (measurements, experiments) depend on the laws according to which we make our measuring instruments. Therefore the results of measurement simply cannot contradict these laws. Where this nevertheless seems to happen it is senseless to assume lack of agreement of the laws with the unobservable 'given'; rather, it is always possible, by formulating appropriate hypotheses, to restore harmony between 'protocol propositions' and the laws originally chosen. To say that physics describes 'reality' or 'the given', is thus, for Dingler too, an inadmissible and senseless way of talking. This moves him, too, to distinguish from the unlimited number of systems that on his view can be constructed, one that in the end has the special property of being 'correct'. The possible misgivings that Dingler's special solution of this problem might provoke need not be examined here. We merely point to the similarity of the problems resulting from 'physicalism' and 'conventionalism' alike.

While the unambiguously formulated view of Neurath just discussed is relatively simple, that of Carnap, explained in some detail in the articles mentioned, seems less so. To this we will now turn.

4. OBTAINING THE PROTOCOL PROPOSITIONS

In Neurath's essay discussed in Section 3 he decidedly rejects Carnap's thesis[13] that protocol propositions are exempt from probation. In his reply, Carnap[14] tries to show that his view differs from Neurath's only in verbal expression. Above all he rejects the suspicion cast on him, that his view must inevitably be supported by appeal to 'ultimate elements' or 'the given', in short that it is 'absolutist'. (Cf. Carnap: 'Über Protokollsätze', p. 228). However, his explanations leave one the impression that he is in two minds.

In searching for protocol propositions two procedures must be noticed above all, according to Carnap. One of them considers protocols outside the system language, the other within it (Carnap, l.c., pp. 216ff, 222ff). With the first procedure, described by Carnap as not indeed incorrect but less perfect, protocol propositions are represented by processes occuring under certain circumstances, that is they are reactions supervening under certain conditions, or as he also says, signals supervening automatically under certain circumstances. We study these regularities of their occurrence, and then, on the basis of translation rules, assign to them expressions of our system language, so that we obtain a lexicon that allows us to translate combinations of signals (that is, protocol propositions) into propositions of the system language, while conversely enabling us to recognize which combinations of signals we must expect on the basis of certain propositions of the system under certain circumstances (Carnap, l.c., p. 216ff). Carnap next tries to show that it is a matter of mere linguistic convention whether we speak of protocol propositions proving themselves or not proving themselves, or would rather say that they are exempt from probation. On that basis his view would differ from Neurath's only in verbal expression. However, let us here anticipate by remarking that in our view Carnap's explanations on this problem reveal a contrast in principle between the two positions.

Protocol propositions as processes (reactions, signals) cannot contradict each other. Only when by means of translation rules we translate them into propositions of the system language, can contradictions arise. This might move us to modify the translation rules or the received proposition of the system in such a way that agreement is achieved. By means of such modifications of the translation rules or of the propositions of the system it is always possible, according to Carnap, "to utilize without contradiction the protocol propositions that happen to be to hand. The protocol propositions remain untouched in this procedure". (Cf. Carnap, l.c., pp. 219, 220). Carnap immediately adds, evidently aiming at Neurath's view: "It is a question of convention whether instead of this procedure we wish to choose another, where in case of incompatibility one might even declare the protocol in question to be 'false' and cancel it" (Carnap, l.c., p. 220). This would satisfy both points of view (Carnap's and Neurath's).

However, it follows from this position that whatever protocol propositions are to hand, agreement or disagreement of the propositions of the system can be regulated independently of the latter by arbitrarily changing the translation rules, agreements and contradictions of the system do not concern the protocol propositions. It is then naturally a matter of arbitrary convention whether in case of inconsistency between propositions of the system the co-ordinated protocol propositions (which lie outside the system language) are to be cancelled as well. For one can deal with them at will, they are 'superfluous entities' in Occam's sense; one can only agree with Neurath when, with regard to this (since altered) view of Carnap, he demands that we ought to speak only of formulating and transforming propositions of the system and, since in any case the only thing that matters is agreement within the system, to leave out as superfluous the protocol propositions external to the system and along with them the rules of translation. In this way we do in fact reach an entirely equivalent but much simpler and less misleading view. There are then only propositions in the system language, of which some are fashioned by means of formulation (that is connection) of certain chosen terms according to syntactic rules and are called 'protocol propositions (within the system language)'. We have aleady met this view of Neurath and thus need not discuss it further. If that is the way Carnap takes the assimilation of his view to Neurath's, then indeed the difference between the two positions is merely one of verbal expression, but in that case all the misgivings urged against Neurath hold equally against Carnap's view. In particular, he too would then have to understand by 'probation' a concept applying in all circumstances, and thus a tautological one. That this however, is precisely what he does not hold seems to me inferable from other passages in his account; but since he himself calls the consideration of protocol propositions "outside the system language" an imperfect procedure, let us turn at once to examining the better procedure, which considers protocol propositions "within the system language" (Carnap, l.c., p. 222ff).

Here too, let it be said from the start, the final result does not impress one as entirely unambiguous. Protocol propositions are reactions or signals. In certain cases the signalling configurations can be words and propositions of the system language, so that here the protocol propositions belong directly to the system language and we are saved the trouble of translation. This entirely possible assumption indeed makes no difference to the sense of the view discussed above: here too the protocol propositions are signals and thus are obtained directly. However, since the signalling configurations are at the same time complexes of signs of the system language, the protocol propositions are inevitably concrete propositions of the system. Testing a proposition of the system here occurs by deriving from it concrete propositions which are then compared with propositions obtained directly (that is, protocol propositions within the system language). Thus we can here say that in verifying propositions of the system we compare them directly with protocol propositions, that is with other propositions of the system. Only we must not forget that one of these was always obtained by derivation and the other directly (by reactions). Thus likewise Carnap (l.c., p. 222, and in the example p. 224).

However, with him, there is the additional point, that there is a special circumstance that allows us to infer equality of status of protocol propositions with the other propositions of the system (l.c., p. 223ff), in which case, as repeatedly explained above, the criterion of adequacy can reside only in formal consistency of the system and the way of obtaining the propositions compared no longer matters. For Carnap holds (following ideas of Karl Popper's) that since protocol propositions are themselves concrete propositions of the system, it is possible that "any concrete proposition of the physical system language may under certain circumstances serve as a protocol proposition" (cf. l.c., p. 224). Since concrete propositions are obtained by derivation from laws, the testing of a law consists in deriving concrete propositions from it – in which one can go back arbitrarily far – and deciding by stipulation where to stop and which concrete propositions to recognize as protocol propositions. Because we thus decide by arbitrary stipulation which concrete propositions are to be called protocols, there can be no protocol propositions with privileged position over the other propositions of the system, for if we decide arbitrarily the lot can fall on any concrete propositions of the system. An absolute privileged position of protocol propositions, as hithero asserted in the 'absolutist' view of the Vienna Circle, thus does not exist; on the contrary, physicalism has succeeded in showing that protocol propositions are relative, using the considerations just indicated above: therefore physicalism may be called the first absolutely relativist epistemology (Carnap, l.c., p. 224, 225, 227ff).

There is an unmistakable contrast between this and Carnap's other view, which we too recognize as correct, that protocol propositions are signalling configurations obtained by reaction (cf. l.c., pp. 222–225). A latent contradiction exists between Carnap's assertion that protocol propositions are relative, that is equal in status with all other propositions of the system (l.c., pp. 224, 225, 227ff) and his other view demonstrated by a concrete example (l.c., pp. 224, 225).

It is indeed correct that in testing protocols or other propositions of the system we derive concrete propositions from those to be tested, a process in which we might 'go back' indefinitely far by enlisting ever more new premisses; but is false that the adequacy of propositions under test is decided by stipulation, by ultimately declaring one of the derived propositions to be a true protocol proposition, that is one that verifies. This view is suggested by Carnap's comments (l.c.); according to it the truth value of protocol propositions depends on a stipulation, and this would lead to the tautological criterion of probation in its worst form. The verifying proposition can never be obtained by derivation nor can any additional stipulation confer on any derived proposition the privilege of verifying. It is correct that we decide by stipulation which derived proposition is to be compared with a protocol proposition obtained by recording a protocol ('observation', reaction): but the proposition with which the derived concrete proposition is compared is always obtained directly by recording a protocol (cf. Carnap, l.c., pp. 224, 225). It is by being directly obtained, whether we call this observation, reaction or whatever, that a proposition is given the privileged position of verifying other propositions, for this directness guarantees that the truth value of the verifying proposition is independent of the truth values of other propositions, and only on that condition can 'probation' of propositions of the system mean more than merely formal consistency. Directly obtained propositions are distinguished by their form, it is their form that enables us to recognize their truth merely by understanding them. This holds equally where we wish to test 'protocol propositions', as stands out clearly in the example discussed by Carnap (l.c., p. 225) even if his mode of expression in places might perhaps provoke misunderstandings. The testing of a 'protocol proposition' becomes necessary when the circumstances under which the reaction occured are not precisely known or are replaced by untested presuppositions (for example reading the position of the hands of a watch presupposing it to be running accurately). Pure protocol propositions (which are only signalling configurations) can thus not be tested at all, their truth or falsehood 'shows' itself. The theoretical hypothetical elements of 'impure' protocol propositions are however testable, and testing 'protocol propositions' consists just in deriving the theoretical presuppositions in the form of concrete systems of propositions and in verifying them. The pure core of each 'protocol proposition', the 'signal', will thus be exhibited with increasing clarity. It may be that such theoretical presuppositions can never be completely eliminated in recording any protocol, but it is certain that they can be gradually and systematically restricted by recording new protocols, which explains why we ascribe greater reliability to one protocol than to some other that rests on less well-tested presuppositions. However, any recording of a protocol is essentially based on being directly obtained by means of reaction ('observation' and so on): this constitutes its core, whatever other presuppositions are also involved. The testing of a protocol proposition P thus consists in deriving concrete propositions from it and comparing them with other protocol propositions Q directly obtained by recording protocols. This too is the sense of Carnap's account at l.c., p. 225, except that I think it would be clearer if in describing the process of verifying protocol propositions he did not simply say that one protocol proposition P is 'reduced' to others Q (which could easily be understood as asserting that P and Q were related a priori by logical deduction) and that all that matters is to ascertain their formal consistency; but instead he had better have said: the concrete propositions derived from P are compared with directly obtained protocol propositions Q. (Confirmation that this last is Carnap's view is given by, among other things, those passages where he says that certain protocol propositions are reduced to others "which describe a calibration carried out by ourselves", or "I have observed this and that setting of the pointer" and the like (l.c., p. 225)).

Every verification is thus based on propositions that were obtained directly by recording of protocols. The view of the physicalists that establishing protocol propositions as a physical reaction can be described by means of spatio-temporal concepts is untenable, for the correctness of this description of the recording of protocols could itself be tested only by recording of protocols. However, we could trust this latter, investigatory recording of a protocol only if we knew that it really was the recording of a protocol, so that our spatio-temporal description of it would be in this case true. Verification of that spatio-temporal description thus presupposes that the description is true. From this it follows that a physical description of the recording of this protocol is impossible. This view is supported by the fact that Carnap holds (with Popper) that only stipulation can decide which concrete propositions of the system we are to recognize as protocol propositions. If the recording of protocol propositions were a process describable by means of physical concepts, then we should not need a stipulation, for in that case we could indicate the empirical criteria of the process and of its results, that is of the protocol propositions. However this may be, the epistemological importance of 'directly' obtaining protocol propositions consists in their not being truth-functions of other propositions; it is this feature that lends them the privilege over all other propositions of verifying, and only then can 'agreement with protocol propositions' mean more than mere formal consistency. To repeat, we believe that from many passages in Carnap's account we can infer that he recognizes this privilege of protocol propositions and therefore is an 'absolutist', even if we also find him making remarks that contradict this. Indeed, we think that the views of Dingler and Neurath are much more closely related than those of Carnap and Neurath.

5. THE 'TRUE' SCIENCE

A further clarification of the problem situation discussed above is afforded by Carnap's explanations[15] in his reply to E. Zilsel.[16]

Of the two questions there discussed, only the second interests us here:

how is the 'true' science distinguished from all other logically possible systems? Confirmation of a system of scientific propositions occurs by means of protocol propositions. Now we may conceive of arbitrarily many 'protocol sets' that are mutually inconsistent but logically equally valid. For each one amongst such sets we may establish scientific systems each internally consistent and confirmed by the respective set (cf. Zilsel, l.c., and Carnap, l.c., p. 179). Our science is one of these possible systems. To what distinguishing characteristics does this system owe its being preferred to all 'imaginary sciences' (that are logically equally valid) and being called the 'true' and 'real' science? Carnap and Zilsel agree that this distinction cannot be one of formal logic. Zilsel goes further and says that the characteristic distinguishing real science cannot be encompassed by any statement, only sham propositions are possible here, pointing to this 'unspeakable' though well-known experience of ours, which forms the background on which 'real' science is based while merely conceivable systems lack it (Zilsel, l.c.). On the other hand, Carnap holds that it is entirely possible to make statements about the distinctive nature of 'real' protocol propositions, so that we are not at all compelled to "admit unspeakable features in science" (Carnap, l.c., p. 179). This distinction is possible in 'descriptive semantics' with the help of concepts from real science. By semantics (or 'logical syntax') Carnap understands the theory of the structure of propositions. In so far as we speak of propositions as physical formations at certain places and times, our statements about them belong to 'descriptive semantics', which is thus a part of real science (Carnap, l.c., p. 178). This is not the place for examining the importance and justification of 'semantics' in general, the less so since Carnap says little about it here, holding out prospects for a fuller account elsewhere.

According to Carnap, the criteria of 'real' protocol propositions and of 'real' science are given by descriptive semantics as follows: "real protocol propositions" is what we call "those statements or written records (as physical formations), that are due to any hunan beings, especially the scientists of our cultural sphere. And by 'real science' we understand the system built by (and to be further developed by) these scientists, so far as it is sufficiently confirmed by those protocol propositions" (Carnap, l.c., p. 180).

What seems to us to make this formulation dubious is that any conceivable protocol set is in any case due to human beings in general and also to scientists of our cultural sphere in particular, as soon as these men formulate, state or write down these sets. Likewise, for each of these sets, the scientists can develop systems of propositions that are internally consistent and sufficiently confirmed by their respective protocol propositions. Accordingly, however. any conceivable protocol propositions and sciences would be 'real'.

However, even leaving aside these difficulties, doubts arise in principle against any descriptively semantic characterisation of the 'real' system of science. Descriptive semantics is a part of 'real' science. In so far as it contains protocol propositions, they are 'real' ones that 'really' distinguish all 'real' protocol propositions and therefore themselves as well. This circularity can also be brought out as follows. Suppose that semantics indicates a property K as distinguishing mark of a real protocol proposition. Then one can assert of a protocol proposition p_1 the proposition p_2: 'p_1 has the property K', only if one already knows that p_2 is itself a 'real' protocol proposition, so that one can assert p_3: 'p_2 has the property K'. For if p_2 were merely a conceivable protocol proposition, p_1 too could not be a 'real' protocol proposition, otherwise we should be able to record the 'real' protocol p_2. If, however, p_1 too is merely conceivable, then either p_2 is false or K is not the distinguishing mark of 'real' protocol propositions. With p_2 and p_3 the same problem recurs, and so forth. Thus within descriptive semantics we reach an infinite regress, which always reveals the senselessness of posing the problem concerned. This infinite regress can be cut short only if, with Zilsel and the view of the Vienna Circle to date, we assume that 'real' protocol propositions, in contrast with merely conceivable ones – although merely 'conceivable protocol propositions' are in my view a non-concept, the result of a nonsensical conceptual construction –, somehow picture the 'unspeakable' or 'given' and their truth 'shows' itself in the very fact of our understanding them: they are distinguished by means of an 'unspeakable experience'.

On this point too we find in Carnap remarks that are not immediately compatible with the view here criticized. Thus on p. 180, l.c., he says that the distinction of real protocol propositions consists in their being notes made by scientists of our cultural sphere; moreover, it is an empirical fact of history that an ever more comprehensive and common scientific system is emerging, the product of scientists working with the tools of modern technology. In the correct formal language this amounts to saying that in descriptive semantics there occur propositions that contain certain con-

cepts (including historical-geographical ones). Now it is childishly easy to bring all these propositions into agreement with any other merely conceivable science, by means of formulating suitable hypotheses and protocol propositions. Thus any conceivable science can be formulated by the scientists of our cultural sphere, and so all conceivable sciences would be equally distinguished. Or are the propositions of descriptive semantics that we have mentioned to be understood in terms of their content? Are they perhaps entitled to a 'privileged position' over all other propositions of the system? Are they perhaps true because they describe certain 'states of affairs'? That would bring us dangerously close to sham propositions, to elucidations. In conclusion, it seems to me a grave error to assert of scientific systems (cf. Carnap, l.c., p. 179) that they might have proved themselves relatively to protocol propositions, but nevertheless be merely conceivable, unreal, 'imaginary sciences'. Here 'probation' is simply understood as formal consistency of the propositions, a confusion that bears within itself the seed of all troubles. In my view, probation is the only distinguishing criterion of 'real' science. Briefly summarized, the contrast between relativist physicalism and the traditional absolutist view of the Vienna Circle consists in this: according to physicalism every proposition of science is a truth-function of other propositions of science, while on the old view of the Vienna Circle there are and must be propositions that are not truth functions of other propositions.
(concluded March 1933).

NOTES

References to *Erkenntnis* are abbreviated 'E', followed by the volume number.
[1] Cf. Schlick, 'Positivismus und Realismus', E3 (1932/33) p. 3.
[2] Ibid., p. 13.
[3] Carnap, 'Die physikalische Sprache als Universalsprache der Wissenschaft', E2 (1931) p. 432.
[4] Neurath, 'Soziologie im Physikalismus', E2 (1931) p. 393.
[5] A precise formulation on this point has, however, not yet been given.
[6] Carnap and Neurath, l.c.
[7] Cf. Schlick, l.c., p. 3.
[8] Neurath, l.c., p. 397.
[9] Carnap, 'Über Protokollsätze', E3 (1932/33) p. 215, and Neurath, 'Protokollsätze', ibid., p. 204.
[10] Cf. Carnap, 'Die physikalische Sprache etc.', E2 (1931) p. 437.

[11] Carnap, 'Über Protokollsätze', E3 (1932/33) p. 215.

[12] This question is also discussed in E. Zilsel, 'Bemerkungen zur Wissenschaftslogik' and in Carnap, 'Erwiderung auf die Aufsätze von E. Zilsel', E3 (1932/33) pp. 143, 177.

[13] Cf. Carnap, 'Die physikalische Sprache etc.', E2 (1931) p. 432.

[14] Cf. Carnap, 'Über Protokollsätze', E3 (1932/33) p. 215.

[15] Carnap, 'Erwiderung auf die Aufsätze von E. Zilsel und K. Dunckert', E3 (1932/33) p. 177.

[16] Zilsel, 'Bemerkungen zur Wissenschaftslogik', E3 (1932/33) p. 143.

CHAPTER III

EMPIRICISM AND PHYSICALISM*

The Physicalists' theory of truth, as formulated by Carnap and Neurath, has been expounded by Dr. Hempel[1] in *Analysis* (vol. 2, no. 4). Within the Viennese Circle Physicalism has led to a division of opinion, which mainly finds its expression[2] in the fact that the Physicalists are reproached by their opponents with having broken with the principles of Empiricism. This charge is denied by Carnap[3] and Neurath[3] in their recently published papers; but I do not consider their counterarguments conclusive, and will endeavour to give my reasons for this opinion. This occasion will also serve to deal with Dr. Hempel's theories.

We shall solve our problem by examining whether the mode of speech of Physicalism is compatible with that of empirical science in its present state.

Statements of empirical science are tested by deducing from them propositions in the form of observation-statements and comparing these with observation-statements that have been actually established.[4] In case the compared statements agree, we speak of 'assertion' of the statements which were to be tested. What happens, however, if the compared statements – to put it according to Carnap, the deduced statement in form of a protocol-statement, and the protocol-statement which is actually made – contradict each other? The Empiricist would say in this case that the proposition obtained by observation has refuted the one to be verified, meaning that the statement subjected to the test is to be altered in such a way that its agreement with the statement gained by observation is brought about. In accordance with this point of view the latter must by no means be altered in order to remove the incompatibility, or else the expression 'the statement obtained by observation has refuted the one to be tested' would obviously be inadmissible.

In opposition to this the Physicalists speak of 'refutation' (and accordingly also of 'assertion') in a way which is incompatible with the above mentioned empiricist mode of speech. Carnap, it is true, is also of opinion (l.c., P. 245), that in case of non-accordance it is the 'actually' made protocol-statement that refutes the deduced statement in form of a protocol-state-

* This chapter is reprinted with permission from *Analysis* 2 (1935) 81–92. *Copyright* © 1935 *by Analysis.*

ment, adding, however, that under certain circymstances it might be permissible, to alter also the 'refuting' statement, which was obtained by observation, in order to remove the incompatibility, for there are no strict rules prohibiting the alteration of any propositions.

Here we only wish to assert, that in opposition to empiricist usage Physicalism means by 'refutation' a symmetrical relation in the sense, that either of two contradictory statements may, as a matter of principle, be altered in the same way.

To be sure, we find it hard to understand why Carnap for all that defininely contrasts two kinds of protocol-statements with each other, viz. the deduced statements in form of protocol-statements, and the actually made protocol-statements. A remark to that effect we find in Carnap's book, p. 244: it is for the observing Physicists to determine, which of the statements which have the form of protocol-statements, and whose form is admitted by syntax, should now really be established as true protocol-statements. We might think that Carnap indeed was thus meeting Empiricism half way. As soon as the truth of protocol-statements is determined by observation, there is no doubt about their being qualified to verify or disprove other statements. But then these observation-statements, when they contradict other propositions of the scientific system, may by no means be declared false, for their truth, according to the above mentioned passage in Carnap's book, has already been determined by observation (i.e. by 'actual' establishment), whereas the truth of the other propositions, which, in order to control them, are being compared with the protocol-statements obtained by observation, is still indefinite and is to be tested by that very comparison. But further on (l.c., p. 245) Carnap, as has been mentioned before, expresses opinions, which make it entirely unintelligible, why he lays so much stress on obtaining protocol-statements by means of observation ('actual' establishment). For if the deductions from a theory which is to be tested contradict the actually made protocol-statements, while we, however, do not wish to relinquish the theory, then, according to Carnap we need only change the statement gained by observation into another protocol-statement in such a way that the incompatibility is removed and the assertion of the theory is attained. Why then introduce the observation? If we are allowed to alter the propositions obtained by observation as it pleases us, in case they contradict some theory, then we might as well have formulated them to begin with in such a way that they assert our theory. This proves that the way in which the Physicalists speak about assertion and refutation renders possible the mode of speech, that just as observation-statements are able to refute a system of hypothesis, a hypothetical system is able to refute statements obtained by observation. This is reminiscent of Hegel's dialectical conception of truth and disagrees, in my opinion, with the empiricist mode of expression.

This opposition between Physicalism and Empiricism becomes still more distinct as soon as linguistic rules themselves are dealt with, which according to the opinion of Empiricists forbid the alteration of certain statements.

Observation-statements such as 'I see red and blue side by side', 'in the yellow I see a dark line', 'I feel pain', 'I hear a loud, high note' etc., admit of no alteration according to Empiricism, and for the following reasons: we formulate statements which have first to be tested, with the reservation that we may possibly have made a mistake regarding them. These are all statements of a hypothetical character, for instance: 'in the adjoining room there is a table and three chairs'. It is by no means impossible that we have made a mistake regarding this proposition. If it results from the test that we have in fact made a mistake, then the statement has to be altered, or perhaps we have to reject it altogether.

Now there are statements of which it is impossible with any sense to say, as far as we use the terminology of our empirical science, that it is possible to make a mistake concerning them. These are observation-statements of the above-mentioned kind. When I make the statement 'I feel pain' somebody may doubt my having spoken the truth but under no circumstances can anyone, acquainted with the terminology of empirical science retort that in saying so I have made a mistake. The language of empirical science, as it is taught to-day, excludes any possibility of using such a mode of speech. In stating 'I feel pain' I may have said something wrong, not because I have made a mistake, but because I had the intention of telling an untruth, i.e. because I wanted to lie. With propositions of that sort I assert either a true thing or an untrue thing, both intentionally; of the third possibility, i.e. that I may have made a mistake in these statements, empirical science consequently can make no sense.

While formulating a proposition such as 'I feel pain' I know already whether I have made a true or a false statement. In this respect a mistake being out of the question, as has been said before, there is no reason for reserving a possible alteration of these statements. They are unalterable, for the possibility of error concerning them does not exist, so there is no ground for the possibility of a further test, the result of which would have to decide about the alteration.

This opinion, which is based on the mode of speech of empirical science and which results from the privileged position of a certain class of propositions, is attacked by the Physicalists. For it violates the fundamental thesis of the parity of *all* statements, according to which all statements, without exception, are alterable as a matter of principle. The Physicalists believe that they are able to prove that also such propositions as 'I see blue', 'I feel pain' etc., may as a matter of principle be tested and altered. It will appear, however, that they succeed in proving this only after having relinquished the rules of Empiricism.

Statements of the kind mentioned above, according to Neurath[5] and Carnap[5], must be interpreted according to extreme behaviourism; so that the statement 'I feel pain' has no other meaning than 'I am displaying some outward attitude or other, as for instance moaning, grimacing etc.; some nervous process or other in my body may be noticed besides.' All these characteristics are undoubtedly testable, and according as they are noted or not the statement 'I feel pain' is retained or altered. In this way it is proved that statements previously designated by us as unalterable can also as a matter of principle be altered.

This proof would stand the test in case the equivalence of the proposition 'I feel pain' with the statement that asserts the existence of the corresponding 'behaviorist' attributes was based on logic. But this is not the case: this equivalence is only empirical. According to our experience it is a fact that, when I assert 'I have pain', the relevant 'behaviorist' data are to be observed in relation to me. But I am justified in making the statement 'I feel pain' and in maintaining it as true even if the corresponding 'behaviorist' data cannot be detected. I do not make a contradictory, but a possibly true statement, if I allege that I suffer pain without any visible sign of pain. And therefore it is not the statement 'I feel pain' – we will call it p – that is verified by the second testing of the characteristics, but always another proposition, which was frequently observed, of the establishment of p itself with its outer characteristics. More about this hereafter.

The incompatibility of the empiricist mode of speech and the extreme behaviorist interpretation of the above mentioned kind of propositions can also be shown in the following way. There is no doubt that there are people who when questioned about the behaviorist criteria of pain would not be able to state them. Is it possible that those people, when complaining about pain, do not mean anything with any sense? Or did people mean anything else when they said they were suffering pain in those times when nerves were not yet discovered and nobody knew that nervous reactions were the empirically decisive criteria of pain, than we do to-day? I consider these consequences absurd. But still less acceptable is the following consequence, resulting from the physicalist point of view. Supposing I complain of head-ache, a Physicalist, having examined me, can not find any 'behaviorist' data (nervous reactions included), that he sought. And he says to me: 'Cheer up, you have made a mistake, you have not got a head-ache!' This statement of the Physicalist has no meaning for the Empiricist. This is the best proof, that the Physicalist breaks with the rules of language of empirical science. An Empiricist in the above mentioned case would possibly reproach me with having told him a lie about my pain, but the rules of his language ex-clude the possibility of calling my statement an error.

From this side the Physicalists attack our unalterable statements with another argument, in order to attain the general acknowledgement of their thesis of the equality of all propositions. We shall see, that this attempt also offends against the rules of the empiricist mode of speech.

According to the Physicalists such statements as 'I feel pain', as far as they are generally intelligible, are not only always translatable into state-ments about 'behaviorist' symptoms, but – consistently generalising the ex-treme behaviorist program – the very character of these symptoms is attrib-uted to them[6]; the result being that if such a statement is considered true we mean that the before mentioned linguistic symptoms have appeared to-gether with certain other exterior characteristics. False we call such state-ments, if the appropriate linguistic symptom appears, not with the charac-teristics known to us from experience, but without them, possibly with cer-tain others. Accordingly such a proposition as 'I feel pain' would be false or true in the same sense as for instance the indications of a clock that is a good or a bad time-keeper, which naturally can always be tested and, according to the results of the test are accepted or altered. So the possibility of alter-ing our statements seems to be evident.

This conception is faced by difficulties, the moment it is tested according to the mode of speech of our empirical science. For there is no doubt that by saying 'I feel pain' some may tell a lie. But shall we ever seriously be able to blame a clock, a barometer or thermometer for having lied to us? At best in a metaphorical sense. If we go by a clock, and miss the train in consequence of its false indication, it is certain that we shall not say the clock has told a lie, just as we do not reproach a barometer 'for having lied to us' if it prophesied fine weather whereas storm and wind have set in. The Empiricist will never speak of lies in such a case, it being against the rules of his terminology. For the same reason it is impossible for him in the above mentioned cases to speak of an 'error' on the part of the instruments. Our unalterable propositions for him are not characteristics arising from a certain empirical lawfulness, but signals which are used according to rules clearly defined beforehand. The Physicalist, on the contrary, must in cases of the above mentioned type reproach the instruments in question with a lie or an error; he is forced to do so if he wishes to obey the rules of the language employed by him. It again becomes obvious that the physicalist's reinterpretation of the statements which we have designated as unalterable offends against the mode of speech of empirical science.

But there is always an objection left for the Physicalists, by means of which they deny that the 'unalterable' statements, even if they are logically possible, are qualified to play a part in science[7]. Even if it is impossible for us to make a mistake in such statements as 'I see orange and yellow side by side', 'I have a head-ache' etc., and their truth does not depend on any test, these statements are for that very reason not intersubjectively controllable, but always only for the one person who has made them. But a statement not verifiable for everybody is not generally comprehensible and consequently does not count in empirical intersubjective science. Thus, if I make the statement 'I feel pain' and refuse the physicalist behaviorist interpretation, it is not intelligible for anybody but me. Even if I am already aware while making them whether I am making a true or a false statement, it is only I myself who am sure of it, and there is no method of verification by means of which the truth of my assertions can be controlled by others.

The 'test' of such statements used in practice, which consists in the investigation whether the corresponding 'behavior' characteristics can be observed, is, as has been mentioned before, no verification of the 'unalterable' statements. From experience we know that if a person complains of a headache we can observe in him certain exterior characteristics or nervous reactions. Further we also know from experience, that the proposition 'I feel pain' was in most cases withdrawn as a lie by the person who had asserted it, whenever these 'behaviorist' characteristics were not to be discovered. Therefore the so-called behaviorist test of an unalterable statement inquires if our supposition that the data referred to are there will be proved true or false, and in case they are not if the person that has made the statement will also withdraw it as a lie. This supposition of ours is verified by the behaviorist test; but logically it is impossible for the statement itself 'I feel pain' to be proved false or true, i.e. be verified, for, whatever the result of the test may be, I may after all have told the truth in this proposition. For the truth of this statement is independent of the behaviorist test. If such statements can therefore not be controlled by other persons, on the other hand the person who has made the statement knows 'directly' i.e. independently of any test, whether he has made a true or a false statement.

But it is upon this that the above mentioned objection of the Physicalists is based. If the truth or falsity of unalterable statements are totally incomprehensible for all others. For I can only understand a statement which I am able to verify. If somebody declares he has a head-ache, and I can not prove whether he has told the truth or not, I can not understand at all what he means by this assertion. This objection of the Physicalists does not work either. For we must distinguish: is it only practically impossible to verify a statement, or is it unthinkable, i.e. logically impossible for us to do so? Generally a statement is only unintelligible if it is logically impossible for everyone to test it. As soon as conditions are imaginable, under which the truth of a statement becomes determinable for everybody, the statement is already generally intelligible, even if the fictitious circumstances do not really exist.

Corresponding imaginable circumstances may be quoted for our unalterable statements. Supposing a blind person, whose optic nerves do not react to physical light-stimuli, always had the colour-experience 'blue', when a seeing person perceived the same colour. It is not unthinkable for us that a blind man should experience colours. We often see colours in our dreams, and surely not under the influence of physical light-stimuli. Now, if the person who sees makes the statement 'I see blue', then the blind person is able to test this proposition directly, i.e. independently of any behaviorist test. It would be out of place to object that it is not the 'same' blue the two are experiencing. For it is a matter of linguistic convention when we speak of an experience as being the 'same'. We can easily agree about the assertion that we are able to say of the two above mentioned persons that they experience the 'same' blue. The chief thing, however, is, that in the supposed case the possibility of 'direct' control (control without a behaviorist test) of an unalterable statement even by another person, would be realized. And these relations can naturally be imagined between any other persons, and not only for the statement 'I see blue' but in the same way for all other unalterable statements. Hereby the logical possibility of verifying these statements is proved, and the consequence is that they are generally quite intelligible, and indeed are understood by everyone. Therefore we are not forced at all to replace the mode of speech of empirical science, which does not permit the alteration of such statements as 'I see blue', 'I feel pain' etc., by the physicalist mode of speech which demands the alterability of all statements, and is incompatible with the former, in order to attain general comprehensible meaning for the above mentioned propositions.

Neurath[8] believes he can give examples, in which the alteration of our unalterable statements also becomes, according to his opinion, indispensable. Hempel[9] gives an example of this sort. It is imaginable that somebody protocols the statement'I see this patch entirely dark-blue and also entirely light-red'. No doubt this statement belongs to our unalterable statements, for all that it is indispensable that it should either be altered or scrapped entirely. For it contains a contradiction. By this Neurath seems to have demonstrated that even the statements, which we declare unalterable are modified under certain circumstances, and that they must consequently be alterable. But if we regard this example more closely, we perceive at once that the truth or falsity of this statement also is independent of any test. It is false in every case. He who formulates this statement knows the moment he is doing so that he has made a false statement, and it is logically out of the question that he is mistaken in this. But where a mistake is out of the question in the making of a statement there is no sense in reserving the alterability of the statement. For in such a case I have either purposely made a true, or purposely made a false statement. There seems to be no sense in introducing the mode of speech that we formulate a false statement purposely and and consciously only for the purpose of declaring it false afterwards, i.e. to reject it. If this also is described as 'alteration' quite another thing is obviously meant here than the alteration of an error. The example given by Hempel can not be an error, whoever formulates it must know, in case he understands the language in which he expresses himself, and is fully acquainted with it, that he has made a false statement. That he can later on describe this proposition as false, i.e. scrap it entirely or partially, is a triviality and absolutely different from the alteration of an error, which never is a triviality. And if we have declared certain statements as unalterable, we did not mean anything but that these statements can never be errors and can therefore not be altered like errors. And in this sense also the statement mentioned by Hempel 'I see this patch entirely dark-blue and also entirely light-red' is *not* alterable. Consequently the other examples of Neurath, being of the same kind, do not prove that our unalterable statements are bound to be altered under certain circumstances.

Finally we want to discuss the reflection also submitted by Hempel[10], by means of which the Physicalists try to meet a difficulty into which they are led by their conception. According to the Physicalists all statements are equal in principle, i.e. they are equally alterable. Consequently the protocol-statements which are to verify a scientific system may be altered, or replaced by other adequate protocol-statements, in case they contradict this system. But then we can imagine an unending number of incompatible sciences for which we can construct the corresponding verifying protocol-statements. How can it be explained, or justified, that we in reality consider as true one single system, viz. the empirical science of to-day, and reject as false all other systems which are incompatible with it, although these are equally verified by protocol-statements constructed and added later?

To be able to appreciate fully in what way the Physicalists endeavour to remove this difficulty, we must mention a special postulate of Physicalism. Carnap[11] has formulated the distinction between the formal and the material mode of speech. The always inexact material mode of speech, according to which e.g. statements express facts, may lead to pseudo-problems. So, for instance, when it is said that empirical statements become verified by their comparison with 'facts'. Such pseudo-formulations according to the Physicalists are excluded automatically as soon as we use the formal mode of speech. Accordingly, we do not say a proposition expresses such and such a fact, but that it contains such and such words or expressions. In many cases this transformation may be advantageous, but we shall see directly that it is impossible for the Physicalists to prove the distinction of our empirical science from all other possible systems in the formal mode of speech. And this logical – therefore insuperable – difficulty just at this point seems to me to be fatal for the physicalist conception. For as a result of this the fundamental physicalist thesis of the parity of all statements becomes invalid.

Concerning the special position of 'true' science, as opposed to the merely possible, Hempel, closely following Carnap, says (p. 57): "The system of protocol statements which we call true, and to which we refer in every day life and science, may only be characterized by the historical fact, that it is the system which is actually adopted by mankind, and especially by the scientists of our culture circle..." Here Hempel is speaking of a certain historical 'fact'. Why does he here, as well as Carnap[12] in the same connexion, employ the material mode of speech? That this is not a mere chance is clearly seen, if we transform Hempel's arguments into the formal mode of speech. Instead of speaking of the 'fact' that the scientists of our cultural circle acknowledge the protocol-statements of our scientific system, we will speak of the statements that register these facts, and which accordingly contain the respective terms. Then Hempel's argument is formulated thus: the system of protocol-statements that we call true is characterized by the quality that certain protocol-statements belong to it, which assert that this very system of statements is acknowledged as true by the scientists of our cultural circle. Here the objection arises of itself that an infinite number of systems of protocol-statements which are not to be contradicted might be quoted, all of which contain those particular statements, which characterize as true the system of protocol-statements of our science, but for the rest are incompatible with this system. And all these imaginable systems have to be considered true exactly in the same sense as our true science. If we consistently obey the physicalist postulate, to employ only the formal mode of speech, the argumentation of the Physicalists themselves leads to the very difficulty which it was to remove. In order to avoid this circle, Carnap and Hempel did not avail themselves of the formal mode of speech, which can therefore scarcely be a matter of chance. But if, for this one argument, it is not admitted that it is translatable into the formal mode of speech, then we are no longer Physicalists. For the statements expressing the above mentioned historical 'facts' – in this case it is impossible to express oneself differently – are distinctive statements, which cannot be verified by comparison with other statements, but perhaps even by comparison with *facts,* and are unalterable. This breaking of the Physicalists with their own theses Hempel believes he can explain by the remark (p. 57, remark 6), that Physicalism was not a 'pure coherence-theory', but a 'restrained coherence-theory of truth'. Strictly he ought to call it an 'inconsistent coherence-theory of truth'.

I believe that I have shown that the physicalist mode of speech is incompatible with that of Empiricism. If we strictly adhere to the physicalist theses, we come on the one hand to formulations which the Empiricist eliminates as being meaningless, and on the other hand, while trying to verify the special position of true science in relation to the merely imaginable one, we enter a circle, a pseudoformulation, that can only be avoided by a return to the empiricist mode of speech, which means giving up the physicalist (formal) mode of speech. As a matter of fact, the Physicalists on this point, as we have seen, become untrue to their own mode of speech. The phrases which they employ here are expressions of the early empiricism. We have acquired our formulations by a linguistic-logical analysis, and should therefore like to call the views represented here *'Logical Empiricism'*.

Vienna, May 1935

NOTES

[1] See C. G. Hempel, 'On the Logical Positivists' Theory of Truth', *Analysis* **2** (1935) 4.
[2] See M. Schlick, 'Über das Fundament der Erkenntnis', *Erkenntnis* **4** (1934) 2 and his recently published paper 'Facts and Propositions' in *Analysis* **2** (1935) 5; and B. Juhos, 'Kritische Bemerkungen etc.', *Erkenntnis* **4** (1934) 6.
[3] R. Carnap, 'Logische Syntax der Sprache', Vienna 1934, p. 243ff.
O. Neurath, 'Radikaler Physikalismus und "wirkliche Welt"', *Erkenntnis* **4** (1934) 5.
[4] R. Carnap, l.c., p. 245.
[5] Neurath, 'Einheitswissenschaft und Psychologie', Vienna 1933.
Carnap, 'Die Aufgabe der Wissenschaftslogik', Vienna 1934, 'Psychologie in physikalischer Sprache', *Erkenntnis* **3** (1932/33) 'Die physikalische Sprache als Universalsprache der Wissenschaft', *Erkenntnis* **2** (1931).
[6] Carnap, 'Psychologie in physikalischer Sprache', *Erkenntnis* **3** (1932/33) 140.
[7] Neurath, 'Radikaler Physikalismus u. "wirkliche Welt"', *Erkenntnis* **4** (1934) 5.
Hempel, *Analysis* **2** (1935) 4, 56.
[8] Neurath, 'Protokollsätze', *Erkenntnis* **3** (1932/33) 209.
[9] Hempel, *Analysis* **2** (1935) 4, 53.
[10] Hempel, *Analysis* **2** (1935) 4, 56–57.
[11] Carnap, 'Logische Syntax', Vienna 1934. See also Hempel, *Analysis* **2** (1935) 4, 54.
[12] Carnap, 'Erwiderung etc.', *Erkenntnis* **3** (1932/33) 179–180.

CHAPTER IV

FORMS OF NEGATION OF EMPIRICAL PROPOSITIONS*

1. FORMS OF NEGATION

Negation may be denoted as the most elementary truth function. One readily grasps that systems which take different views on the negation of propositions must issue in quite different kinds of theorems. The deep contrast between logicists and intuitionists is well-known, and their controversy about the foundations of arithmetic turns not least on the question of what rules to obey in using the negation sign.

The logicists hold that the truth value of not-p depends only on the truth value of p. Not-p is thus a truth function of only one variable: $\sim p =_{df} F(p)$. From the truth or falsity of p follows the falsity or truth of not-p.

The intuitionists, on the other hand, declare that not-p is a truth function of (generally) several variables. We are to assert not-p only if we can indicate at least one true proposition q different from p and not-p (that is, not equivalent to either), which must moreover be incompatible with p. Here, the truth value of not-p thus depends not only on p but also on the truth value of at least one other proposition q: $\sim p =_{df} F(p, q)$. For example, in order to prove the falsity of the proposition not-p, let us say "the number π nowhere contains the sequence of digits from 0 to 9", it would on this view not be enough to prove this to be incompatible with the axioms of arithmetic (that is, logistically speaking, the truth of p), but we must indicate, by means of a proposition q, where precisely in π a sequence of this kind occurs.

On this view it remains an open question whether for each negation at least one q, not equivalent to p or to $\sim p$, must belong to the argument values of the truth function not-p. We shall not here pursue the relevant discussions of the mathematicians. What interests us is which of these two forms of negation is customarily used in the real sciences, that is, with empirical propositions. At first sight it might seem as though this could only be the logicist form of negation, and this certainly applies in many cases.

* Translated from the German original 'Negationsforman empirischer Sätze', first published in *Erkenntnis* 6 (1936) 41–45.

However, there are empirical propositions – and precisely these have a very important role – that are negated exclusively in a sense that is at least cognate with the intuitionist view, in which particular of their use they differ from all other propositions.

2. 'COMPLETE' AND 'INCOMPLETE' CONTRADICTIONS

Let us first explain a number of concepts. By truth function (tf) we understand a proposition whose truth value can be determined if the truth values of some other propositions are known, which latter we call the arguments of the tf; according to their number we speak of a truth function of one or more variables. The relation of arguments to tf is in general many-one. To a definite combination of truth values of the arguments there always corresponds only one truth value of the function, while to one value of the function there may correspond several combinations of the argument values. This relation becomes one-one for truth functions of only one variable, for example logicist negation, where from the truth of not-p it only ever follows that p is false. If now in a system there arises such a contradiction in the logical sense, it can be mended only by cancelling one of the propositions, just because the truth value of either depends only on that of the other. In such a case we shall speak of a 'logicist' or 'complete' contradiction.

It is obvious that 'contradiction' and 'incompatibility' change their meaning if the negation sign is used in a sense other than the logicist one. If for example the truth value of not-p could not be inferred even when the truth of p in a certain system leads to a 'contradiction', this last must here mean something other than a logicist or complete contradiction. The truth value of 'not-p' in this case no longer depends only on the truth value of p, but would require for its determination our knowing the truth value of at least one other proposition, as explained by the example of intuitionism in Section 1. On this view negation is thus a truth function with more than one argument. It would be advisable in such cases to use a sign different from that for logicist negation ($\sim$), perhaps the intuitionist $\ulcorner$. However, it is important that in so doing we are already speaking of 'contradiction' before the truth value of not-p is fully decided. Thus intuitionists, for example, say that from the contradiction of a proposition p with the propositions of arithmetic it does not in general follow that not-p is true (because we do not yet know the truth values of propositions q, r and so on, that are arguments of the tf not-p in such a case). Thus we here speak of 'contradiction' although not all the free places of the tf not-p are occupied. In contrast with complete, logicist contradiction, let us call the present one 'incomplete'. An incomplete contradiction assumes the character of a complete one, if the incompatibility continues to exist when all free places of the tf not-p are filled by appropriate argument values. If, however, an incomplete contradiction vanishes in such a case, it proves itself to have been merely 'apparent'. In an anologous sense we speak of complete, incomplete or merely apparent incompatibility.

From this it becomes clear what is to be understood by the 'removal' of a contradiction. Complete contradiction can be removed only by arbitrarily dropping one of the two contradictory propositions. It is here impossible that an error was made and that no contradiction was involved. By contrast, this is always possible with incomplete contradictions; and precisely because here one might always learn that a contradiction had been hastily and 'erroneously' assumed (that is, incomplete contradiction may turn out to be apparent), it is inadmissible to set this possibility forcibly aside by striking out a proposition. Thus a complete contradiction is removed by arbitrary cancellation, while with incomplete ones we must first fill the free places of the 'contradictory' truth functions and only then will it be decided whether the contradiction is after all complete and we may cancel, or whether the incomplete contradiction dissolves into an apparent one.

3. THE OBTAINING AND VERIFYING OF PROPOSITIONS

With propositions that are not analytic, two points must be watched: the obtaining and the verifying of them. The two are so closely linked that one might easily believe oneself to be concerned with only one single factor governing propositions. However one recognizes that there are here two different things after all, as soon as we take into account how differently they are connected in different propositions.

With certain propositions, the obtaining and verifying of them are necessarily separated in time. Such propositions are always obtained by logical operations in the most general sense, even if it is merely by collocation of signs according to logical rules. Following Carnap, let us call these propositions system propositions. The way they are obtained by logical operations

further points to how they can be decomposed, transformed, in short, reduced to other propositions. Their truth value can always be inferred from
the truth values of other propositions; in other words, system propositions
are always truth functions of other propositions. Let us provisionally assume that the reduction always issues in propositions whose truth or falsity
is already known or of which we otherwise know how we must determine
their truth value. Then we can say that in obtaining system propositions we
are also shown the way in which they can be verified, which always
amounts to comparing of propositions with each other (to see whether they
are compatible). If the way to verify a proposition is known to us, then we
understand the proposition. Accordingly we understand system propositions before we have verified them, before we know whether they are true or
false.

Besides system propositions there are, however, propositions in which
obtaining and verifying them coincide, namely those not obtained by logical operations. They are never obtained by transforming other propositions nor therefore reducible to them. Their truth values are in principle not
inferrable from those of other propositions; in other words, they are not
truth functions of other propositions. Let us call such propositions 'constatations'. The term 'protocol' would be quite suitable too, but the physicalists have recently given it a sense according to which 'protocol statements'
must also be counted amongst truth functions, so that there is no further
difference between protocols and systems propositions.

Against this view, we understand by constatations propositions that are
in principle different from system propositions. Constatations are true or
false independently of whatever truth value belongs to any other propositions. According to the initial terminology of the Vienna Circle one might
call constatations statements about the 'given', picturing the structure of
the 'given' and being true if they have the same structure as the 'given'.
What is certain is that constatations are not verified in terms of comparing
propositions for compatibility. Indeed, when verifying constatations we
not only do not refer to any other proposition but in a sense one can say
that constatations need not be verified at all, since at the moment when
we obtain constatations we already know their truth value as well,
for obtaining and verifying such propositions coincide; that is, at the
moment when we understand them we already know whether they are true
or false.

4. NEGATION OF CONSTATATIONS

However, at this point a difficulty arises. What happens when we negate a constatation? Negation is a logical operation and we said earlier that constatations are not obtained by logical operations. Is 'not-p' then no longer a constatation, even if 'p' is? Moreover, it surely seems that we can infer from a constatation's truth value that of its negation and vice versa. Hence with constatations, too, the negation sign is used in the logicist sense and every constatation must be taken as a tf of its negation. Therefore it seems that there can be no propositions that are not truth functions. It is probably these reflections that finally moved physicalists to deny any difference between 'protocols' and system propositions. The difficulties to which their extreme position leads I have discussed in an essay in *Erkenntnis* 4, p. 397[1], and I will here make some additional remarks.[2] However let us first examine what are the conditions for 'constatations in negated form' to appear.

It can be shown that if the negation of constatations is logicist in kind, so that the negation of a constatation p is a truth function of p alone, then it is impossible that there should be any propositions that are not truth functions. For in that case incompatible constatations p_1 and p_2 are necessarily reducible to the form of a contradiction, say $p_1 . \sim p_1$. From this it follows that p_1 itself is a tf of p_2 and the assertion that constatations are not tf's of other propositions false. Constatations too could then always be reduced to other propositions by means of logical operations, a process that it would be arbitrary to inhibit. Thus the main thesis of physicalism would remain correct, namely that the verification of any proposition occurs by comparison of propositions with each other to ascertain consistency. To what circular formulations this view leads I have discussed in the essay cited above. Having thus clarified one of the paths that led to physicalism, we now must discover the source of error which in our opinion sustains the erroneous physicalist-relativist view.

In descriptions that merely constate ('protocolize') what we observe, what strikes us is that no negations – that is, propositions of the form 'not-p' – occur amongst the descriptive propositions. This would be rather strange if constatations could be of the form 'not-p' as well. A simple example, however, shows that it is no accident if 'negated constatations' never occur in descriptions. If, for example, 'not-p' means 'this is not red', this assertion presupposes at least the holding of one constatation q (let us say 'this is blue') that we must be able to indicate and that is incompatible with *p* but not equivalent with 'not-*p*' and therefore cannot itself have the form of a negated proposition. It is now obvious at once that, since not-*p* can be asserted only on the basis of *q*, the description will contain only *q* while not-*p* is quite irrelevant to it and can occur only in a theoretical system constructed by logical operations starting from the description containing *q* and *p*. From this we can see that the negation of a constatation cannot be a truth function of only one variable.

Since *q* is incompatible with *p* and not equivalent to not-*p*, no formal logical operation can yield *q* from *p* and not-*p*. This is indeed the main contrast between the logicist and intuitionist forms of negation. However, while in mathematics the source from which we obtain proposition *q* – which is of course an argument of the tf not-*p* – may be problematic, in the case of non-analytic propositions we know a suitable non-problematic source, namely experience. We thus recognize a certain similarity between intuitionist negation and the negation of constatations. Let us examine the latter somewhat further.

If *p* is a constatation, then not-*p* has been recognized to be a function of at least one other proposition *q* incompatible with *p* but not equivalent to not-*p*. In intuitionist mathematics *p* counts as proved false if we succeed in indicating a suitable true proposition *q*. If for example the proposition 'in the number π a sequence of digits 0 to 9 begins at the 1,000th decimal place' were to turn out true, this would prove the falsity of the proposition *p* 'the number π contains no sequence ot the digits 0 to 9' and not-*p* would hold. In this respect negations of constatations behave differently, even if first appearances speak against it.

Let *p* once more stand for 'this is red' and *q* for 'this is blue', which is incompatible with *p* but not equivalent to not-*p*. Then it seems as if with *q* true, *p* is proved false. However, this need not be so. For if a third proposition *r*, let us say '*p* was found at 7 o'clock and *q* at 8 o'clock' is true, then we cannot infer from *q*'s being true that *p* is false. It is a matter of principle that the 'incompatibility' of constatations is never – in contrast with contradictions in logistic systems – 'complete', requiring the cancellation of one of the mutually contradictory propositions, but that we can always indicate other propositions which, if true, show the incompatibility of the original ones to be only apparent; and this possibility is unbounded.

From this insight it follows that contradictions between constatations are always incomplete. Normally we reduce contradictions to the form p & not-p. Since for constatations contradictions are always incomplete, it follows that the negation of p (or of any constatation) is always a tf of several variables.

Moreover, it is important that contradictions between constatations can never become complete: however many free places we fill with argument values, it always remains possible in principle to find new propositions (argument values) and thereby recognize that the contradiction concerned is only apparent. At the same time an incompatibility between constatations can never, with absolute certainty, be proved to be merely apparent, for it always remains possible that propositions are found which, when put into the free places of not-p, will show the incomplete contradiction to exist after all. This is what is meant by the often stated but imprecise saying that empirical propositions – whose truth values are based on constatations – can neither be completely verified nor completely falsified. However, the circumstances discussed could not exist if negations of constatations were tf's of only finitely many variables. For if they were, then after filling the finite number of free places it would have to be possible to decide with absolute certainty that a contradiction between constatations is complete or apparent. Since, as we saw, this is in principle impossible, and in any given case we can without limit indicate further propositions on which the truth value of a given negated constatation (not-p) depends, these rules for treating constatations and their negations entitle us to assume that the latter are tf's of infinitely many variables. Thus the truth value of the negation of a constatation always depends on (p and) the truth or falsity of an unbounded number of propositions none of which are equivalent to each other or to p or not-p.

To clarify this very important circumstance concerning constatations and their negation, let us summarize: our initial thesis that constatations are not tf's of other propositions is proved by the fact that the 'incompatibility' of constatations is always in principle an incomplete contradiction, and never one in the logicist sense. The negation of constatations accordingly is not logicist (as for system propositions) but, like intuitionist negation, a tf of several variables, indeed of infinitely many.

This contrast with system propositions marks the special position of constatations. The truth of a constatation shows itself in the structural sameness of proposition and 'the given'. If a constatation is 'false' – that is, negated –, we no longer have a constatation but a system proposition. The negation of a constatation is therefore no longer verified like a constatation but like any other system proposition, by reduction to other propositions. Because of the infinitely many arguments, such a verification of negated constatations must of course remain in principle incomplete.

Let us further remark that attempts to verify system propositions (by reduction to other propositions) always lead to propositions that are compared as to compatibility with constatations, whose truth is guaranteed independently of any reduction by the way they are obtained. Complete contradictions can be removed only by cancelling one of the propositions in contradiction, as we said. Incompatibilities between constatations are never complete and to try to remove them by this same method would be inadmissible, as mentioned before. For the removal of a contradiction must, so to speak, be 'adequate' to its origin, that is, to the way the propositions concerned were obtained. Formal (complete) contradictions arise through arbitrary stipulations or formulations that are inconsistent and may therefore be removed by arbitrary changes of propositions, that is cancellation. With constatations, however, incompatibilities never become complete and arise independently of any arbitrariness or logical operations, so that we must never remove them by arbitrary cancellation, but by trying to introduce new propositions whose truth values are decided by new constatations not arbitrarily obtained. It is a peculiarity of constatations that this is always possible.

Our language is so arranged that we cannot see from the external form of a proposition whether it is a constatation or a system proposition. For any proposition obtained by observation there is always a duplicate obtainable from other propositions by logical operations, such as deduction. There are good reasons for this. That we can obtain duplicate propositions by different methods must be presupposed if we are to be able to 'trace back' system propositions to constatations. For from the system propositions to be verified we derive propositions that are duplicates of constatations with which they may then be compared. The point of verifying system propositions is to ascertain agreement between propositions obtained independently of each other and by quite different means (by logical operations or by observation).

5. INCOMPATIBILITY OF 'PROTOCOL STATEMENTS'

The view expounded above contrasts with that of the physicalists. As we have said, their orientation starts from the thesis that verification always consists in comparing propositions with each other, and thus comes to proclaim that all propositions have equal status. The only criterion of truth remaining on this view is logicist consistency. The same result is obtained – as our enquiry showed – if we assume that the negation of any proposition whatever is a tf of only one variable, thus being logicist in form. Removal of a formal logicist contradiction occurs by arbitrary cancellation of one of the contradictory propositions. This equal status of propositions is lost if the negation of certain propositions is not logicist in form and contradictions between such propositions no longer arise from arbitrary inconsistent stipulations, nor therefore may be removed by arbitrary cancellation. The group of such propositions would then seen to be privileged over all other propositions, in the sense that they are exempt from cancellation in case of contradiction. These propositions would thus constitute a (constantly growing) basic reservoir of unmodifiable statements authoritative for the propositions of systems built up on it. In constatations we have actually discovered propositions whose negations are tf's not of one but infinitely many variables. The always incomplete contradictions between such propositions arise not from stipulations nor therefore may they be removed by cancellations.

In contrast, the physicalists, if they wish to aknowledge all the consequences of their position, are forced to prove that the above-mentioned possibility of using incompatible constatations consistently is not a matter of principle, so that one could indicate cases where removal of a contradiction unavoidably requires cancellation of a constatation; which would at the same time prove – though perhaps less recognizably so, because of the untransparent structure of the propositions concerned – that negations of constatations as of all other propositions, do have the logicist form after all and contradictions between constatations are always complete.

This, I think, is what Neurath tries to prove by means of the example he gives in *Erkenntnis* 3, p. 209ff. He there attacks an older (and since altered) view of Carnap's, according to which protocol propositions stand in no need of probation and can be used consistently even when 'incompatible' and therefore must never be cancelled. Carnap there (*Erkenntnis* 2, p.

432ff) still takes 'protocol proposition' roughly in the sense of our 'constatation'. Neurath comments (l.c.):

the fate of being cancelled can strike a protocol proposition too. No proposition can claim a 'noli me tangere', which Carnap decrees for protocol propositions. Take a particularly drastic example: suppose we know a scholar named Kalon who can write simultaneously with both hands. Now let him write with his left hand: 'Kalon's protocol at 3.17 pm [Kalon's speech thought at 3.16 pm and 30 seconds was (at 3.16 pm the only thing in the room was a table perceived by Kalon)]', while with his right hand he is writing: 'Kalon's protocol at 3.17 pm [Kalon's speech thought at 3.16 pm and 30 seconds was (at 3.16 pm the only thing in the room was a bird perceived by Kalon)]'.

Neurath thinks that not both of these two 'protocol propositions' can be used in one and the same system, and that at least one of them must be cancelled as false. We shall not investigate whether propositions of the form of Neurath's examples may be called 'protocol propositions', rather we assume with Neurath that propositions occuring in protocols can be represented as such peculiar propositional incapsulations. Nor will we consider the following no doubt noteworthy objection that might be raised against Neurath's argument from a logicist point of view. For we might a priori accuse the person who simultaneously records two contradictory propositions, without considering their content, of not understanding the language he is speaking, so that what he has noted down are merely senseless sequences of signals but not propositions. Or if we do not wish to deny him the ability to make sensible records, we should have to explain that he speaks a language unintelligible to us. In either case it would be impossible to communicate with one who is in the habit of making such peculiar records.

However, as we have said, we do not wish to use these objections, though they are certainly not altogether unjustified from a logicist point of view; rather, let us ask ourselves whether the incompatibility of the two propositions in Neurath's example can be removed only by arbitrary cancelling one of them.

On looking more closely we see that this view is erroneous. In particular from a radical physicalist point of view the incompatibility of Neurath's two protocol propositions can easily be removed otherwise than by cancelling. We need only 'formulate' certain other propositions whose validity shows up the incompatibility of the two protocol propositions as merely apparent, so that they might be used consistently. Thus it does not appear from the conditions Neurath gives for the taking of these protocols, nor indeed from the propositions themselves, whether the writing down of them actually occurred at the time mentioned in them, that is, immediately after Kalon's obeservations, or whether Kalon first made a mental note of his observations (that is, Neurath's protocol propositions) to write them down later. If we assume this we are enabled further to assume that Kalon for example records with his left hand observations separated by 24 and 48 hours and so on from those simultaneously recorded with his right hand. On these assumptions, which are quite compatible with Neurath's conditions, we can already see that the two protocol propositions can be used consistently. However, if one wishes to refrain from these assumptions and the noting down of the protocol propositions is to occur immediately after observation, this still by no means excludes that we might remove the incompatibility, which is indeed an incomplete one. For the conditions mentioned by Neurath are compatible with the assumption that Kalon in taking the protocols was in a position from which he could look at two rooms simultaneously, recording with his left hand what he observes in one and with the right hand what he observed in the other. The possibility of consistenly using both protocol propositions on such an assumption is immediately clear. The assumption that both propositions express the same thing, only in two different languages, is also possible. We must, however, not imagine that by excluding this least-mentioned assumption as well, the incompatibility between Neurath's protocol propositions becomes complete and removable only by cancelling one of the propositions. However precisely we fix the circumstances under which the protocols are taken – and Neurath obviously thinks that by doing this as precisely as possible he can make consistent use impossible, that is, prove the incompatibility to be complete –, we can still in principle make infinitely many assumptions enabling us to range the incompatible protocol propositions consistently into a given system. (It is of course inadmissible to stipulate by definition that any possibility of consistently using the incompatible protocol propositions is to count as excluded. This would turn protocol propositions themselves into stipulations, depriving them of their character and meaning as protocols).

On the basis of these unlimited possibilities we recognize that the negations of constatations have the preculiar property of being tf's of infinitely many variables. We therefore think that Neurath's example does not prove that the incompatibility of protocol propositions has the character of a complete contradiction in the logicist sense, removable only by cancelling one of the propositions in contradiction.

This peculiarity of the form of negated constatations moreover appears to justify their special status which the physicalists wrongly dispute.

POSTSCRIPT

The above essay was completed in February 1934, and handed to the editor. Meanwhile *Erkenntnis* has published two discussions that touch on the subject of my essay in several points.

Thus M. Schlick in 'Über das Fundament der Erkenntnis', (*Erkenntnis* 4, 2) uses the term 'constatation' for propositions that are partly used according to similar rules as those propositions I have here also called "constatations".

Closer still is the affinity of my arguments with the views put forward by K. Dürr in 'Die Bedeutung der Negation, Grundzüge der empirischen Logik' (*Erkenntnis,* 5, 4).

This paper appeared almost eighteen months after I had submitted mine, so that I could obviously not refer to it. With Dürr I hold that the negation sign liked with constatations (which he calls 'original' propositions) is used according to special rules. He too thinks that constatations (or original propositions) are the only empirical propositions that are non-hypothetical and cease to be so as soon as they are negated, when they cease to be constatations but become a certain function of constatations. Certain differences between our views can of course not be concealed. Thus Dürr defines the concept of the 'true' in a way that makes it impossible to call his 'original propositions 'true', though they might be 'false' for him under certain conditions (pp. 217–219). Here our views differ, as also when he calls the logical system of the logicists a 'metaphysical' logic, because the rules according to which they use the negation sign often do not admit calculus treatment of empirical (and above all 'original') propositions with the help of the logicist system of signs. In contrast with 'metaphysical' logic, he calls his own suggested system 'empirical' logic (because it is suitable for the treatment of empirical propositions). This opposition seems to me infelicitous: firstly, because the two logical systems differ only by different formal rules of use for their signs, which is a matter purely internal to logic; and secondly, because the system of logistics is in many cases eminently suitable for the treatment of empirical propositions, indeed, for physical calculations we often enlist parts of mathematics that are readily based on mathematical logic, while the use of empirical logic would here be likely to encounter difficulties of principle.

NOTES

[1] This volume, Chapter II, p. 16 above.
[2] Cf. also the author's 'Empiricism and Physicalism', *Analysis* **2,** No. 6 (1935), 81-92; this volume, p. 36.

CHAPTER V

PRINCIPLES OF LOGICAL EMPIRICISM*

1. REALITY CONCEPTS, METAPHYSICAL AND EMPIRICAL

In criticizing the logic of philosophical realism, philosophers have gradually developed positions whose interrelations may be described as follows.

Metaphysical realism distinguishes two or more 'realities', perhaps best
called 'appearance' and 'reality' proper; the former is the world of sense,
real in its own way, but taken to be transitory, changeable, subjectively dependent and so on, which cannot be the case for reality proper, the un-
changeable world of things-in-themselves.

This kind of dual account has been rejected both by empiricist and by
positivist critics. For them, statements about the real world or real events
are tested by empirical facts (sensations, perceptions, feelings and the like)
that belong to the world of appearance. What cannot in principle so be
tested, cannot be a statement. To make sense, we must therefore speak only
of one reality. This new position we call 'empirical realism'.

The form of empiricism to be outlined here is based on logico-linguistic
investigations and emphasizes the distinction between 'language' (logic)
and 'reality': empirical statements are taken to assert something about
'facts', which is to be verified ultimately by comparing the statements (or
their implications) with 'reality'. This view we have called 'logical empiricism'. At the International Congress for Unity of Science (Paris 1935), this
term was used indifferently for all the various points of view represented
there, and some of them were incompatible as the reports show[1]. By contrast, we shall compare it with the 'physicalist' theory of science put forward by Carnap and Neurath[2].

'Physicalism' has lately tended to go even further than empirical realism.
The physicalists hold that to speak of language and its statements, as opposed to facts and reality, can easily mislead us into constructing metaphysical pseudo-problems. Consider the following example: an observer
who wishes to test the truth of statements by comparing them with reality
can do so only in terms of 'his own reality', namely his own experiences; this

* An earlier translation of this article appeared in *Mind* 46 (1937) 320–346. *Copyright* © 1937
by Mind.

excludes intersubjective verification, our statements will be unintelligible to others and a universally intelligible empirical science would be impossible. We must therefore express the matter differently and speak only of language but not of a 'reality' as opposed to language. According to the physicalists, a trace of 'metaphysical realism' remains wherever this distinction survives[3]. This structure does not apply to 'logical empiricism', as we shall show presently. 'Reality' can mean only the world of experience: we need not accept a multiplicity of realities. In several articles I have criticized the physicalist theory of science in terms of logical empiricism[4]. Here, I shall follow the opposite course: first, we shall describe the logical empiricist position and then present objections to it by the 'relativists', who do not contrast 'language' with 'reality' but speak only of language and statements.

2. INVARIANT SYSTEMS OF STATEMENTS

Statements are derivable from physical theories roughly confirmed by experience, and may be classified into those that do and those that do not agree with observation, and those not yet testable by observation because of practical difficulties.

If a physical theory T_1 is given up in favour of another, T_2, because, as the phrase goes, it 'gives a better description of the events', then the two theories are related as follows: all statements derived from T_1 and confirmed by observation must be translatable into statements that can be derived from T_2; failing this we have not given up T_1 to describe the physical occurrences, but use it in addition to T_2. Moreover, it must be possible to deduce from T_2 statements that agree with observation but cannot be transformed into statements that have been derived from T_1. Statements thus derived from T_2 may be incompatible with certain statements derived from T_1, but it is also possible that neither they nor their negations contradict T_1. At all events, the system of statements deduced from T_1 and confirmed by observation must be transformable into a sub-class of the statements derived from T_2 and likewise confirmed. Similar if T_2 is superseded by T_3, and so on. The statements that are derived from a T_n and agree with observation therefore constitute an invariant group K_n, and from any of the T_{n+i} that supersede T_n in the development of theory, it must always be possible to deduce a group of statements isomorphic with K_n.

However varied the language used in the individual theories T_n, involving important changes in linguistic signs and rules, always remember that T_n must imply a system of propositions isomorphically related to a complex of statements derived from T_{n-1} and confirmed by observation. When formulating a new theory we are therefore not entirely free to choose our language and axioms, since certain parts of the new theory must be translatable into parts of the theory that it replaces, and conversely: the new theories and their language must allow a transformation that leaves the observationally confirmed statement systems of the old theory invariant. This fact, vital in the construction of theories, is one of the reasons why logical empiricism distinguishes between language and reality.

When one tries to speak only in terms of language and its formal characteristics, it soon becomes clear that we can give no reason why the physical theories that supersede each other as described above must be connected by invariants. Formally, everything in language is arbitrary: we fix the rules with a view to the ways the signs are used, and transformation rules for complexes of signs are chosen arbitrarily, as indeed are the axioms. As regards validity, analytic statements are either consistent, or contradictory, or analytically non-determinable. (Other degrees of validity, such as are found in so-called many-valued logics, need not be considered here.) Since all statements derived from a certain theory are formally equally valid, we cannot determine by means of formal criteria which partial system of statements should remain invariant when they are translated into other theories. The logical empiricist does not make the mistake of trying to find such a formal criterion. For him, the statements derived from a certain theory are compared with 'reality'. those that agree with it must be represented by a group of statements isomorphically related to the corresponding group in any new theory.

The physicalists, on the other hand, here prefer not to speak of 'reality', 'facts' and so on, and their reason is this: as they develop, physical theories often change so radically that different theories would have to be taken to correspond to different realities; in giving up one theory for another we should have to say that the former did after all not describe 'true' reality, as when we give up Newtonian mechanics in favour of quantum mechanics. For Newton, reality consists of particles in definite places with definite momenta, while quantum theory does not admit talk of such 'realities' in the case of micro-processes. If then we try to correlate language with reality, we are forced to speak of many 'realities', one 'true' and the others 'seeming' or 'less true'; in short, to use expressions that are entirely metaphysical.

The claim that because physical theories differ in language we must recognize different realities rests on a misunderstanding of how laws of nature are used in physics. Physical laws are represented by functions. Values derived from such functions never describe observable and measurable quantities exactly: every calculated value Z and the corresponding methods of measurement have a range of inaccuracy R, and all measured values that fall within this range are taken to agree with Z. (Such a range exists also for the magnitudes that are inserted for the variables of functions. They characterize the so-called initial state; thus we likewise make errors of measurement when trying to determine or test these states.) The range R indicates the inaccuracy of measurement, the 'limits of error', which depend on our methods of measurement (instruments, and so on). One can imagine circumstances where measured and calculated values are so related as logically to exclude the use of formulae (and the physical concepts occuring in them): in such cases we can no longer speak of 'natural laws'.

It may happen that a formula yields differences in value so slight that we lack a sufficiently refined method of observation to test them. In that case the attendant error ranges are so large as always to include many values calculated from that formula. Such hypothetical 'single processes' in principle escape testing, unless we can construct sufficiently discriminating instruments. Failing this, it would contravene the customary terminology of physics to go on calling such formulae 'physical laws'.

An example from modern physics will make this clearer. Nothing observable corresponds to the formulae for individual de Broglie waves: only the mean squared amplitude of the waves, interpreted as the probability of a corpuscular distribution in a certain region of space, can be tested by observation. However, the several individual waves with their different phases are by no means exactly defined by the mean squared amplitude. What can be calculated from the formulae is a set of waves, all equally possible, but there are no criteria of observation (and, since the velocities figuring in the functions exceed that of light, no conceivable measurements) that could tell us which particular individual waves (from amongst the calculated ones) are to be correlated with the physical initial state of the observed distribution. Putting it differently, many mathematically deduced initial states correspond to each observable inital state. The former are 'unreal', they cannot be determined, measured or tested by observation. If de Broglie's theory nevertheless ascribes certain attributes (such as velocity, direction, phase-differences) to each individual wave, this cannot amount to asserting physical laws or making statements about facts.

Heisenberg has put forward a theory to cover the same field as de Broglie's theory but dealing essentially with quantities that can be verified by observation. Concepts such as 'individual wave' or similar unverifiable and hypothetical 'processes' do not occur in Heisenberg's formulation. Yet physicists do not assert that the two theories describe two different 'realities'. On the contrary, they accept as a matter of course that both theories describe the same events, precisely the same reality, the more readily so as in this case the two theories are 'mutually translatable': all statements derived from the one and confirmed by observation can be expressed in terms of the other, and conversely. What interests us here is that where de Broglie formulae are unrelated to any observable facts and do not correspond to any Heisenberg formulae, physicists speak of the former as purely formal linguistic peculiarities; such as might for instance be found in comparing German with Latin. The former contains both definite and indefinite articles, the latter does not: just because this is only a difference in linguistic signs, we can assert that empirical statements expressed in German describe the same 'reality' as their Latin translations.

Here we are in no doubt as to contrasting language with reality, and asserting that both theories describe the same reality. What, however, are we to say when two theories are related as follows? A number of statements are deducible from theory T_1 and confirmed by observation. Then T_1 is superseded by a theory T_2 that gives greater precision to the values derived from T_1 and confirmed empirically; besides, T_2 enables us to derive additional statements that conform with reality but are not deducible from T_1. Newton's mechanics and quantum mechanics may serve as an example.

Here, it seems, we cannot say that the two theories describe the 'same' reality. The mechanics of Newton speaks of masses and their space-time relationships. These laws were at first considered valid not only for dimensions of a certain magnitude but for particles (masses) and space-time quantities of any size whatever. This view was, however, disproved as soon as atomic processes became accessible to observation and measurement. Thus, for Newton, continuous change in the distance between masses produces continuous change in their mutual attractive force, in their accelerations and so on, and conversely. In atomic processes, continuous variation in distance between corpuscles is ruled out in certain cases, where we can have only discrete changes in energy (by finite amounts) producing discrete changes in energy levels.

Even greater difficulties seem to arise from the fact that Newtonian and quantum mechanics differ in the way they use terms like 'mass', 'particle', 'corpuscle'. In classical mechanics, bodies have a certain 'mass'; every body is in a definite place at a definite time and has a definite momentum. These attributes were taken as verifiable, measurable quantities. At the same time, it was assumed that these attributes were always measurable whatever the order of magnitude, and that micro-processes could therefore be described in terms of mass and corpuscles of the same properties as bodies in macro-processes. More accurate instruments of observation revealed not only that Newton's mechanics were inapplicable to corpuscular processes. but also that there was no use for phrases like 'corpuscles at a definite space-time location and having a definite momentum'. Moreover, quantum mechanics claims to be valid not only for micro-processes: some of its laws are applicable to macro-processes. The deviation of these laws from classical mechanics leaves measurement unaffected, but does produce calculable mathematical differences.

If nevertheless we wish to cling to the contrast between language and reality, how can we avoid saying, in cases such as these, that classical mechanics and quantum mechanics describe 'two different realities'? Both theories are based on experiment and observation, therefore both describe 'reality'. On the other hand, the more recent theory rules out the use of classical terms in certain cases; once quantum theory had been established, results that concerned micro-conditions and relied on Newtonian calculations could not even be called wrong: they simply no longer made sense. If, then, we are to speak of reality at all, must we not distinguish between several 'different' realities, or say that one theory represents a 'true' reality and the other an 'untrue' one?

These objections to logical empiricist formulations in my view rest on misinterpreting the syntactic rules for using laws of nature. More specifically, if we replace Newtonian by quantum mechanics, does this invalidate statements derived from classical theory but sill confirmed by observation? Certainly not. These laws and statements let us call them N_i are still recognized by physics. On the other hand, there must be statements Q_i deducible from quantum mechanics and confirmed by the same observations as the N_i, otherwise one could not give up Newton's theory in favour of quantum mechanics. How are the Q_i and N_i related? Do they denote different realities, even though their validity is tested by the same observations? In analysing the relation between Q_i and N_i we actually provide the answer to the essential part of our question. For the fact that from quantum mechanics we can deduce certain statements not obtainable from classical theory is irrelevant to the logical question involved here, namely whether, in comparing the two theories, we are forced to speak about 'different realities', as described above. The classical theory makes no statements about the facts in question. Likewise irrelevant are those cases in which classical mechanics has given us statements that can be tested but have proved to be wrong, whereas statements in quantum mechanics for the corresponding facts do agree with observation. Here one can simply say: each theory in its own way describes the same reality, but the calssical account is wrong.

What, then, about the statements N_i and Q_i? By hypothesis, both types of statement have been confirmed by experience, and indeed the same measurements, observations and experiments confirm a N_i as well as the corresponding Q_i. Does this circumstance by itself give us the right to say that both sets of statements describe the same facts, the 'same reality'? One might object on the ground that the two theories use different concepts and that this linguistic difference persists even where two corresponding statements are confirmed and tested by the same observations. Yet clearly mere differences in language cannot force us to say that the two theories refer to two different realities. If in every case two corresponding statements can be confirmed and tested only by the same observations, then the one must be a translation of the other; that is, both agree with the same facts. According to our assumption, it must be possible (as is indeed the case) to translate the statements in Newton's mechanics that are confirmed by observations into statements of quantum mechanics having exactly the same meaning. Hence the two theories refer to the same facts. Here, too, we need not speak of 'different' realities denoted by the terminologies of the two theories.

Another important case is the following. It may happen that certain N_i are satisfactorily confirmed by observations with an inaccuracy range R_1. Thus, if a value indicated in an N_i deviates from a value obtained by observation, the difference lies within the range R_1, which means that we cannot test the deviation by means of instruments and the like used in the observation. What if the means of observation are improved so that now the N_i are tested by new observations that have a smaller inaccuracy range R_2? The differences between calculated and newly observed values may of course still fall within the accuracy range R_2. Let us, however, assume that this is not the case as regards the values given in the N_i. Therefore the N_i are confirmed by the old observations with inaccuracy range R_1, but they deviate from the new observations with the smaller inaccuracy range R_2. As we have seen, there must in any case be statements Q_i that are confirmed by the observations with error range R_1. The values indicated in the Q_i need however not be mathematically identical with those figuring in the N_i, so long as the differences between calculated and measured values all lie in the range R_1 (observations with smaller error ranges being excluded by hypothesis); if they do, then the N_i and the Q_i count as physically equivalent. Relatively to such observations, the Q_i are merely translations of the N_i, and conversely. Let us now further assume that the Q_i differ from the N_i as follows. We investigate the deviation of the statements N_i and Q_i from the observations with the smaller error range R_2. Suppose we then find that the differences between the values mathematically deduced from the Q_i and the observed values all fall within R_2, whereas some of the analogous differences between the values of the N_i and the observed values are found to lie outside this range. Relatively to these finer observations the Q_i can no longer be reckoned as merely translations of the N_i. Nevertheless, we can still deduce, from the Q_i, statements that are physically equivalent to the N_i. Rules of transformation can be given for so changing the Q_i as to yield statements of this sort. Such, for instance, are the rules for 'approximation' or for 'neglecting' certain quantities. They allow us to eliminate magnitudes of a certain kind from precise laws (Q_i) or to substitute for them other 'approximately' equal magnitudes, so as to give us the less exact laws (N_i). Of the latter we then say that they are 'valid as a first (second, and so on) approximation'. What corresponds physically to this method is the arbitrary assumption that we shall consider only those observations whose error ranges do not fall below a certain magnitude.

So much for the way propositions belonging to different theories are related. I hope it will by now be clear that we are not forced to speak of two different realities, notwithstanding linguistic differences between the two theories. That between the propositions of classical and quantum mechanics there are just such relations as between our N_i and Q_i is shown by the fact that we can deduce classical mechanics from quantum mechanics by means of the 'approximation' rules ('principle of correspondence') and that this has actually been done.

Classical mechanics retains its validity for observations whose error ranges have a certain size. Its statements about such observations agree in their meaning exactly with those statements of quantum mechanics that concern the same observations: that part of classical mechanics which agrees with observation is taken over as invariant into quantum mechanics. If at any time quantum mechanics should be in turn displaced by another theory, the latter too would have to contain these invariants (and of course others besides). Theories that displace one another being thus bound together by invariants not eliminable by conventional rules of language is one of the reasons why logical empiricists distinguish between language and reality. It is therefore quite unnecessary that 'different theories must describe different realities', as will by now have become evident. The formulae of quantum mechanics, for instance, can be transformed into those of classical mechanics for a certain domain of magnitudes. It is incorrect to think that quantum mechanics absolutely rules out talk about bodies with definite simultaneous position and momentum co-ordinates. This is forbidden only for certain kinds of observation, where the error ranges fall below a certain order of size. From Newton's mechanics no statements can be derived concerning these new experimental conditions. The laws and statements of the older theory have meaning only as regards coarser observations, where quantum mechanics likewise permits one to speak of bodies that are at a certain place and have a certain mass and velocity. To say that reality is described by physical theories thus seems to be entirely justified, and there is no reason to distinguish between 'two or more' realities. On the other hand, to wish not to speak of 'reality' at all, but only of statements and linguistic expressions, seems inexpedient (other doubts set aside) since theories that displace one another are linked by invariants, a fact independent of any conventions of language.

3. NON-BEHAVIOURISTIC VERIFICATION

Logical empiricists have additional reasons for calling physics (empirical science) 'a description of reality'; these are provided by the logico-linguistic rules governing the scientific use of certain empirical statements.

Statements, especially certain empirical ones, are said to be verifiable only by being compared with reality. What are the rules for using these statements? The empiricist construes the signs we adopt to form or present statements as signals to be used according to certain established rules. If we have formed a statement p from signs according to these rules, or presented p as a signal, there are two ways in which this sign complex can be related to truth, falsehood, or the decidability of p. Either one can decide whether p is true, false or analytically undecidable, merely by means of the rules for using the signs; or one cannot. If the latter, the formation rules for p at best tell us when it is logically possible to decide its being true or false; but to determine whether p is to be called true or false, knowing the linguistic rules and signs is not enough. This applies to all empirical statements.

The empiricist respects these conditions when he says: "to decide whether an empirical statement should be called true or false, it is not enough to understand the linguistic signs and the rules for using them; rather, the statement must be compared with 'reality'". Of course, empirical statements are often verified by referring back to other empirical statements and thus by comparing them with other statements and not with 'reality'. However, the empiricist holds that these references back cannot go on for ever; in the end we reach statements that can be verified only by comparison with 'reality'. The rules for using such statements do not allow them to be false, merely because they contradict (or agree with) some other empirical statements. (We leave open the question whether there may be statements that can be verified by comparing them partly with reality and partly with other statements. Incidentally, such propositions can be divided into an hypothetical and a non-hypothetical part.)

In virtue of what other features of those rules can certain statements be verified only by comparison with 'reality'? Let us call such statements 'K-statements'. The rules that stipulate how a statement p is to be used must always specify under what circumstances p is to be called true or false. Logical empiricists have stated such rules for K-statements in various ways that we shall now analyse.

First, K-statements may be called true (false) only when they themselves do (do not) agree with reality. Second, if somebody knows the rules according to which a K-statement is used (that is, understands what he means when he states p), then in stating p he must always know at the same time whether he has made a true or a false statement. In other words, a K-statement must never be called an error. Each of these two rules determines one of the necessary conditions under which K-statements are to be true (or false). How are they related? Let us reformulate their meaning.

If a statement p may be called true (false) only when it is (is not) itself in agreement with reality, then, clearly, p can be true (false) no matter whether it is compatible or incompatible with other statements. Such a statement p cannot be logically equivalent to an hypothesis, since that is always additionally testable by comparing statements deduced from it with others already accepted, and the result of such comparisons may suffice to mark an hypotheses as verified or falsified. For K-statements this is ruled out, but they may be 'empirically equivalent' to an hypothesis. 'Empirical equivalence' means that experience shows us that we commonly regard both statements as true, or false, together.

We have said that anyone who knows the rules for using a K-statement p and then states p must also know whether his assertion is true or false, which is reminiscent of the rules governing non-empirical statements. However, the difference in principle must not be overlooked. If we understand the rules for forming and using an analytic statement, we can deduce from them whether that statement is consistent, contradictory or undeterminable. These formal questions are rather unimportant when we consider the rules for using empirical statements. We can assume that rules used in forming and applying empirical statements (which includes K-statements) do not lead to absurdities, or that we can test this from the rules alone, should we wish to do so. However, the rules for using a K-statement alone cannot tell us whether if asserted it is also true. One of these rules is indeed that a K-statement must not be called an error – this much they do tell us; in other words, anyone making this kind of statement with a complete grasp of the language he is using must know whether he has made a true or a false statement. Likewise, 'error' cannot be mentioned as regards analytic statements: at most one might make 'mistakes in speaking' or 'in writing', that is 'mistakes in performance' (which can of course happen with K-statements too); which is something quite different. It will by now be clear that the rules for forming K-statements are only superficially analogous to those for forming analytic statements.

Statements that can be erroneous are hypothetical. As stated above, logical empiricist state two conditions for using K-statements, and these regulate the relation between hypothetical and non-hypothetical statements. On the one hand K-statements are not testable (verifiable) by comparison with other statements, a process applicable only to hypotheses. On the other hand, a K-statement cannot be called erroneous: only hypotheses can prove to be errors. Therefore the rules for using K-statements stipulate that such statements cannot be called hypotheses. It is meaningless to say that K-statements are more or less probable, or cannot be fully verified, and so on. A K-statement is either true or false, and it is verified the moment it is made. It cannot be accepted as true (false) and later turn out to be false (true) and then be modified as an error. It can be made only deliberately as true or as false and, if the latter, can be rejected only as being a lie.

This account might provoke the following criticism. A K-statement, we have said, is true or false according to whether it agrees or disagrees with reality. One might therefore object that a certain testing process is used for K-statements too, and that this makes sense only if the truth or falsehood of a statement is merely hypothetical. However, this criticism arises from imprecise formulation: to verify a K-statement is not to test it after the event; to obtain it we compare it with reality. Assertion and verification are here identical, the rules for obtaining and applying such statements exclude their being tested later. (Trying to find out whether somebody in making a K-statement has uttered a lie is not verifying.) Since obtaining and verifying these statements is one and the same thing (namely, comparing them with reality), anyone making a K-statement knows whether he has made a true statement or a false one, which is how we put it earlier. Which statements in empirical science are K in type? Let us first eliminate those statements that are verified in ways incompatible with the rules for K-statements as cited above.

Certain empirical statements imply a number of other less general statements that can thus be deduced without the help of any further empirical statements. If we take a simple example like 'this thing here is a table', we can derive many other statements from it merely by analysing the concepts; for instance statements concerning the existence of parts of the table, their function and so on. These statements can likewise be tested for truth; let us assume that, no matter how this test is carried out, some of these statements are not confirmed. Suppose, for instance, we were to deduce that the top of the table functions as a support, but on testing we observe that objects placed on it fall to the ground, without our noticing any change in the table. We can now either maintain our statement 'this thing here is a table' by introducing new hypotheses, or modify the statement and say that it was only a hallucination and that no table is to be found at the place mentioned. In any case we see that such a statement cannot be K in type, since not only was it verified by being compared with 'reality' but also other statements were deduced from it and the verification of the original statement depends on the truth of the deduced statements. Furthermore, on making the statement 'this thing here is a table' one does not know at once whether it is true or false, since error is not excluded. These conditions always prevail where an empirical statement implies others of its kind and the latter can be derived from the former by logical analysis alone.

One characteristic of K-statements must therefore be that no other statements are implicit in them and deducible from them alone. Thus, when a K-statement is related to other empirical statements (leaving aside merely tautological transpositions or a negation of the K-statement) then this relation can only be empirical. Logically we can therefore always maintain or reject (as a lie) a K-statement without making use of auxiliary hypotheses, no matter if other statements are verified or not. If nevertheless we sometimes seem to 'test' the truth of a K-statement by means of other statements, it is only a matter of expediency, forced on us by fortuitous empirical features of the verifying process. If we are in doubt whether somebody's K-statement is a lie or not, since he alone can verify his K-statement, we must enlist the help of the 'empirical' equivalence between that K-statement and other empirical statements. The latter are then tested and depending on the results we either accept or reject the K-statement just as if the empirical equivalence were a logical relation. That this is not a verification of K-statements we have underlined earlier.

Some empirical statements do not imply further empirical statements. ('Implication' based on the law of excluded middle or on the law of contradiction is obviously not applicable here.) These are statements that express 'experiences' in such a way that they can be verified only by non-behaviouristic methods. Logical exmpiricism, unlike physicalism, demands that corresponding rules for using statements be established.

The physicalists, with their so-called 'behaviouristic thesis', demand that all psychological statements be translatable into statements about space-time processes: in the case of psychology, about behaviouristic processes, in the most general sense of the term. According to this view, a statement about mental processes must always be verifiable by means of behaviouristic data concerning processes that take place in the human body and in the nervous system. The psychological statement is to mean no more than a statement about such processes. This simply means that every psychological statement must be logically equivalent to a set of statements about space-time processes, and therefore these latter statements should be analytically implied by the former. On this view it would follow that all empirical statements are mere hypotheses; K-statements would be logically impossible.

The logical empiricist rejects this behaviouristic thesis. On his view, certain psychological statements are used according to rules that exclude their logically depending on other statements that refer to behaviouristic data. For example, if somebody answers the question 'what do you see?' with 'I see blue', this answer is to be logically independent of any possible statement about space-time processes and in particular about behaviouristic data. Such a statement implies no other empirical statements, nor can other statements be derived from it by logical analysis alone. This does not contradict the fact that in science we often construct 'empirical' implications between psychological statements of the type just mentioned and other empirical statements. For example, we know from experience that the statement 'I see blue' 'implies' that certain physical measurements produce such and such results, or that certain processes could be observed in the optical nerves or in the brain of the person who made the statement, and so on. Here the empiricist emphasizes that the K-statement 'I see blue' can be true or false quite independently of the truth or falsehood of the statements which imply it or which it implies.

Similarly the logical empiricist, who characterizes K-statements (such as 'I see blue', 'I feel pain') as follows. Whoever makes such a statement knows whether he has stated a truth or a falsehood: about this he cannot be mistaken. Obtaining and verifying these statements is one single process. Verification is non-behaviouristic, and consists in comparing the statements with 'experience'. Their meaning is not that of the behaviouristic statements that are usually regarded as 'empirically equivalent' to them.

The physicalists hold that if we actually adopted these rules in dealing with empirical statements, logico-linguistic difficulties would arise: if he who makes a statement such as 'I feel pain' cannot be mistaken (since to verify it he need not observe any space-time processes), then he alone can understand it; to others it would be unintelligible. In science, however, such statements are of no interest: scientific statements must be intersubjectively intelligible (verifiable)[5], so that statements such as 'I feel pain' are there tested only behaviouristically and can have no meaning other than that described in the corresponding behaviouristic statements.

In several articles[6] I have tried to show that we can conceive of circumstances in which statements such as 'I see blue' are verifiable by others as well, using a non-behaviouristic method. That this is logically possible entitles us to ascribe to these K-statements an intersubjectively intelligible meaning, using them under rules that deviate from the rules that govern the use of statements (always hypothetical) about space-time processes.

Another physicalist notion connected with this should be mentioned here. According to physicalism, a series of signs such as 'I see blue' has a two-fold character. In the first place, the series can be construed as merely a symptomatic reaction of the person uttering the statement, in which case the series is not a statement but rather an indication, somewhat like a pointer reading on a barometer. If, as we always can, we describe this linguistic reaction, along with the conditions under which it occurs, in terms of space and time, we simply get the meaning of the sequence of signs 'I see blue'; the description is logically equivalent to the meaning of the series considered as a statement. The statement itself is then obviously a hypothesis, stating such things as 'my optical nerves are struck by light-rays of such and such wavelength, whereupon my speech organs perform certain movements, and so on'. When tested, this statement can of course turn out to be false.

This view is entirely contrary to that of the logical empiricists. In the articles cited above, I have mentioned some of the points on which our views differ. Here I shall confine myself to some additional remarks. On the physicalist view, the sign complex 'I see blue', being a linguistic reaction, belongs to that series of space-time processes which is described by the statement 'I see blue'. Call this statement p, and the sign for it 'p'. Then the meaning of p differs according to whether 'p' is spoken or written down, since the signs of the statement constitute a different process in the two cases, while belonging in both to the sequence of events that p describes. Sometimes p may even assert no more than that 'p' occurs. If I assert 'I see blue' and this sign complex is the only space-time process to be observed on testing the assertion, then the physicalist could say that in that setting the statement merely says that the signs were uttered in a specified way. If we accept this view, we can no longer use the sign 'p' as a signal under arbitrary rules, nor understand it as such. This we could do only if the causal determination of the signal (sentence), which of course always exists, is not connected with the sequence of events described; that is, between the two there must be no causal link involving the laws of nature. This shows the physicalist view to be incompatible with the logico-linguistic principle that language is always a system of signs used under rules that are arbitrarily fixed. Physicalism regards language as a physical process. It is thus conceivable that merely by observing the above sign 'p' occurring in a single instance, the statement p should be tested and verified. In certain cases we can indeed exclude any other method of verification. In contrast, logical empiricists retain, for the above p as for all other propositions, the logico-linguistic principle that the signs of language are signals used under rules arbitrarily determined, so that language is always a system of logical relations and never a physical process. It must therefore always be conceivable that in a formally unobjectionable language a statement p can be presented by means of a sign 'p' that does not belong to the processes stated in p. On the physicalist view there are statements (such as 'I feel pain', our K-statements) for which this cannot be so. For logical empiricism it must at least be logically possible to present the statement p by means of a sign 'p' that is not itself relevant to the verification of p. If I make the statement 'I am speaking in a loud tone of voice', that statement can indeed be verified by observing the sign for the statement; but I might just as easily have written it down, and then the signs would be useless for the verification, although the statement would not have acquired another meaning. By consciously ignoring the above logico-linguistic principle we may of course construct statements that make an assertion about the signs used for presenting them, for example, 'this sentence is in English'. In English this statement is true, but translated into another language it is false; therefore it is not translatable. Such a statement could not be expressed in a language that involves only formal logico-linguistic principles.

With K-statements it is more difficult to see that the signs for statements function purely as signals, because they are verified by non-behaviouristic methods and in practice cannot be verified intersubjectively. However, there are various circumstances to show that in psychology, as well as in daily life, the signs for these statements are used purely as signals.

If a person is heard to say 'I feel pain', the rules for using this statement are such that whatever behaviouristic data are observed, they do not determine whether he has told the truth or not. This already involves the assumption that the signs presenting the statement are not amongst the facts expressed by it; a circumstance further confirmed when we note that the statement 'I feel pain' can be presented by signs of the most varied kinds without changing its meaning. It makes no difference whether the statement is spoken or written, formulated in German or English; which could not be so if the sign for the statement were part of the behaviouristic reactions that alone are alleged to be described by the statement.

In sum, we may characterize K-statements thus: they are not hypotheses (that is, they can never be called erroneous); whoever makes such statements also knows whether he has told a truth or a falsehood; they cannot be made with the reservation that they might be false and may have to be modified later; they can never be verified by being compared with other statements but only by being compared with 'reality', that is, by means of non-behaviouristic methods. The consequences are as follows.

The rules for using K-statements in the language of everyday life, in psychology or in other empirical sciences, do not allow us to assume that all empirical statements are hypothetical in character. Of statements in scientific systems we may say that the statement S is an hypothesis when considered in relation to other, less general statements derived from it, but at the same time we can also call it a non-hypothetical statement in relation to more general statements that refer back to S and can be justified by it. This two-fold status is forbidden by the rules for using K-statements. Less general sentences cannot be derived from a K-statement, nor can it be established or refuted by being compared with other statements. Given the rules for using K-statements, there can be no other statements towards which the former might be hypothetical. In contrast, all statements of a theoretical system are hypothetical in relation to K-statements. By the same rules, we cannot say that all empirical statements (which includes K-statements) are only more or less probable. A K-statement is never tentative, it always deliberately states something as true or as false.

As we have said, K-statements are verifiable only by non-behaviouristic methods. I have shown elsewhere[7] that even where a K-statement might be verified by persons who did not make it, this too can be done only by means of non-behaviouristic data. Hence the applicability of the physical space-time language as the only (universal) language of science rests on certain conditions. If experience were to establish a complete parallelism between non-behaviouristic processes and certain corresponding physical processes, then a purely space-time description would in general suffice us.

Even then, however, we could not dispense with the insight that the corresponding physical statements are only empirically equivalent to certain K-statements, and there might be cases in which we must give up linguistic uniformity in favour of a non-behaviouristic mode of expression. Thus, if we observed that we always assert 'I see blue' when certain processes take place in our brain, it would become scientifically acceptable to use instead of that statement the physical statement 'such and such a process is taking place in my brain'. This way of talking could however be used only with the reservation that under certain conditions the K-statement might remain true even if not 'confirmed' by space-time data.

However, the universal language in terms of space and time would certainly be inadequate where we have not observed an empirical equivalence between K-statements and the corresponding physical statements. We cannot now determine whether at some future time we may be able to pursue empirical science by means of space-time statements alone (subject to the reservation just mentioned). Yet even if science could not logically give up its two-fold language, it would be wrong to conclude that we cannot then make meaningful and at the same time generally intelligible statements about those processes that cannot be described by means of space-time concepts. As previously mentioned, this is what the physicalists do assert, and I have cited my attempts to show that our K-statements might be verified by other persons in a non-behaviouristic manner. Grant that possibility and the statements are already intersubjectively intelligible, whether the conditions of verification are realized or not.

The physicalists, however, go further still and attempt to prove that even for the speaker a statement such as 'I feel pain' is meaningful only if it is interpreted as a statement about behaviouristic data. This view rests on the surely mistaken idea that such statements can be tested later; that is, they believe that obtaining the sentence and verifying it are two separate processes. This is not so for K-statements, being indeed incompatible with the rules for using them, as I have explained in various places[8].

It is therefore wrong simply to assert that statements verifiable only by non-behaviouristic methods are unimportant for science. Such statements can be intersubjectively intelligible, witness the case of K-statements. Therefore they must be given a role in the system of all empirical statements, that is, in empirical science. What this role is will become evident when we examine the relation between K-statements and the 'invariants' discussed in Section 2 above.

4. *K*-STATEMENTS AND INVARIANT SYSTEMS OF STATEMENTS

The empiricist distinction between language and reality comes into play mainly in two situations. Firstly, as soon as we say that 'physical theories describe reality', this refers to a certain relationship (not entirely describable in formal terms) between theories following and superseding one another in the course of scientific development. Every physical theory contains as one of its parts a system of statements that are confirmed by observation. As we have said, this confirmation and the corresponding observations always lie within certain limits of error. If one physical theory is given up in favour of another, then the new theory takes over from its predecessor at least that portion (although perhaps expressed in different terms) which agrees with observation, as just explained. If the new theory contains none or only part of that portion, then the old theory has not been given up for the new, but both are used in describing reality (for example, wave theories and corpuscular theories in the description of certain atomic processes). Thus, every physical theory contains an 'invariant' system of statements that is taken over into all subsequent theories. There is no formal criterion by which the invariant part could be singled out from a theory. Agreement with observation is the only decisive factor here.

The second case where the logical empiricist distinguishes between language and reality is in regard to those statements that can be verified only non-behaviouristically. *K*-statements are so used that they cannot be verified by being compared with other statements. The concepts that figure in such statements cannot be analysed further, so that logical transformation rules alone cannot yield any new and testable statements. It is a step in the wrong direction to wish to analyse and transform the statement 'I feel pain' as follows: 'I' means 'this man with certain physical qualities', 'pain' means 'such and such processes in the nervous system'. On this basis, the statement 'I feel pain' could be translated into the statement 'such and such nerve processes are taking place in the body of this man'. This translation is wrong, because in the statement 'I feel pain' the word 'I' has no independent meaning, does not indicate an object; it becomes meaningful only within the total context of the statement. Furthermore, the word 'pain' does not mean processes in the body. Thus, in agreement with our view, *K*-statements can be verified only in a non-behaviouristic manner, by comparison with reality. Moreover, that is how they are obtained: obtaining them and verifying them are one and the same process.

In regard of the 'invariants' we spoke of a system of statements agreeing with 'reality', whereas in the case of K-statements the individual statements are compared with 'reality'. How are K-statements and the invariant system of statements related?

First, we note that whenever empirical statements are tested by subsequent observation, one always reaches a point where one must ascertain whether certain K-statements are compatible with statements deduced from the sentences to be tested. Thus, when we say that certain systems of statements are confirmed by observation, we find here too that confirmation is based on certain K-statements. The derived statements, which can ultimately be tested only by comparison with K-statements, can always be put into the form 'If I do such and such, then I shall obtain such and such a K-statement'. We shall call such sentences 'V-statements'. Verification is then achieved by my performing the acts prescribed by the V-statement, which will reveal whether I obtain the K-statement predicted or some other K-statement. From an empirical system of statements S, many V-statements can be deduced. If some of these are confirmed by K-statements, then a system of statements S' (which is part of S) can be constructed for the group of confirmed V-statements. This can be done in such a way that, given the accuracy limits of observation, only the confirmed V-statements can be deduced from S', which is then an invariant part of the system S. This elucidates the connection between saying 'a theory is confirmed by observation, agrees with reality' and saying 'certain statements are directly compared with observations, with reality, and thereby verified'. K-statements are directly compared with reality, in that they are verified in a non-behaviouristic way. Systems of statements agree with reality if the V-statements deduced from them are confirmed by K-statements.

It might here be objected that we can always 'imagine new observations' that could compel us to modify or even to cancel an invariant system of statements; and that it is therefore not true that certain parts of such systems must be taken over unchanged into physical theories that supersede their predecessors. However, we must first define what is here meant by possible 'new' observations. If a certain empirical regularity has been observed and represented by an invariant system of statements S', this means that certain K-statements have confirmed the V-statements deduced from S'. 'New observations' cannot mean that under identical conditions of observation, for example with the same experimental arrangement, we can obtain, in random fashion, K-statements now falling within the limits of observational inaccuracy, and now falling outside these limits. Otherwise the empirical regularity could not have been observed in the first place, nor could the invariant system of statements have been constructed. (Where we can observe only statistical regularities, there can be invariants just as with non-statistical laws. In these case, however, a V-statement is such that it can be confirmed only by several K-statements but not by one alone.)

If 'new observational results' that might lead us to give up an invariant system refers to K-statements obtainable by improved means of observation, we saw in Section 2 that new theories based on results so obtained never exclude as false such old theories as were confirmed by less exact observations. The latter remain in the new theories as 'limiting cases'; that is, that part of the old theory which was confirmed by certain K-statements is always deducible from the new system if we 'neglect' certain terms. Refining the observational process means reducing the limits of inaccuracy by changing the conditions of observation. Under these new conditions, new K-statements can be obtained, but this is irrelevant to the verified 'invariant' sub-system of the old theory, whose statements are meaningless with regard to the new limits of inaccuracy and assert nothing about the new observations. From a theory we can deduce certain V-statements that specify the conditions of observation under which K-statements should be obtained, and only the K-statements of observations so obtained are relevant to confirming or refuting that theory.

Logically, therefore, new observations (that is, K-statements obtained under new conditions) can never make us give up or modify an invariant system of statements confirmed by certain K-statements. There is thus a connection between the 'unalterability' of K-statements and the 'invariance' of confirmed systems of statements. Since K-statements are used under rules that forbid us to call them erroneous, they can be neither altered nor cancelled (except trivially, if they are lies). From systems of statements we can deduce V-statements, which assert that certain K-statements would be obtained under certain conditions. If I follow the V-statements and so obtain the K-statements, this confirms the former and the system from which they are derived. A V-statement can be mistaken and is shown to be so if under the conditions specified in it the predicted K-statement is not obtained. A system of statements whose V-statements are confirmed by the corresponding K-statements becomes 'invariant'; that is, every system describing the facts concerned must include the invariant system. The latter must be deducible from the new system, either as a limiting case by omission of certain orders of magnitude or according to appropriate rules of transformation.

We have, I think, shown that logical empiricists are not trying to be metaphysical when they distinguish between language and reality. On the contrary, the distinction refers only to certain rules for using statements and ways of speaking. Since we have investigated the relevant rule for empirical science, we are justified in calling our position 'logical empiricism'. We have analysed the syntactic rule for applying those statements and ways of speaking that are particularly characteristic of empirical science. If, as the physicalists do, we reject the syntatic pecularities that distinguish those statements from the rest of scientific statements, thereby turning all scientific statements into hypotheses, then we reach a position that, although based on logical and linguistic analysis, is incompatible with certain ways of speaking in empirical science. What they reject is the characteristic feature of empirical description, namely the two-fold use of empirical statements. Given the difficulties arising from this rejection, it seems to me that the position of physicalists like Carnap and Neurath is too flimsily based to pass for empiricist.

NOTES

[1] See *Erkenntnis* **5** (1935) 6 and the lectures published by Herman & Co., Paris, 1936.

[2] I have previously used the term 'logical empiricism' with precisely this meaning, in my essay 'Empiricism and Physicalism', *Analysis* **2** (1935) 6 (see this vol. p. 36).

[3] Cf. O. Neurath in *Erkenntnis* **3** (1932/33) 2 and **3** (1932/33) 3; and also **4** (1934) 5 and **5** (1935) 5.

[4] Cf. Juhos, in *Erkenntnis* **4** (1934) 6 and **6** (1935) 1 and in *Analysis* **2** (1935) 6 and **3** (1936) 5.

[5] On these problems Carnap has recently met his adversaries half way (cf. 'Testability and Meaning', *Philosophy of Science* **4**, 1, pp. 10-11). He no longer considers it impossible to construct an intersubjectively intelligible language by appropriate combination of several languages that are only subjectively intelligible.

[6] Cf. Juhos, essays in *Analysis* **2** (1935) 6 and **3** (1936) 5, and in *Revue de Synthèse* (Oct. 1936).

[7] See the author's papers in *Analysis* **2** (1935) 6 and **3** (1936) 5; also an article in *Revue de Synthèse* **12**, 2, 'Discussion logique de certaines expressions psychologiques'.

[8] See Note 7 above and the author's essays in *Erkenntnis* **4** (1934) 6 and **6** (1936) 1.

CHAPTER VI

THE METHOD OF EPISTEMOLOGICAL ANALYSIS*

1. ANALYSIS OF FORM AND OF CONTENT

Each individual science has developed for itself a method by which it obtains its knowledge. Every piece of knowledge must be statable or representable by a proposition. Propositions are linguistic expressions. Only if a language, with certain chosen signs and rules of·use for them is given, can propositions be set up and expressed by linguistic signs. The kinds of such signs and rules at the same time determine the kinds and forms of proposition representable in that language. What kind of propositions can be represented in the language of our sciences and what kind are present in our system of scientific propositions, these are the basic questions of recent epistemology.

Kant had already placed at the beginning of his epistemology the question what are the properties of the propositions that occur in our sciences. In examining the propositions of mathematics and the natural sciences as to what their character in respect of validity is, with the help of what method we obtain them and what are the relations between the concepts occurring in the propositions, Kant already used the method of epistemological analysis in a modern sense.

Following the Kantian characterization of kinds of scientific propositions by special epistemological properties (analytic, synthetic, *a priori, a posteriori*), the individual sciences and epistemology have since greatly sharpened and enriched the methods for logically analysing and marking the forms of scientific language, a development of over a hundred years. Today we can say that we dispose of a highly developed analytic method in epistemology, fit to determine what is the epistemological character and sense of propositions and concepts in the individual sciences. In particular, the method lends itself for testing whether these propositions are correctly formed or whether they are senseless because of logically inadmissible uses of linguistic signs and expressions. Moreover this method can be fruitful in enquiries in the special sciences, by bringing out new possibilities of con-

* Translated from the German original 'Die erkenntnisanalytische Methode', first published in *Zeitschrift für philosophische Forschung* **6** (1951) 42–54.

cept and theory formation, or pointing out obscurities, if any, in modes of scientific expression.

This analytic method of epistomology applies two main procedures for investigating the individual sciences as to their foundations, basic concepts and propositions and modes of expression; it applies those same procedures for examining the modes of expression and formulations of problems in philosophy. These two procedures may be denoted as that of form and content respectively.

Formal analysis examines linguistic expressions occurring in an individual science, in philosophy or in daily life – whether propositions, individual names, characterizations, concepts of objects, properties or relations – independently of their meaning (and therefore independently of the objects they denote), as to the rules by which these expressions are formed from linguistic signs, the linguistic system to which they belong and the rules to be used for forming further expressions. The questions how far the system of propositions can be axiomatized, and what conditions a logically impeccable system of axioms must satisfy (independence of the axioms, consistency of the system and so on), these too are problems of formal analysis. As the basis of formal analysis we may take the question: what stipulations of logical grammar characterize the expressions (propositions, concepts) of a science and to which, if any, linguistic system do they belong?

In contrast, the analysis of content examines the relation of linguistic expressions in individual sciences, philosophy and daily life to the objects they denote, especially by what assignment rules the meaning of linguistic signs and complexes of signs is fixed. Recent logic usually denotes questions of this kind as 'semantics'. To this we need not object, only we must not imagine that the analysis of content, if called 'semantic' is confined to examining definite systems of rules that establish relations between precisely defined systems of linguistic signs to an equally precisely defined system of objects. For this is the form of semantic theories currently being put forward above all in America.[1] The general analysis of content in epistemology does indeed examine such systems as well; but the linguistic expressions to be investigated often do not belong to any uniformly formalized language nor can sufficiently precise determinations of form be given for the objects denoted to permit the meaning of the expressions concerned to be represented as a relation between two logically exact systems. Thus, the means of analysis of content are and must be richer than the formalized method of logical theoretical semantics.

For the analysis of content we may take as basic the question: about what objects do the expressions, above all the propositions of a science, speak? From the answer to this and cognate questions – as also from the answer to the questions of formal analysis – certain criteria emerge for the epistemological characterization of the linguistic expressions used, and it may turn out that according to the various criteria ascertained we must distinguish different meanings in a single expression.

Ascertaining the unambiguous sense of linguistic expressions (signs, concepts, propositions and systems of propositions) can be taken as one of the main goals of the analytic method of epistemology. This goal is served above all by a precise examination and determination of the methods by which we obtain propositions in the individual sciences. Here we often have to analyse both forms and content. From the way propositions are obtained we can infer not only their epistemological character but also the conceptual connections that are to be in question when we describe the objects concerned.

All these investigations lead us to exhibit on the one hand the logical structure of the linguistic expressions used, and on the other those elements of the total domain of objects to be described on which we try to map the structure of the expressions. The analysis of linguistic forms and their assignment to the objects to be described mostly shows us further possibilities of concept and theory formation in the domains concerned, which may lead to fruitful development in the enquiries in individual sciences. It is an essential feature of the analytic method of epistemology that it always proceeds hand in hand with enquiry in the individual sciences, that is, it only ever recognizes linguistic expressions as unambiguously determined in sense if they satisfy those criteria that are regarded as binding for the methods and statements of the individual sciences. Thus viewed, the analytic method recognizes only one kind of knowledge, namely that obtained through the methods of individual sciences.

2. FORMAL ANALYSIS OF THE CONCEPT OF EXISTENCE

A good example of formal analysis of a linguistic expression is the critique of the concept of existence. The comparison of existential propositions in the sciences on one hand and in philosophy on the other brought to notice even purely externally a difference in propositional form or in the way the 'concepts of existence' were marked. Wherever an object is asserted to exist – whether an empirical object or a concept purely analytically defined – this always happens with indication of properties that are to belong to the object or of relations it is to have to other objects. For example: 'there is an object that has the property red'; 'there exists a natural number related to the number two by "smaller than"'.

In contrast, in philosophy there are existential propositions in which existence is explicitly asserted of an object independently of all possible properties or relations. The proposition 'there is a thing in itself' in Kant's philosophy – the 'thing in itself' being indeed intended as devoid of properties, forms and relations to anything else – is a particularly clear example of the 'philosophic' use of the concept of existence. Metaphysics usually calls this philosophic concept 'absolute being' and contrasts it with the individual scientific concept of 'relative being' mentioned above.

A further noteworthy distinguishing mark of the two kinds of existential propositions consists in the fact that the propositions in which a relative existence is asserted are always testable (whether by empirical or by logical analytic methods), while for propositions asserting absolute existence there is in principle no claim that its validity could be tested.

It was the logistical symbolic language established by Frege and Russell that made it possible to represent the logical structure of existential propositions and therefore formally to analyse the concepts of existence. The decisive insight here consisted in the recognition of the logical connection between existential propositions and so-called universal propositions. For it turned out that an existential proposition can always be represented as a universal proposition negated in a determinate way. The proposition 'there are things that are red' is equivalent to the negated universal proposition 'not all things are not red'. From this insight it follows that the linguistic signs by which universal and existential propositons are marked must have the same epistemological character. Indeed, symbolic logic has succeeded in finding two forms of linguistic sign that unambiguously mark universal and existential propositions and show that their rules of use have the same epistemological character and that the two kinds of proposition can be transformed into each other as mentioned above. These characterizing symbols are the 'sentential operators'. The language of symbolic logic distinguishes universal and existential operators that have no independent meaning but express a meaningful proposition only when combined with propositional functions (expressions with variables). Translated into words, the two operators are: 'for all x' and 'there is an x such that' (for example 'there is an x that has such and such a property').

From the form of existential operators and the rules for using them it is obvious at once that these expressions cannot be sensible propositions on their own. Sentence-like expressions such as 'there is something or other', 'there is an object' and the like are no more than independently used existential operators deliberately left unconnected with any sequent propositional function asserting a property or relation for 'something'. This deliberate detachment of the existential operator from predicate or relation functions appears even more clearly in the sham proposition 'there are things in themselves for which no properties or relations can be asserted'. The sense of all such 'propositions' that assert 'absolute existences' is thus that of the independently used existential operator, in words: 'there is an object such that...' From the structure of this expression it is clear that it is not a complete proposition with a meaning. If it is nevertheless used as such – as occurs in any assertion of 'absolute' existence –, we violate the logical rules laid down for the existential operator, and this must always lead to meaningless linguistic formulations. Thus it becomes intelligible why the theses of metaphysical systems can never stand up to logical criticism: for they always contain, in more or less concealed form, assertions of 'absolute' existences independent of any properties or relations.

A formal analysis of the metaphysical method would show that in its theses it always uses linguistic expressions in such a way that the logical linguistic rules holding for them in scientific and in everyday language are in principle violated. In doing this, metaphysics pursues a certain purpose and for the same reason omits to give for its expressions (whether taken from ordinary languages or newly invented) rules of use sufficiently consistent in order to fix their sense unambiguously. To investigate these metaphysical purposes is however a task for the analysis of content.

The further formal analysis of the concept of (relative) existence must take into account that our sciences contain both 'analytic' and 'synthetic' existential propositions. Accordingly we must distinguish the concept of 'analytic' or 'logical' existence from that of 'synthetic' existence. The former is an existence by definition, that is, the 'objects' that analytic propositions say exist only ever do so with respect to those linguistic systems (within those languages) in which the stipulations defining these objects (always concepts) occur. Such 'logical existence' belongs for example to arithmetic and numbers, on the logicist interpretation. In contrast, 'synthetic existence' belongs to all objects that exist independently of all stipulations, that is, outside language. Thus the existence of empirical objects is synthetic, and so, on the intuitionist view, is that of arithmetic and numbers.

3. CONTENT ANALYSIS OF THE CONCEPT OF CAUSALITY

The assertions of quantum physics presented epistemology with the task of ascertaining the sense of the various physical propositions and modes of expression touching on causality. This analysis can of course be only one of content, for by causality we denote factual states, that is elements of reality. We merely need to discover, precisely by analysis of content, what kind of real elements is denoted by this expression.

What character is to be ascribed to the principle of causality had already been clarified by epistemology. In conventionalism it is interpreted as a convention, as a prescription always to seek out a regular connection between the events to be described or to assume such a connection. It is a question of expediency whether we decide to adopt such a convention. Physical enquiry will have to decide whether following this prescription will always lead to the simplest possible description of sequences of events.

However, we might also construe the principle of causality as an empirical proposition in the sense of empiricism, in which case its validity can be tested by observation. If this were to show that within certain domains no regularity can be ascertained in the way events run, it would refute the universal validity of the principle.

When the analysis of knowledge had thus clarified the epistemological position of the principle of causality in the system of propositions of empirical science, it seemed natural to indicate the content of the causal concept in general by the following marks. If for the description of events, the principle of causality demands the setting up of laws (or is valid where we observe regularities in the run of events), then we might generally indicate the meaning of that concept as 'regularity of the run of events'. Such a definition was given by M. Schlick,[2] who sees the decisive criterion of causality in the ability of laws of nature, to make predictions possible: if with the help of such a law we can predict what event will follow a certain other event and this is confirmed by observation, then we usually speak of a causal connection, no matter what the form of the law from which the prediction was derived. Now we certainly do recognize the coming true of a prediction obtained by means of known regularities as a criterion for causality, but if we were to accept Schlick's definition as final we should drift into a strange position as regards certain modes of expression concerning causal connections in recent physics.

If causality simply means physical regularity, then by 'chance' we must understand absence of law. However, modern physics knows of a definite kind of laws – namely the statistical – that designate as 'lawless' and therefore a matter of 'chance' the series of events which they describe, but nevertheless allow us to predict their occurrence (up to a certain degree). Events described by such laws would thus be a matter of 'chance' and 'causally conditioned' at the same time. Yet chance and causal dependence are contradictory determinations. Since, however, physics makes testable predictions by means of propositions about processes whose course is thus marked both by chance and causality, these apparently contradictory concepts certainly have an unobjectionable sense. It is the task of further content analysis to find out which different elements of a real run of events are denoted by the physical expressions 'chance' and 'causality'.

If instead of 'chance' and 'causally conditioned' we use 'lawless' and 'regular' respectively, events describable only with the help of statistical laws would have to be denoted as both lawless and regular. The contradiction can be removed only if we distinguish two ways of being regular, and also of being lawless. Indeed, physics emphasizes that these strange and apparently contradictory properties belong only to such courses of events as can be described only with the help of statistical laws, and that that form of law must be distinguished from the form of the other laws occurring in physics.

Accordingly we must distinguish in physics two formally different kinds of regularity and therefore also of causality[3], to which of course must correspond two forms of lawlessness or chance. In this way there need be no contradiction if for certain processes we assert the validity only of laws of the one form, while at the same time designating the processes as lawless with respect to the second form of law.

The analysis of the two forms of physical law mentioned reveals the following formal difference in the formulae representing the laws. Physics describes events with the help of spatio-temporal quantities. In the formulae representing laws of nature the space-time co-ordinates can appear explicitly as constants or implicitly as variables. We must thus distinguish two forms of laws: those which describe a sequence of events with the help of explicit co-ordinates we shall call 'first-order laws' and those using implicit co-ordinates 'second-order laws'. (Physical laws, which always describe a multiplicity of sequences of events, are almost without exception mixed forms of first and second-order laws. With regard to sequences that they describe by means of constants they are then first-order; and with regard to events whose order they render by a relation of variables, second order.)

As is easily seen, second-order laws always presuppose the holding of first-order ones. We merely need to substitute individual values for the variables of second-order laws to obtain propositions stating first-order regularities of the kind 'an event of kind A is always followed by one of kind B' (A and B being marked by constants). On the other hand, the holding of first-order laws does not presuppose the holding of second-order ones. For in a sequence R of courses of events, we may have observed their regularities and stated them in a series of first order laws ('A_n is followed by B_n ') without being able to represent R by a relation of only implicit co-ordinates (which can be done only when R shows a continuous order of its elements).

From this conceptual difference between first and second-order laws there follows an important difference as to form of the predictions derivable from these laws. From a first-order law we can only ever predict the repetition of sequences of events of the same kind as those explicitly denoted by that law: these we shall call 'first-order predictions'. In contrast, from a second-order law we can always also predict sequences not explicitly denoted in the law and therefore not used in obtaining it and indeed perhaps never observed at all: these we shall call 'second-order predictions'.

If now 'causality' means 'regularity of events' and we speak of causality wherever we can predict the occurence of events with the help of a law of nature, the question arises which of the two orders of empirical regularity in physics is to be identified with the concept of causality. Moreover, when are we to speak of the existence of causality: when the events can be foretold by first-order predictions, or only if second-order ones are required? These questions could of course be answered by an arbitrary decision. If, however, we wish to reproduce the concept of causality in the sense in which physics usually speaks of 'causality', then we must examine how physics links its concept of causality with the two orders of physical regularity.

Here we see a hint in the way physics marks events that can be described only with the help of statistical laws. According to their meaning, statistical laws are first-order (which does not exclude their being stated along with second-order ones in physics, in which case the descriptive law is of mixed form). If the logical account is correct, such a law only ever states probabilities of events, obtained by measurement or counting, so that it can predict only that events of the kind concerned will occur with the probabilities explicitly assigned to them in the law. Such predictions bear all the signs of first-order predictions. If we have thus recognized that statistical laws are first-order, it is no longer a contradiction if processes describable only with the help of statistical laws are in physics called 'regular' and 'lawless' at once: for they are regular in the first-order sense and lawless in the absence of a statable second-order regularity, so that they cannot be described by second-order laws.

This example shows that corresponding to the two forms of law physics likewise distinguishes two forms of causality. Sequences of events describable only by first-order laws, whose occurrence is thus foretellable only in first-order predictions, can thus be accorded a form of causal conditioning that we shall call 'first-order causality'. If, on the other hand, a sequence of events is describable by a second-order law which makes no explicit reference to the individual kinds of events whose sequence it describes, so that the events are foretellable by second-order predictions, we can accord the sequence a form of causal conditioning that we shall call 'second-order causality'.

It is immediately obvious that wherever we can assert the existence of second-order causality we can do so also for first-order causality, but not, in general, conversely, as follows from the already mentioned relation between first- and second-order regularities. A law that for example describes a planetary orbit in explicit from, marking the planet's solar distance and the shape of its orbit by numerical values, allows us to predict only that the path will be repeated as explicitly stated in the law, that is as always hitherto observed under the same circumstances. This is a first-order prediction derived from a first-order law. From Newton's law of gravitation, on the other hand, we can derive the relative motions of any two masses at any distance, and therefore also the orbits of all planets round the sun. In all these cases events are predicted that are neither explicitly mentioned nor characterized numerically in Newton's law. Thus we have here second-order predictions derived from a second-order law. Further examination of the criteria distinguishing first- from second-order causality would show that second-order causality or regularity can be asserted only ˙for sequences of events whose order is 'continuous'. Regularly recurring discontinuities ('anomalies') in an otherwise continuous order or courses of events can never be foretold in second-order predictions.[4]

Corresponding to the two orders of causality we must likewise distinguish two forms of 'chance'. Second-order causality excludes any form of chance, while first-order causality is compatible with a certain form of it. If anomalies and statistical processes which are of course describable by first-order laws and foretellable by first-order predictions, can nevertheless be designated as 'lawless' ('uncaused'), we know that this is to be understood as the impossibility of describing such courses of events with the help of implicit co-ordinates, that is, second-order laws. Given the latter's form, anomalies and statistical precesses are to be denoted as 'lawless', that is, a matter of chance. Let us call this form of chance 'first-order'. We see that first-order chance and mere first-order causality co-exist. First-order chance exists only where we can assert the existence of first but not of second-order causality. In courses of events not describable even by first-order laws so that their occurrence cannot be predicted in any way and thus neither kind of causality can be asserted, in that case let us speak of 'absolute' or 'second-order' chance.

Now we can understand the disquiet of physicists about the insight that processes in micro-domains can be described at best with the help of statistical laws (first-order laws), that is, we can at best assert first-order causality in these domains. From this it follows that such processes can be foretold only by first-order predictions, which means that here we can predict only the repetition of counted or measured frequencies of events under recurrence of the same circumstances. Kinds of events not used for obtaining first-order laws in micro-domains (no other laws are possible in them) and perhaps never yet observed, simply cannot be predicted in these domains, precisely because of the holding of only first-order causality. This does of course mean a considerable restriction on the goals of physical knowledge.

It would be easy to show how the distinction between two concepts of causality, here obtained by analysis of content, affects further concept formations and formulations of problems linked with the concept of causality. We merely mention the problem of free will, which must surely now be formulated in a new sense, seeing that there are two orders of causality the first of which can co-exist with first-order chance.

From the above examples of analysis of form and content we believe we can recognize in what way the analytic method of epistemology leads to an understanding of the sense of scientific as well as everyday concept formations and statements, which may perhaps manifest itself in the formation of new concepts and theories and therefore generally in an increase of our knowledge.

NOTES

[1] Cf. Carnap, R., *Introduction to Semantics,* Chicago 1946.

[2] Cf. Schlick, M., 'Die Kausalität in der gegenwärtigen Physik', *Die Naturwissenschaften,* 1931; also in his *Gesammelte Aufsätze,* Vienna 1938.

[3] Cf. Juhos, B., 'Praktische und physikalische Kausalität', *Kant Studien* **39** (1934), 2; *Erkenntnisformen in Natur- und Geisteswissenschaften,* Leipzig 1940; 'Theorie empirischer Sätze', *Archiv für Rechts- und Sozialphilosophie* **37** (1945), 1; *Die Erkenntnis und ihre Leistung,* Vienna 1950.

[4] For the way in which physics used to deal with anomalies and by what methods it tries to make them disappear, see Juhos, B., *Erkenntnisformen in Natur- und Geistenwissenschaften,* Leipzig 1940; and the last two references in Note 3 above.

CHAPTER VII

PROBABILITY INFERENCES AS SYNTACTIC INFERENTIAL FORMS*

1. RELATIONS OF CONSEQUENCE, DEDUCTIVE AND PROBABLE

The inference rules used in the linguistic systems of modern logic, that is, in calculi, are deductive throughout. To derive conclusions from premisses we there use schemata from which we can read off necessary and sufficient syntactic marks of deductive inferential forms. Denoting the premisses by p_1 and p_2 and the conclusion by s, then in deductive inference the premisses are always linked conjunctively, that is they constitute a 'conjunctive class of propositions' $p_1 \& p_2$.[1] Moreover their relation to the conclusion is implication, $p_1 \& p_2 \to s$.

It follows that a deduction is valid (binding, sound), or the formula representing it is logically true, only if the conclusion s itself appears in the conjunctive premiss class. This condition is fulfilled if s is identical with one or several of the premisses ($p_1 \& p_2 \to p_1$), or with a conjunctive part of a premiss (for example if $p_1 \equiv s \& r$, then $p_1 \& p_2 \to s$), or the negation of s is incompatible with a premiss (for example, if $p_2 \equiv \sim (p_1 \& \sim s)$, then $p_1 \& p_2 \to s$).

This property of deductive inference forms, that the conclusion figures amongst the conjoined premisses, is also expressed by saying that the conclusion is already asserted in the premisses. The relation between premisses and conclusion is here always implication, so that any deductive dependence between propositions can be represented by it.

The purely syntactic schema for the relation of conjunctive classes to their elements ($p_1 \& p_2 \to p_1$) suggests the question how we are to represent the corresponding relation for disjunctively joined propositions: how, to take the simplest example, is $p_1 \vee p_2$ related to p_1? It is obvious that the dependence cannot (or at least not in general) be represented in a 'binding' way by implication: the formula that would here represent an implicational connection would not in general be logically true.

If, following Carnap, we call a class of disjunctively joined propositions a

* Translated from the German original 'Wahrscheinlichkeitsschlüsse als syntaktische Schlussformen', first published in *Studium Generale* **6** (1953) 206–214.

'disjunctive' class, we can reformulate the question thus: how is a disjunctive class of propositions related to its elements? As with deductive inference, where the premisses form a conjunctive class containing the conclusion related to the premisses by what is called a relation of 'consequence' or 'derivation', let us denote disjunctive classes and their elements in their mutual relations as 'premisses' and 'conclusions' respectively. What then is meant by relations of 'consequence' or 'derivation' between a disjunctive class and one of its elements, that is, between disjunctively linked 'premisses' and a syntactically allied 'conclusion'? Let us write the two schemata as follows:

$$(p_1 \mathbin{\&} p_2) \mathbin{/} p_1 \dots \text{(I)} \qquad\qquad (p_1 \vee p_2) \mathbin{/} p_1 \dots \text{(II)}$$

Schema (I) represents a deduction and we know that the premisses are related to the conclusion by implication. What is the corresponding relation for schema (II)? If here too we were to take implication as the relational link, we should by no means always obtain a logically binding inference, since the formula $(p_1 \vee p_2) \to p_1$ is not logically true (tautological): we could deny it without obtaining a logically false proposition. Thus implication in this case turns out to be an inadequate (non-'binding') connection. Here we must observe that our calculi do not dispose of suitable connections to express this relation between disjunctive classes and their elements. New stipulations are required to provide means for representing this extended domain of forms accurately and completely.

Let us begin by introducing a general nomenclature for the different connections between propositional classes and their elements: the relation between conjunctive classes and their elements is to be called 'deductive consequence', and the corresponding relation for disjunctive classes 'probable consequence'.

To represent a deductive consequence we use implication, for probable consequence we have as yet no suitable device. In order to discover one we must study the logical properties of this relation of consequence.

Deduction is characterized by the fact that a proposition and its negation cannot both be conclusions of consistent premisses. In contrast, this possibility is not in general excluded for problable consequences. If for example there are black and white balls in a container and p_1 and p_2 record the drawing of a white and black ball respectively, then both p_1 and $\sim p_1$ are in probable consequential relation to the disjunctive class $p_1 \vee p_2$. Hence several conclusions deduced from the same conjoined premisses (the same conjunctive class) can always themselves be conjoined and conjointly are a deductive consequence of the premisses. Not in general so for several probable consequences of a disjunctive class (disjunctively linked premisses). Except for extreme cases, the position is rather that in by far the majority of instances the individual 'probable' inferences from the same premisses are disjunctively exclusive and incompatible. Once again we have two different syntactic schemata to illustrate the two different relations of consequence. Let a conjunctive class and a disjunctive class be denoted respectively by $[K]$ and $[D]$; then $[K] / s_1 \& s_2 \& \ldots \& s_n$, $[D] / s_1 \vee s_2 \vee \ldots \vee s_n$ represent these contrasting properties of the two kinds of consequence, where the disjunction sign '$\vee$' is mostly (and above all in the logically decisive cases) to be taken as the exclusive or ('either-or').

Besides this syntactic difference of the two inference forms there is a further important distinguishing mark. All deductive conclusions from a conjunctive class are uniformly so related by implication. In contrast, the probable inference of 'conclusions' from disjunctive classes can be of different kinds. Not only can the different conclusions follow from the same premisses 'with different probabilities', there is also the logical possibility of stipulating different kinds of probable consequence for a disjunctive class and a probable conclusion from it. For we know that in probability calculus and in experience where we often use probable inferences, we can variously define probable consequence between a premiss class and its conclusions.

In roulette, for example, we can stipulate the same *a priori* probability for each of the possible numbers from 0 to 32. However, we might just as well stipulate the average frequency of a number's occurrence as its probability value. These two procedures need by no means produce the same probability for the same number. Alongside these two procedures for assigning probabilities, however, there obviously can be infinitely many others. In the above example we might stipulate that each of the numbers in the sequence shall have half the probability of its precursor and so on. Probability values can be assigned to individual cases within a class in infinitely many ways, to each of which syntactically corresponds the choice of a specific type of probable consequence relation from the infinite set of possible such relations between a disjunctive class and the consequences derivable from it 'with probability'. In the case of roulette, the disjunctive class comprises the propositions 'the ball will come to rest on 0 or 1 or ... 32', and the conclusions derivable 'with probability' are 'the ball will come to rest on 0' and so on.

The fact that in probable inference a value of probable derivation is assigned to the conclusion with respect to the disjunctive premiss class, enables us to characterize the syntactic character of probable inference more precisely: it is always a function whose arguments are the premisses of a disjunctive class and the conclusion. If we apply probable inference in practical experience or in probability calculus, the premisses indicate the empirical or presupposed data on the basis of which we are setting up the conclusion as 'hypothesis'. The relation of probable consequence, chosen in the form of a particular function in any given case, then assigns to the hypothesis set up a determinate value of probable derivation.

As we have said, there are infinitely many functions that can represent relations of probable consequence, and in any individual case we are at liberty to choose any one of these. One of them in particular is used in a priori probabilities. Empirical statistics, on the other hand, usually adopts a different function to represent probable consequence relations amongst its propositions. Here, the values of the function are defined as 'average frequencies' and treated accordingly. Alongside these two there are infinitely many other functions that can be chosen for the purpose, as previously mentioned.

We may then ask whether we have any right to speak of 'consequential relation' where we are free to choose what dependence relation is to count as 'valid'. It is precisely because the relation of deductive consequence (always represented by implication) is uniform that all deductive systems show the same syntactic features. If, however, we have the choice of infinitely many 'relations of consequence', we obtain propositional systems of quite different syntactic form, depending on which 'consequential relation' is chosen as holding between the propositions. In that case it obviously makes sense to speak of 'consequential relations' only if we succeed in discovering common properties marking all these possible relations of dependence between propositions. Failing this, we are not justified in putting probable relations of consequence alongside deductive ones as equally syntactic relations of dependence.

2. THE SYSTEM OF RELATIONS OF PROBABLE CONSEQUENCE

Recent attempts to define the concept of 'probability' have always led to the result that in each case some particular more or less precisely specified relations of probable consequence was chosen and its special features were declared to be the criteria of probability in general. Thus L. Wittgenstein[2] defined probability as a 'consequential relation' between propositions, provided that the cases in which an atomic proposition can take on the thruth values 'true' or 'false' are 'equi-probable'. This means that he is using the '*a priori*' relation alone, stipulating it as holding between the propositions. In contrast, R. v. Mises[3] understands by probability 'relative frequency' (although the way he determines the concept is not consistent), thus choosing as consequential relation between propositions the one that empirical statistics usually adopts in deriving its propositions. H. Reichenbach[4] opts for a similar 'consequential relation', but characterized with more logical precision. A more subtle kind of relation is defined by A. Wald[5], whose function brings elements of *a priori* and statistical probability into a certain relation and thereby assigns the individual values of the function (probable derivation values) to the conclusions.

For practical use now this, now that definition will be more expedient, each indicates a determinate form of probable consequence relation for the propositions. In practical cases, experience and observation then decide which type of relation is most adequate for the empirically observed dependences between propositions in that field. As to logical syntax, however, all the infinitely many types are equally justified, so that the question arises whether they have common features enabling us to treat them systematically as a whole or whether each type must be characterized independently of all the rest, in which case of course we could not justify calling them 'relations of probable consequence' in general.

Now R. Carnap[6] has proved that all possible forms of probable inference stand in a certain syntactic logical relation so that they can be uniformly and systematically treated together. He starts from the insight that 'relations of probable consequence' (he calls them 'inductive inferences', which seems to me inexpedient because misleading) can be represented by functional dependences whose arguments are the premisses and conclusion of the propositional connection concerned. Denote the premiss class by e and the conclusion by h, then, as we know, there are infinitely many relations of probable consequence that can connect e with h. Depending on which of these we choose, the premisses give the conclusion (or hypothesis) different probabilities. Representing such a relation generally by a function $c(e, h)$

with e and h as arguments, we must accordingly distinguish an infinity of such functions, say $c_1 (e, h)$, $c_2(e, h)$, ... $c_n (e, h)$, ... The values of any such c-functions are the probability values we must assign to the hypothesis h starting from premises e. Depending on which c-function or probable consequence relation we choose for the derivation, different probability values may result for h. Is there a common syntactic property of the infinity of c-functions that might allow us to treat them systematically as a whole?

Carnap[7] actually succeeded in proving that a comprehensive and regular relation can be indicated for the series of c-functions, that is, our relations of probable consequence. Besides the series of c-functions (probation functions) Carnap defines a second series of e-functions (estimate functions), which have the same syntactic character as the former. The c-functions represent forms of inference that we apply in setting up hypotheses, while e-functions represent probable inference forms that we tend to adopt for discovering ('estimating') the 'real' value of quantities that have been measured imprecisely. There is no logical syntactic difference between the two types of function, so that they can be transformed into each other, and Carnap indicates a general function G from which they can be derived as special cases. It is sufficient here to consider the syntactic features only of c-functions or of the general function from which these can be derived.[8] The c-functions can be arranged in a continuous series in such a way that each may be viewed as a special form of a general function. This series extends between two extreme forms for limiting cases. One of these gives probability values as a ratio only of empirically ascertainable factors on which probability is defined as depending, namely the total number s of observed (counted) cases, and the number s_i of those cases that actually occurred and whose probability is to be determined. One extreme form of the c-functions is thus represented by the statistical probability s_i/s. The other extreme form defines probability as a ratio of only logical a priori features, by which sets and subsets of cases can often be characterized. These are the total number k of 'all possible' cases and the number w of 'favourable' cases that may be indicated prior to any experience (observation). Then the second extreme form of the c-functions is represented by the formula w/k of *a priori* probability calculus. All other possible forms of the c-functions (our relations of probable consequence) then form, as Carnap was able to show, a continuous series between these two. For we can indicate a general function of the arguments k, w, s and also a parameter λ taking all real values from 0 to ∞ , each giving a determinate c-function. This general function set up by Carnap is $G(k, w, s, s_i) = (s_i + \lambda\ w/k) / (s + \lambda)$. For $\lambda = 0$ it takes on the form s_i/s of 'statistical probability' and for $\lambda = \infty$ it tends to the limit w/k of 'a priori probability'. Carnap further showed that the special probability theorems and principles used both in theory and in practice can if properly formulated be taken as special forms of G marked by a special value of λ . We leave undecided (as Carnap does) whether there might not be other probability functions not belonging to G. Still, his system is a uniform syntactic treatment of the forms of probable inference currently used in experience and in probability calculus. This system therefore allows us to examine the syntactic features of these forms more closely.

For a start, a discussion of the continuous series of such forms shows that the infinitely many logically equal possibilities of arbitrarily choosing a particular form for a given case amounts to choosing an arbitrary real value for λ between 0 and ∞. It turns out that for certain values of λ we obtain forms that not only deviate from those commonly used in statistics or pure mathematics, but may even be diametrically opposed to these principles. This too is quite in tune with the logical possibility of stipulating arbitrary relations of probable consequence between given propositions. In empirical descriptions one almost always uses forms of consequential relation (c-functions) that agree in their derived probability value with the requirements of the so-called 'principle of induction'. This principle is not a syntactic rule but a content rule imported from without, and used as principle of choice for the most suitable form of probable inference in the determination of probabilities in empirical practice. In general, the 'principle of induction' says that an event B that hitherto always followed an event A in the same way under the same circumstances will do so again if those circumstances recur. Applied to the probability of an event to be expected, this means average frequency of its occurrence, where the probability thus determined is to count the better secured the greater the number of observed series having given that same relative frequency for that case.

By following the 'principle of induction' we are ultimately always led to the inferential form that uses 'statistical probability' as its criterion; but for the simplest possible way of maintaining agreement with the principle it may turn out to be expedient under different empirical conditions to adopt different forms of probable inference. In games of chance under the usual conditions, it is the formulae of probability calculus, that is, the choice of a

priori probability, that show themselves as the simplest way of securing agreement with the principle of induction, while in determining probabilities of mortality it is the use of purely statistical methods. Carnap mentions other cases in which determinate connections of *a priori* and statistical probabilities, that is, values of λ between the limits, show themselves most adequate for the principle.

However, the principle of induction (as well as other principles of choice used in empirical descriptions) is a rule imported from without that cannot be logically or syntactically justified. Its validity is based, as emphasized already by Hume, on a belief which, though practically useful, cannot be rationally established. In view of the syntactic λ-system of probable consequential relations this manifests itself in that λ can take on values resulting in c-functions (probable consequence relations and therefore probability values) by no means agreeing with the principle of induction. Indeed, we can even choose c-functions that would make the probable expectations for a kind of event the smaller, the greater the empirically ascertained relative probability of their occurrence. These would be forms of probable inference that are the opposite of those used in individual sciences.

From this insight it follows firstly that we must regard probable consequence relations or inference forms as a purely syntactic domain of forms, represented perhaps by Carnap's λ-system of probability functions; and secondly, that the forms used in the sciences and in everyday practice constitute only a small subset of the total possible domain of such forms. Since for practical empirical purposes forms are chosen according to non-syntactic principles of content (particularly the 'principle of induction') imported from without, which do not touch the logico-syntactic character of relations of probable consequence, it is in our opinion unjustified to denote probable inferences and probability logic generally as 'inductive' inferences and logic respectively. These terms used in this sense by Carnap we consider to be misleading and therefore inexpedient.

The syntactic character of the system of forms of probable inference shows that we are here dealing with an extension of the domain of syntactic inference forms. In calculi sufficiently rich in elements we therefore must now include rules of probable derivation alongside the deductive ones like substitution, transformation, inference schemata and so on; that is, we must indicate the determinate forms of probable inference (perhaps by specifying individual c-functions) that are to be applicable and valid in a calculus. For the choice of such forms of probable inference for a calculus, the only relevant considerations are syntactic ones, such as consistency, completeness and the like.

The system of infinitely many different forms of probable inference in relation to the application of a subset of them in empirical enquiry shows a certain analogy to the likewise syntactic set of infinitely many possible geometries in relation to the 'physical' geometry of empirical space. Physical geometry is chosen for the syntactic domain of infinitely many logically possible geometries in accordance with non-syntactic considerations of practical expediency (probation, simplicity of description and the like) in just the same way as the forms of probable inference empirically used are chosen from the syntactic λ-system.

3. 'PROBABLE CONSEQUENCE' AND 'LOGICAL TRUTH'

If, as our investigation has shown, forms of probable inference take their place alongside deductive ones as new syntactic relations of derivation, the question automatically arises whether there might not be certain relations between the two syntactic domains of forms. It is indeed striking that relations of probable consequence, as previously mentioned, assume the character of deductive relations in extreme cases. This happens for instance when in an inference from a disjunctive class $[D]$ the conclusion enumerates as possible cases all the elements of that class. In all such cases the 'relation of probable consequence' between premisses and conclusion may be represented by implication. This always yields a logically true proposition (a tautology), so that the relation has become deductive. Syntactically, this can be represented by $(p_1 \lor p_2) \to (p_1 \lor p_2)$.

Replacing implication by $\sim$ and $\lor$, this turns into $\sim A \lor A$ and further into $A \lor \sim A$, a form of proposition used in logicist calculi as criterion for deciding the 'logical truth' ('universal validity', 'tautological character') of the formulae.[10] This suggests a connection between the concept of 'relation of probable consequence' and the concepts 'relation of deductive consequence' and 'logical truth'.

One syntactic criterion of all relations of deductive consequence is that formulae expressing such a relation between two sub-formulae are always tautological and therefore directly reducible to the form $A \lor \sim A$, or transformable into a conjunction of sub-formulae each of which can be equivalently replaced by such a form. In the latter case the total conjunction can again be replaced by a propositional form $A \vee \sim A$. Hence any formula expressing a deductive relation of consequences between sub-formulae is equivalent to the form $A \vee \sim A$, that is $A \rightarrow A$. This is, however, also one of the extreme forms of probable inference relations, as shown above. The latter cannot in general be reduced to the form $A \rightarrow A$ but assume it only in syntactically extreme cases. Thus it turns out that forms of probable derivation are the more general inferential form, of which the deductive ones must be regarded as a sub-class or as special forms (syntactically extreme forms).

From this insight as to the syntactic connections between probable and deductive relations of derivation the following further consequences arise. Investigations on the concept of 'truth' in modern logic have sometimes pointed to the epistemological difference between 'logical' and 'synthetic' (empirical) truth.[11] Decisions as to the logical truth of formulae in calculi relative to all possible domains of individuals are ultimately always made with the help of the critical propositional form $A \vee \sim A$. In view of this sole syntactic criterion 'logical truth' may be defined as a syntactic property of the formulae concerned.[12] Definitions of this kind have the necessary distinguishing feature that a logically true formulae must be divisible into at least two parts; that is, every such formula must always be a connection between two sub-formulae. A look at the forms of probable relations of consequence enables us to range the special syntactic form, required for the connection between these two sub-formulae in order that the total formulae should be logically true, into a definite domain of logical syntactic connections. Since deciding whether a formula is logically true ultimately involves reducing it to the form $A \vee \sim A$, that is $A \rightarrow A$, and this is the syntactically critical form for any deductive relation of consequence, we may define 'logical truth' generally thus: a formula (proposition) is 'logically true' if it expresses a deductive relation of consequence between sub-formulae (partial propositions). For any tautology we must therefore be able to show that it states such a relation.

Within the total system of relations of probable consequence logically true formulae are thus represented as expressing a special form of these relations. The recognition that logically true formulae as necessarily expressing deductive consequence between at least two sub-formulae allows us to indicate distinguishing features for characterizing logically true statements on the one hand and synthetically true ones on the other. Given a statement P, we must first examine whether with suitable syntactic representation it turns out to have several parts or not, that is, whether it states a consequence relation or not. If the former, then we must further examine whether P asserts or denies a deductive consequence relation, or not. If it does, we know that P must be syntactically transformable into $A \to A$ or $\sim(A \to A)$ respectively. Only in that case is P logically true (or logically false). Since the deductive character of a consequence relation is decidable, if at all, only by syntactic means, the same therefore holds of logical truth and falsity.

If, on the contrary, a polymerous P states a non-deductive consequence relation (that is, a relation of probable consequence in the narrower sense), then, if P consists of a combination of signs that have meaningful content, we must be dealing with a synthetically true or synthetically false proposition. Likewise if P turns out to be of one piece and therefore does not state a consequence relation (for example a property of an individual or a relation between individuals): such statements of content can be only synthetically true or false. This follows directly from the fact that logically true (or false) propositions always assert (or deny) a deductive consequence relation and any such must relate at least two parts. Synthetic truth or falsity of statements can never be decided with the help of syntactic criteria alone, for this we always need to consider semantic content, that is, the objects denoted by those statements. This is the basis of the view expounded in my articles mentioned in Note 9, that a uniform concept of truth is not sufficient for science (contrary to Carnap's view[13]), but that we must define 'logical truth' as a syntactic concept and 'synthetic truth' as a semantic one.

Summing up, we may say that logically true (or false) statements always express (or deny) deductive consequence relations so that these statements must always have several parts, while synthetically true (or false) statements are either of one piece or, if of several, express non-deductive consequence relations, that is, relations of probable consequence in the narrower sense.

NOTES

[1] The distinction between 'conjunctive' and 'disjunctive' classes of propositions was first defined in the discussion of other logical syntactic problems in Carnap, R., *Formalization of Logic*, Chicago 1943.

[2] Cf. Wittgenstein, L., *Tractatus Logico-philosophicus*, London 1922.

[3] Cf. v. Mises, R., *Wahrscheinlichkeit, Statistik und Wahrheit,* first ed., Vienna 1936.

[4] Cf. Reichenbach, H., 'Axiomatik der Wahrscheinlichkeitsrechnung', *Math. Zeitschrift* (1932); *Wahrscheinlichkeitslehre,* Leyden 1935.

[5] Cf. Wald, A., *Statistical Decision Functions,* New York 1950.

[6] Cf. Carnap, R., *The Continuum of Inductive Methods,* Chicago 1952.

[7] Cf. ibid., p. 25ff.

[8] For a detailed analysis of c-functions, see Carnap, R., *Logical Foundations of Probability,* New York 1950.

[9] Cf. the arguments of this section with Juhos, B., 'Die Wahrheit wissenschaftlicher Sätze und die Methoden ihrer Bestimmung', *Methodos* 4/13 (1952), 19–38; 'Die Voraussetzungen der logischen Wahrheit in den Höheren Kalkülen', *Methodos* 5/17 (1953) 31–43.

[10] Cf. the references in Note 9 above.

[11] This difference I try to exhibit in the first item in Note 9 above. The article contains further references on this topic.

[12] This view of mine, maintained in the articles in Note 9 above and in my book *Die Erkenntnis und ihre Leistung,* Vienna 1950, is opposed to the definitions of truth given by Carnap and by Tarski. These latter try to define logical truth, too, as a semantic concept; but this, as I try to show in the articles mentioned, is incompatible with the procedures used in calculi to decide logical truth. Cf. Carnap, R., *Introduction to Semantics,* Chicago 1942; Tarski, A., 'Der Wahrheitsbegriff in den formalisierten Sprachen', *Studia Phil.* I (1936).

[13] Cf. Carnap, R., l.c.

CHAPTER VIII

'POSITIVE' AND 'NEGATIVE' USE OF STATEMENTS*

1. TRUTH VALUE AND CONTENT

Propositions whose truth value is formally decidable (those that are decidably logically true or logically false) are usually said to be 'empty'. This has been explained in ways that are certainly appropriate even if they do not completely characterize these propositions. Thus, logical propositions are said to be empty because they are true or false whatever is the case; that is, they do not tell us anything about what is or is not the case. Therefore formally decidable logically true or false propositions cannot convey anything. This becomes obvious in propositions like 'it is raining or it is not raining' and 'it is raining and it is not raining'.

If we accept this way of taking the content of propositions, then meaningful information can be conveyed only by propositions for which there must be specific conditions that make them true as well as ones that make them false. This is sometimes put by saying that meaningful propositions can sensibly assume either truth value; that is, it makes sense to think of them as true or as false without change in their meaning. This is easily illustrated in the proposition 'it is raining'.

In what follows we shall consider only propositions with a meaning in this sense. What they state must evidently be distinguished from the truth or falsity of the statement. We might compare this with the distinction between the absolute value of a number on one hand and its positive or negative value on the other. $+2$ and -2 have the same absolute value $|2|$. Thus the true proposition 'it is raining' and the false proposition 'it is raining' have the same meaning, that is, they both state 'it is raining'. In modern semantics the meaning of a sentence 'p', belonging to it whether it is true or false, is the proposition p that it denotes. From this distinction between the sentence 'p' which is true or false and the proposition p it denotes in either case, some remarkable logical consequences follow.

* Translated from the German original 'Der "positive" und der "negative" Aussagengebrauch', first published in *Studium Generale* **9** (1956) 79–85.

2. CONDITIONS TO BE SATISFIED BY ANY
DEFINITION OF TRUTH

It is important to be clear on the fact that the proposition 'p is false' does not necessarily mean the same as '$\sim p$' ('not-p'). As we shall see, customary linguistic systems use '$\sim p$' in the sense of 'p is false' on the basis of a special stipulation, non-observance of which may lead to paradoxes.

Because the content of p (the proposition p) is independent of its truth value (of truth or falsity of p), p cannot say anything about its own truth or falsity: a statement p, whether sentence or proposition, can never contain indications as to its own truth or falsity. From this it follows that every statement of the form 'A is true', where A may be a sentence or a proposition, can have sense only if within A the term 'true' (or 'not true', that is, false) does not figure at all or not in the same (or not in the contradictory) sense as in the statement. In whatever way we then define the concepts 'truth' and 'falsity', the definiens must express a statement that is meaningful independently of the concepts to be defined.

Reflecting further that there can be statement like '"A is true" is true', '"A is true" is false' and so on in unlimited sequence of orders, suggests that it is expedient to define 'truth' and 'falsity' recursively. This is taken into account by the so-called 'semantic' definition of truth if suitably interpreted[1]; that is, we can give this definition recursive form by additional stipulations. On this definition the proposition p asserts the same thing as the statement 'the sentence "p" is true'. This may be interpeted as requiring every statement of form 'A is true' to have a constituent in statement form but not containing the expressions 'true' and 'false', or not in the same sense as in 'A is true'. It is readily seen that only if thus construed is the definition '"p" is true' $\equiv_{df} p$ recursive in form. Therefore, if we accept the above semantic definition of truth, there must be an ultimate form of statement A_0 figuring as constituent in statements like 'A_0 is true (or false)', but not containing the expressions 'true' (or 'false'). For simplicity we shall henceforth replace A_0 by p_0.

The expressions p_0 satisfying the above conditions can be propositions or sentences. Proposition is the name we give to the content of sentences. If a proposition p and the sentence 'p' denoting it are both p_0-expressions they contain no expressions 'true' or 'false', that is, nothing that denotes truth values. It is expedient to describe such propositions and sentences as having 'zero-order statement form', so that we can in this sense speak of zero-order propositions or sentences.

According to the semantic definition of truth, only sentences can be true or false: 'a sentence "p" is true if and only if p'. At the same time, modern epistemology defines truth as 'truth of propositions'. This definition is called the 'absolute definition of truth'[2], in contrast to the 'semantic' one. According to this a proposition too can be true or false, independently of whether there is or is not a sentence 'p' denoting (stating) the proposition p. It is therefore difficult to represent the absolute definition symbolically. Its sense is best clarified by initially contrasting it with the semantic one. On the latter definition the truth or falsity of a sentence 'p' presupposes that there is at least one linguistic system in which 'p' can be represented. This may be a verbal language like German or English, or a symbolic one like mathematics, or some other system of signs for which logico-linguistic rules of use are given. If there is no such linguistic system, as was the case for instance before there were human beings, there can be no true or false sentences, nor, according to the semantic definition of truth, does it make sense to wish to speak of truth or falsity at all.

As against this, the definition of 'absolute' truth requires that all possible content, that is, any possible proposition, should be capable of truth or falsity, whether or not there are linguistic systems whose sentences can denote the propositions. Before there were human beings, when there were no linguistic systems nor sentences, according to the absolute definition of truth the proposition 'the moon is spherical' was true and the proposition 'the moon is square' false.

Both definitions of truth lead to difficulties that cannot easily be removed. If the semantic definition is applied to the truth of logically analytic sentences, it follows that 'logical truth' comes to depend on the adventitious empirical existence of linguistic systems. Moreover, it becomes problematic what we are now to understand by 'proposition'. This becomes even more urgent in view of how the holders of the semantic theory of truth construe their definition: a sentence 'p' is to be semantically true if it agrees with the proposition p. What is to be understood by a 'proposition' here, and in what is 'agreement' to consist? If the true sentence 'p' agrees with the propostion it denotes, but the false proposition 'p' does not, then according to the semantic definition, agreement can consist only in the relation between denoting sentence and denoted proposition; but then the true sentence 'p'

must denote a proposition different from that denoted by the false sentence
'p': disagreement of false 'p' with p means that false 'p' and true 'p' agree
with or denote different propositions. In that case, however, the same sentence 'p' would have different meanings depending on whether it was true or
false; which contradicts the usual way of marking empirical sentences,
whose meaning remains the same in either case.

Difficulties of another kind are raised by the definition of 'absolute'
truth. According to this, propositions (possible contents of statements) are
true or false independently of whether there are any linguistic systems by
means of which the proposition can be represented by sentences. Thus true
or false are not linguistic expressions but 'logical constructs' that in semantics are usually called 'entities'. In the above example, the proposition 'the
moon is spherical' is to have been true even before there were beings en-
dowed with language and thus before there were sentences that might have
represented the proposition. However, earlier still there was not even a
moon and it is conceivable that the moon became spherical only gradually.
At those times the proposition 'the moon is spherical' is to have been false,
so that the truth value of propositions turns out to depend on empirical
events without these logical entities being themselves empirical realities.
Yet the theory of absolute truth requires that every proposition must be
true or false. It must therefore remain quite undetermined how the truth or
falsity of a proposition is to be decided. All scientific procedures for deciding the truth value of sentences, statements, communications, in short all
forms of expressions that can be true or false, presuppose that the 'expressions' to be examined are available in some linguistic form. If this is not the
case or even impossible in principle (as with propositions that are to exist as
entities when there are no linguistic systems) then it is in principle impossible to indicate a decision procedure for the truth value of such 'logical constructs' (possible contents of statements). This is epistemologically disturbing, even if certain logicians make light of it.

We must note that according to their definitions 'semantic' and 'absolute
truth' are different and mutually independent concepts. Any attempt to define one in terms of the other either deprives one of them of its independent
meaning and makes it superfluous or leads to inconsistent definitions.[3]
That the two are incompatible if viewed as defining the same concept 'truth'
is clear from their respective presuppositions alone. From the semantic
truth of a sentence we can logically infer that there is a linguistic system in

which that sentence is represented, while no such inference is possible from the absolute truth of a proposition. We need not here enlarge on objections to these two definitions nor examine whether a uniform definition of truth might not be found. Rather, we are interested in the common conditions that both (and any other possible) definitions of truth must satisfy.

One condition holding here is above all that mentioned above, namely that every statement of the form 'p is true' must contain a sub-expression of statement form not containing symbols for the predicated truth value or its negation. This condition compels us to distinguish a sequence of orders of statement forms and truth values. Forms of zero-order do not contain sub-expressions denoting any truth value. Denoting these forms by p_0, we can describe a truth value predicated of a p_0 as a 'first-order truth or falsity'. If an expression p contains symbols for nth order truth values, then of such a p we can state only an $(n+1)$ order truth value. This recursive condition must be satisfied by any definition of truth, including the 'semantic' and 'absolute'. The form of definition used in either case, $T(p) \equiv {}_{df} p$ ('"p is true" is equivalent by definition to p')[4] can, in our opinion, be recursively defined. We must choose a recursive form since otherwise the definitions of truth can lead to paradoxes, as we shall see. From the recursive form of definition it follows that in any statement of the form 'p is true' continued decomposition must in the end reveal a constituent of form p_0.

3. 'POSITIVE' AND 'NEGATIVE' USE OF STATEMENTS

In order to recognize how the form of definition $T(p) \equiv {}_{df} p$ is to be specified recursively, we must give a clearer account of the sense of truth value ascriptions such as 'p is true', 'p is false' and so on. If a statement p is always also to convey 'p is true' and this meaning is assigned to p by stipulation (by definition), it is not a necessity of thought that p should have this meaning. This manifests itself in the fact that we can think of the statement of p as true or false without change in its meaning, provided it is not logical and analytic, but synthetic and empirical. Therefore instead of stipulating $T(p) \equiv {}_{df} p$ we can stipulate something else. That this actually occurs in many cases we shall presently explain.

In practice, the stipulation mentioned signifies a linguistic rule of use for the way we are to understand communications. If somebody tells us that p, then we are to take it as true (understand it as a true message). Let us call this rule adopted almost everywhere in science as well as in everyday use of language a 'positive use of statements'.

In practical instances it may happen that we convey a message p while not following the rules of positive use: here p is supposed not to be taken as a true message. This is the case for instance if we tell somebody the truth in a mock-ironic manner. Suppose a painter shows us pictures and asks us for our opinion, but we find them bad, it is precisely by saying 'splendid pictures' in an ironic tone, that we can tell him what we think. The ironic tone is to convey that in this case the message p is to be understood as 'p is false' that is $F(p) \equiv_{df} p$. This stipulation is logically just as possible as $T(p) \equiv_{df} p$. It is a matter of arbitrary choice according to which of the two we wish to convey something in a given case. In linguistic usage, the stipulation $F(p) \equiv_{df} p$ signifies the rule that a statement p is to be understood as 'p is false', a rule of use that is certainly possible; we shall call it the 'negative use of statements'.

In itself the stipulation $F(p) \equiv_{df} p$ no more contains a contradiction than the much better known and usual definition $T(p) \equiv_{df} p$. Both must be made more precise by being stated in recursive form. A contradiction is obtained if we use one and the same statement positively and negatively at once. The two definitions are of course incompatible.

4. NEGATION AND FALSITY

As mentioned above, the negation 'not' and the expression 'is false' must be kept apart in their meanings. This holds for all linguistic systems, two-valued ones included, these last being the only ones concerning us here. (It is of course possible in setting up a linguistic system to define negation as meaning 'is false'; if we do, then we are not using the negation sign in the usual sense of propositional calculus but in some restricted way, about which more later.)

It is appropriate to recall that we may use statements negatively, in order to clarify the difference between 'not-p' ($\sim p$) and 'p is false'. We need mention only that in propositional calculus $\sim p \equiv p$ is always a contradiction; while the definition 'p is false' $\equiv_{df} p$ is not in itself inconsistent (it leads to contradiction only if used at the same time as 'p is true' $\equiv_{df} p$).

Construed as a rule, the latter expresses negative use. That it is consistently usable is shown not only by the mentioned examples of 'ironic messages' but also by occasional games in which negative use is deliberately included in the rules but under exclusion of the negation sign.

An example of this last kind occurs in bidding at bridge. Here a player must reveal his hand to his partner, while observing certain restrictive rules, amongst them the veto on any kind of negation sign. The only permissible bids are of the form 'my hand corresponds to "one club"' and so on. In some cases one must tell partner that one is void or not strong enough in a suit: it is part of the logical attraction of this bidding game that negation signs must not be used for such messages. There are various logico-linguistic rules for conveying them. Sometimes it is enough to pass, which conveys abstention from any bid including the one desired by partner. In other cases one can make a bid incompatible with the expected one, for example by calling a different suit, from which partner infers that the bidder's hand lacks support in the suit required. At times, however, one cannot use these rules because they would mislead partner. It may then be possible to convey this lack by means of negative use, bidding the suit in which appropriate support is lacking. Here the statement p (say, 'my hand corresponds to "two clubs"') is conveyed in the sense of 'p is false'.

There are several causes for the undoubted reluctance to allow negative use as logically equally possible along with positive use: since the latter has been used throughout almost all linguistic systems, it has become a habit of thought that a message p can be understood only as 'p is true'. We thus feel the stipulation $T(p) \equiv_{df} p$ to be a necessity of thought and believe that any change in it must lead to contradiction. We therefore further feel that p and 'p is true' must mean the same, and that the above stipulation is based on a tautological equivalence. However, it is precisely this extreme consequence that shows itself to be untenable: for if p and 'p is true' necessarily meant the same it would be impossible to think that an un-negated empirical statement is false, whereas it is a mark of empirical statements that we can think of them as true or false sentences without change of meaning. Moreover, it would follow that in understanding an empirical statement p we must always grasp that it is true into the bargain (as with logically true sentences); but this runs counter to the epistemological character of empirical sentences. The habit of thought that identifies the meanings of p and 'p is true' is thus real but logically unjustified. That is why the stipulation $T(p) \equiv_{df} p$ is logically possible but by no means based on a necessary (tautological) equivalence.

Once we have seen this, another equally ingrained habit of thought turns
out to be equally unjustified in logic. The view that 'not-p' and 'p is false' are
identical in meaning is widespread. It may be understood as a logical consequence as the above mentioned habit, for if we negate the two sides of 'p is
true' $\equiv p$ construed as a tautology, the equivalence is maintained and we
obtain the tautology 'p is not true (that is, false)' $\equiv \sim p$, which states the
meanings of 'not' and 'is false' to be the same. If, however we recognize the
former equivalence as a possible stipulation that is not tautological, it follows that the same is true for the latter, so that logically we may stipulate 'p
is true' $\equiv_{df} p$ as much as 'p is false' $\equiv_{df} p$.

That negation and falsity do not mean the same can be seen from other
examples too. In general, the function of negation is to delimit various domains from each other. In negating concepts (for example, non-smoker)
this is easy to understand. Thus classes and their complements are marked
from each other by negation. Applied to statements, negation to begin with
means only that they are to be classified into two domains. These latter may
be construed as the classes of true and not-true statements respectively, but
this is by no means the only, that is, necessary, way of interpreting the classification. Above all, we may reserve the un-negated class for false statements instead of true, so that non-false (true) statements would go into the
negated class. In that case $\sim p$ would mean 'p is true' and we might feel inclined to regard negation as necessarily identical in meaning with truth.
However, one might classify statements quite differently and independently of their truth values, where the negation sign can be used in the same
classifying way as in the case where the classes are the true and the false. If
statements are classified into finite and non-finite (transfinite) we actually
do not use the negation sign to mark the elements of the negated class, but
in principle there is no reason why we should not, although it could no
doubt lead to the inappropriate view that negation and transfiniteness were
identical in meaning.

When contrasting the definitions of 'semantic' and 'absolute' truth, we
explained that it is not easy to indicate what the concepts 'true' and 'false'
mean. This still imcompletely clarified sense of truth values may further the
belief that 'not' and 'is false' have the same meaning. It may therefore be
fruitful to indicate a classification of statements with the help of the negation sign, but with predicates completely clarified in content as distinguishing feature. For example, the sentences in a book might be printed partly in

black and partly in red ink, these being the only two colours, so that we can classify sentences into black and non-black, denoting them by p and not-p respectively. However, non-black here means red, so that one might feel tempted to take 'not' as meaning the same as 'is red'.

If, in this way we recognize that p does not have to mean the same as 'p is true', nor $\sim p$ the same as 'p is false', it becomes intelligible that rules of positive and negative use are equally possible logically. However, one might here object that for any given linguistic system only one of the two stipulations is admissible troughout, since 'p is true' $\equiv_{df} p$ and 'p is false' $\equiv_{df} p$ are mutually incompatible. This objection is valid as soon as we are dealing with a closed linguistic system, of the kind aimed at by exact science. However, the language of everyday speech and any other exclusively practical messages do not make stipulations to be used throughout. Rather, the rules that are set up are only conditionally valid. For example, under certain empirical conditions linguistic expressions are to be used and understood according to the rules of positive use and under other such conditions according to those of negative use. No attempt is made to give a systematic total articulation of practical languages and connections between statements. Depending on whether a statement is to be understood as a true or as a false message at a given moment, we use statements positively or negatively, adapting to empirical circumstances, as can be seen clearly from the example of normal against mock-ironic messages or from the use of statements in special languages. That positive and negative use are mutually incompatible is of import for practical communication only where we test the consistency of a connection between statements made (for example reports), where we must of course know for each statement whether it is to be understood according to the rules of positive or negative use.

5. CONFUSION OF POSITIVE WITH NEGATIVE USE[5]

A special kind of semantic paradox arises from the simultaneous application of positive and negative use to statements, that is, mixing up the two possible kinds of use.

Suppose a person A at time t makes the statement p, in symbols $S(A, t, p)$, no matter whether p stands for a sentence or a proposition, since our argument will apply equally to semantic and absolute truth. If at t, A asserts the single statement p, symbolic logic marks this as follows.

Using $F(p)$ for 'p is false', the expression

$$F[(\iota p)\ S\ (A, \iota, p)] \tag{1}$$

means 'the one and only statement that A made at ι is false'.
Denote the expression in (1) by E, so that

$$E =_{df} F[\ (\iota p)\ S\ (A,\ \iota,\ p)] \tag{2}$$

Now let A assert E at ι, so that we have the identity $[(\iota p)\ S\ (A,\ \iota, p)] =_{df} E$, which substitued in (2) gives

$$E =_{df}\ F(E) \tag{3}$$

The meaning of (3) is simply that E is to be used and understood according to the rules of negative use. If we do not take notice of this rule fixed for E and simultaneously use it in the positive way that has become a habit of thought, we thereby stipulate $E \equiv_{df}$ 'E is true', or in symbols

$$E \equiv_{df} T(E) \tag{4}$$

From (3) and (4) we obtain the contradiction $F(E) \equiv_{df} T(E)$. Note that neither (3) nor (4) is in itself alone inconsistent. Contradiction occurs only when both stipulations are simultaneously applied to the same statement E. The well-known paradoxical sentence 'this sentence is false' has this form of a semantic contradiction. By 'this sentence' we are here to understand the characterization 'the one and only sentence written in this specific position on the sheet of paper'. The statement that this particular single statement is false is then of form (1), but that statement (let it be called E) states its own falsity; that is, E states that it is equivalent to 'E is false', or $E \equiv F(E)$ which has the form of (3) and states that E is to be understood according to the rules of negative use. If now we also adopt the rule that with any statement p it is asserted that p is true, this amounts to recognizing $E \equiv T(E)$ as valid, which is form (4) and with the previous formula gives $F(E) \equiv T(E)$: from the falsity of E follows its truth and conversely.

6. THE RECURSIVE DEFINITION OF THE CONCEPT OF TRUTH

In the last example of a paradoxical sentence the concept of truth is used in a sense that is not compatible with the recursive definiton of the truth value concepts. This inadmissible use of those concepts is logically connected with the above mentioned simultaneous application of positive and negative use to the same statements. Designating a statement not containing expressions for truth values by p_0 as before, and the sentential function 'x is true' by $T(x)$, it is expedient to define truth in the form $p_1 =_{df} T(p_0), p_{n+1}$

$=_{df} T(p_n)$. This definition[6] presupposes for any truth value statement ($T(p)$, $F(p)$, $T[T(p)]$ and so on) that we can indicate a sub-expression p_0 having the form of a statement but not containing expressions that denote truth values. For instance, the sentence 'it is true that it is snowing' has the form $p_1 \equiv T(p_0)$. The sub-expression 'it is snowing' has the form p_0. The sentence 'it is false that the sentence "the earth is spherical" has always been true' is of the form $p_2 \equiv F(p_1) \equiv F[T(p_0)]$, with 'The earth is spherical' as p_0.

According to the recursive definition, the sentence 'this sentence is false' would have to have in it a sub-expression in statement form but not containing any term denoting a truth value; that is, the sub-expression must be of form p_0, but there is none such in the sentence. The sub-expression 'this sentence' is not of statement form but denotes an object which is itself of course a statement, namely 'this sentence is false'. This last, however, contains the expression 'false' which denotes a truth value, so that the original sub-expression is not of form p_0. Therefore in the original sentence and in the paradoxes arising from it, the concepts of truth value are used in a way that is incompatible with the recursive definition of truth.

If the sentence 'this sentence is false' had in it a sub-expression of form p_0, then we could not derive the paradox from formulae (1) – (4) in the previous section. The following example will make clear that, conversely, the non-occurrence of a sub-expression p_0 in a truth value statement always allows us to use the schema (1) – (4), or a logically analogous one, to apply positive and negative use simultaneously to that statement and thereby generate a paradox.

Hitherto it was thought that a paradox could be derived only from the sentence 'this sentence is false', but not from the sentence 'this sentence is true'. This view is based on the erroneous presupposition that positive use is a necessity of thought. That negative use is possible is not even considered. Suppose, then, that we use the statemenents of a linguistic system L according to the rule that every message p in L is to be understood as 'p is false', and $\sim p$ as 'p is not false', that is, 'p is true'. If in L we write down the sentence 'this sentence is true' and call it E, then E asserts its own truth, or $E=$ (E is true).

This can be interpreted as the rule of positive use almost always adopted in our usual linguistic systems. However, E belongs to L, for which we have chosen by convention that the rule of negative use is to hold, so that any statement p is to be equivalent to 'p is false', and therefore $E \equiv$ (E is false).

From these last two relations there arises a paradox which is of exactly the same logical semantic character as the paradox usually demonstrated as arising from the sentence 'this sentence is false'. In the sentence 'this sentence is true' too there is no expression of form p_0. The concept 'true' in this sentence is therefore not used in the sense of a recursive definition of truth. This alone enables us to give E the meaning 'E is true' and at the same time apply negative use to it, which generates the paradox.

NOTES

[1] Cf. Tarski, A., 'Der Wahrheitsbegriff in den formalisierten Sprachen', *Studia Philosophica* **1** (1936), from whom the semantic definition is taken over and generalized by Carnap, R., *Introduction to Semantics,* Chicago 1942 and 1946. A definition of truth that has the same significance is given by Carnap, R., *Symbolische Logik,* Vienna, 1954, p. 89ff, although it is not there represented in symbols.

[2] The contrast of the 'semantic' and 'absolute' definitions of truth figures already in Carnap, R., *Introduction to Semantics,* Chicago 1942. The two are not as yet clearly distinguished there. On the one hand, he tries to define the absolute by the semantic concept, while on the other he speaks of the 'absolute' truth of propositions as independent of the existence of linguistic systems, if I understand him aright. In his *Symbolische Logik* (l.c.) he deals only with semantic truth. A. Pap regards the semantic definition as inadequate. I know his view mainly trough personal communications. He believes he can give an exact and satisfactory symbolic representation of the absolute definition of truth.

[3] Both follow from the definition of 'semantic' and 'absolute' truth given in Carnap, R., *Introduction to Semantics,* Chicago 1942 and 1946.

[4] That the semantic definition puts the first p into quotation marks to indicate that we are to understand by it the semantically signifying sentence 'p', is immaterial for the conditions that all truth definitions must satisfy and that are of interest to us here.

[5] Cf. the arguments of this section with Juhos, B., *Elemente der neueren Logik,* Vienna, 1954, p. 219ff.

[6] Strictly speaking, the symbols T in the defining equations should be distinguished by indices. For what follows this is, however, unimportant.

CHAPTER IX

THE NEW FORM OF EMPIRICAL KNOWLEDGE*

1. THE 'EMPIRICAL CONTINUOUS' FORM OF KNOWLEDGE OF
CLASSICAL PHYSICS

Empirical science aims at recognizing and describing the connections of phenomena in the world of experience. Description is a representation of these connections by means of concepts. For this we require two kinds of concepts, namely those by which we mark phenomena and those by which we convey their relations.

In the exact empirical sciences the phenomena to be described (there called 'states' and 'changes of state') are marked by metrical quantities or changes in these. The regular relations of phenomena, for example their regular sequence in time, are represented by functional relations between metrical quantities or changes in these. The two kinds of concepts used for describing how phenomena hang together are therefore metrical quantities or functional relations between these. In modern logic, functions with one argument are called 'properties' or 'one-place predicates', and functions with two or more arguments 'relations' or 'many-place predicates'.

In empirical description what matters is therefore to look for many-place predicates that express the relations between the metrical quantities marking the phenomena. The most general form of such relations, stated in laws of nature, concerns the way changes in different quantities depend on each other. For example: the path covered by a body changes with 'time' according to such and such a function.

Amongst the changes that may occur in states and in the metrical quantities marking them, special importance attaches to 'differential' changes. To the mathematical form of a differential change in a metrical quantity there corresponds in empirical description an arbitrarily small but measurable change of state. For the form of description (and therefore of knowledge) in classical physics it is characteristic that it presupposes that we can in principle sufficiently mark and completely describe all states and changes in them by means of metrical quantities and differential changes of them.

---

* Translated from the German original 'Die neue Form der empirischen Erkenntnis', first published in *Archiv für Philosophie* 8 (1958) 110–128.

From this presupposition there follows a quite determinate form of laws of nature by which classical physics seeks to describe how phenomena are connected. A state is said to be sufficiently marked if characterizing quantities are assigned to it in such a way that with the help of the relevant laws of nature we can derive testable predictions. The elementary metrical quantities used by exact science for marking phenomena (states) are 'length', 'time' and 'mass'. All the rest (like 'velocity', 'acceleration', 'momentum', 'energy', 'action' and so on) are defined as relations of the three elementary quantities. In order sufficiently to mark a state at least two measured values are needed (for example a position and a momentum marking a body's state at a point in time).

If now we presuppose that the sufficient marking of states can be achieved only with the help of measurable values and that further all changes in state (and especially arbitrarily small ones) are amenable to measurement, allowing representation by corresponding (and perhaps differential) changes in the metrical quantities, it follows that the content of the descriptive differential laws is the mutual dependence of differential changes in the various metrical quantities. Differential (that is, arbitrarily small) changes in metrical quantities and therefore in states are to be in principle amenable to measurement, according to the presuppositions of classical physics. Such changes are also customarily described as occurring in arbitrarily small or proximate spatio-temporal domains. Successive changes of state in space-time domains are also called transmissions or transfers of action.

Laws that describe transmissions of action (changes of state) in arbitrarily proximate spatio-temporal domains we call 'laws of proximate action'. Actions have a speed of propagation. It is part of the presuppositions for laws of proximate action that empirical actions travel at finite speeds only. If we assume an infinite speed of propagation (as classical mechanics does for gravitation) we can no longer sensibly speak about changes in state in arbitrarily proximate domains, since changes appear everywhere 'instantly', that is, independently of time. 'Propagation of changes in space-time domains' now becomes an empty concept. If nevertheless the laws for action at a distance (namely the transmission of actions at infinite speed) appear in the form of differential laws in which metrical quantities are differentiated with respect to time as well, what may be derived from these laws regarding changes in state in neighbouring space-time domains is empty, because action at a distance is time-independent. Such derivations do not denote changes in state ascertainable by measurement.

Thus we may regard laws of action at a distance as degenerate laws of proximate action: for they have the external form of the latter, which make statements about nothing but relations of dependence between differential changes in various metrical quantities. However, while in the case of genuine laws of proximate action such statements always count as testable by measurement of the corresponding arbitrarily small changes in states or measured values, in the case of laws of action at a distance it is in principle impossible to speak of measurable differential time changes in certain metrical quantities. In view of the identity of form of the two kinds of laws in classical physics and the above-mentioned difference in their contents, laws of action at a distance are often called 'laws of pseudo-proximate action'.

What causes the degeneracy is the presupposition that there are transfers of action at infinite speeds, which transcends experience. Leaving aside this assumption, which certainly lies beyond all conditions of measurability, we can characterize the form of knowledge in classical physics as regarding its object (namely states and changes of these in their regular connections) as sufficiently and completely describable by quantities whose individual values and their changes can be captured by measurement. An essential mark of the differential laws that represent this form of knowledge consists in the fact that in them metrical quantities are always differentiated only with respect to other such quantities. This presupposes that every metrical quantity is continuous and that its continuous changes are measurable. In a certain sense we therefore can denote the form of knowledge of classical physics as 'empirical continuous'.

2. UNCERTAINTY DOMAINS AS PREREQUISITE FOR THE EMPIRICAL-FICTITIOUS FORM OF KNOWLEDGE

In relativity theory, we give up the assumption, which transcends experience, that there are actions transmitted at infinite speed of propagation. What is here made into the basic principle of empirical knowledge is the insight that measurement always means the establishing of a transfer of action (chain of action) between measuring instrument and the phenomena to be measured. If these transfers only ever travel at finite speed, and here there is even a finite maximum speed for the propagation of action, they will affect the form of possible laws of nature under those conditions and along with it the general form of empirical knowledge.

If we are dealing with the propagation of actions at speeds that cannot be neglected in comparison with the speed of the signals used for measuring, then this latter is simply one of the phenomena whose regular connections are to be ascertained by measurement in the given case. Thus, in such cases the speed of signals (which is a metrical quantity characterizing the measuring instrument) is one of those metrical quantities between which the law of nature to be determined states a functional relation of dependence; therefore that speed belongs to the set of arguments of the function to be established, that is, to the object of knowledge.

This form of laws means something new as against the classical forms insofar as it now seems that laws of action at a distance are impossible. All other metrical quantities depend on the signalling speed occurring in the formulae in such a way that for transmissions above that speed the descriptive functions become meaningless.

Logically more important still is the circumstance that from this finite limit for transmission speeds a relation between phenomena may become derivable that can in principle not be set up by means of classical physics and thus signifies a new form of concepts and leads to a new form of knowledge. From the fact that there is a finite maximum transmission speed it is known to follow that under determined conditions no relations of action can exist between phenomena. In such a case we say that the phenomena have the relation of indeterminate temporal sequence or action.

The conceptual definition of an 'uncertainty domain' for phenomena is not possible in the system of concepts of classical physics. If the relation of indeterminate action exists between the measuring instrument (observing system) and the state to be measured, we cannot set up a transmission for the purpose of measuring. Thus, the state to be measured must remain undetermined as to the values of measurement that mark it. Yet phenomena that are indeterminate as to action with respect to the observing system are by no means mere unreal ad hoc entities: they are empirically real in the same sense as other phenomena to which we can ascribe individual characterizing values by measuring procedures. Moreover, one can indicate or determine a spatio-temporal order for phenomena in uncertainty domains, enabling us to obtain further predictions. However, this requires a new logical step determining the new form of empirical knowledge so obtained.[1]

For phenomena in the relation of indeterminate temporal sequence or action, a spatio-temporal order can be fixed only by stipulation that we can determine which amongst a set of phenomena in indeterminate temporal sequence are to count as simultaneous, earlier or later. Since because the states are in a relation of indeterminate action it is in principle impossible to measure (that is, neither ascertain nor test) the 'values of measurement' arbitrarily assigned to the individual states (phenomena), these stipulated 'values' are fictitious. This characteristic then obviously belongs to the whole system of values stipulated to mark the phenomena occurring in uncertainty domains; that is, the space-time metric thus chosen is fictitious. If phenomena or states are in principle inaccessible to measurement, what, we might ask, could be the point in marking them by arbitrarily assigned values for which we stipulate an arbitrary spatio-temporal order? Surely, empirical description rests on observed (measured) values and is to be testable by observation (measurement). If now phenomena in uncertainty domains are marked by fictitious values (and there is no other way of doing this under the conditions mentioned), anybody can choose a fictitious system of values or assign arbitrary individual values to phenomena and fix functional relations between them. Yet this is surely not an objective scientific description of the course of phenomena that anybody can test, nor therefore can it (when sufficiently confirmed) hold for everybody.

To this we must reply that the description of phenomena in uncertainty domains does not stop at assigning fictitious values to states. Rather, the proposition expressing these fictitious markings are linked with observational propositions in which values obtained by measurement are assigned to other phenomena. We can then use the fictitious ascriptions and the observational propositions as premises from which testable predictions may be derivable. Clear examples for this are the predictions derived in relativity theory from time relations determined by measurement and ascription of fictitious simultaneity.

If predictions obtained with the help of fictitious assignment of values are sufficiently confirmed by observation, we call the chosen fictitious metric expedient; otherwise (if the derived predictions largely fail to be confirmed) we reject the fictitious system of values as inexpedient and replace it by a new system or merely modify it, which again amounts to an arbitrary stipulation, since the new values assigned to phenomena are again inaccessible to measurement; the fictitious character of values assigned to phenomena in uncertainty domains persists.

Empirical description with the help of fictitious values (always manda-
tory with uncertainty domains if we are to obtain testable predictions) no
longer satisfies the conditions and marks of classical empirical continuous
forms of knowledge, for that requires states describable exclusively by metrical quantities whose individual values and differential changes in them
are measurable. This implies a method of marking phenomena by values
that are obtained by measurement alone or are in principle testable: this
method ceases being applicable as soon as uncertainty domains arise, no
matter why they supervene, as we shall see. In the case of relativistic uncertainty of action or temporal sequence, they arise because of the finite limit
to the speed of propagation of action, in quantum physics because of something else. In every case, however, the only order we can stipulate for phenomena that to the observer lie in uncertainty domains is fictitious, and this
we do by choosing a fictitious value system and therefore by assigning fictitious individual values to individual phenomena. In the relativistic metric
the choice is by stipulating that electromagnetic waves should have the
same constant velocity c in all directions independently of the observer's
state of motion and further that all mechanical and electromagnetic laws
are to be invariant to translational transformations. From this we can deduce (calculate) what individual values (say, of time) are to be assigned to
the individual phenomena in uncertainty domains. These relativistic value
assignments are known to be in principle neither obtainable nor testable by
measurement, so that like the chosen metric itself they are fictitious. To this
corresponds the fact that the relativistic metric is one amongst many possible ones, chosen only for being expedient, which will be the case if it permits
a description that is as simple as possible (using a minimum of auxiliary hypotheses) and at the same time allows predictions that are as far-reaching as
possible. If the fictitious relativistic metric failed to satisfy these criteria,
that is, if it were inexpedient, then it would have to be replaced by another
metric that would, however, be again fictitious.

In view of this logical step of assigning fictitious values to phenomena
in uncertainty domains, that is, stipulating a fictitious order that when
combined with observational propositions leads to the derivation of testable predictions, it seems appropriate to give the name 'empirical fictitious' to the form of knowledge gained with the help of this new method.
The new form differs from the empirical continuous one of classical physics

in that it is in principle impossible always sufficiently to mark the phenomena (states) with the help only of metrical quantities, that is, quantities whose individual values and differential changes can in principle be captured by measurement. On the contrary, according to the conditions for the empirical fictitious form of knowledge, in order to mark phenomena in uncertainty domains sufficiently, one must use fictitious values along with measured ones. However, this is done by complementing the system of observational values by a fictitious value system, or by ranging the former into the more comprehensive latter. This method of recent empirical enquiry we shall call the 'method of fictitious predicates'.[2] We find it applied in an especially clear manner in quantum physics.

3. PROBABILITY DESCRIPTION, A SPECIAL CASE OF THE EMPIRICAL FICTITIOUS FORM OF KNOWLEDGE

The fictitious values used in relativity theory for fixing a space-time order for phenomena in uncertainty domains as to action or temporal sequence have external form of measured values as used in physics for describing phenomena. If for example we stipulate fictitious simultaneity for phenomena in uncertain temporal sequence, then we ascribe to them fictitious time values that have the same external form as measured ones. The fictitious values thus introduced are treated within the system of descriptive propositions by the same rules and operations as the observed ones (for example in derivations); in particular they are stipulated to be continuous and differentiable, as must be the case for all metrical quantities on the basis of the presuppositions of classical physics.

That is why the laws of proximate action in relativity theory do not differ from the corresponding laws of classical physics: in both, metrical quantities are always differentiated with respect to other such quantities. There is of course a difference of content between classical differential expressions and certain relativistic ones: the former express the way arbitrarily small and in principle measurable changes in a metrical quantity depend on analogous changes in other metrical quantities, while the latter may express relations of dependence between differential changes of quantities some of which are fictitious and in principle not measurable either as to individual values or as to changes in these. One might then speak of 'fictitious' differential expressions, whose importance within physical description consists in fixing a relation between the order stipulated for phenomena in uncertainty domains and the order that can be marked by observed values. By thus functionally relating fictitious to observed values it becomes possible to derive new predictions that may be tested by measurement (observation). In contrast with relativity theory, where introduced fictitious values have the same external form as measured ones, quantum physics in describing phenomena uses fictitious values to define quantities whose form is logically quite different from that of metrical physical quantities. These new quantities are 'probability quantities'.[3]

The appearance of uncertainty domains in relativistic measurement is due to the impossibility, under certain conditions, of transfers of action between phenomena and therefore of measuring procedures to determine measured values marking those phenomena. There simply are no measuring procedures that under these conditions would allow us to capture either individual values of the metrical quantities concerned or changes in them (for example the quantity 'time').

However, it is conceivable that under certain conditions there may well be measuring procedures for ascertaining these values, but with limited accuracy only. The unavoidable inaccuracy can be indicated in the form of a value range for the metrical quantities concerned. These 'uncertainty ranges' are then the uncertainty domains within which we can in principle not determine anything by measurement as regards individual characterizing values or changes in them.

Now it might be thought that here too one should be able to stipulate fictitious values for the uncertainty domains so that one exact fititious individual value is assigned to each metrical quantity within these domains. This would correspond to the procedure used in relativity theory, whereby phenomena in domains of uncertain temporal sequence have such exact fictitious individual values assigned to them. However, experience shows that this form of the empirical fictitious method is not expedient in quantum physics and fails to lead to empirically useful predictions. For when the experiments are performed it turns out that ascribing of exact fictitious individual values to states in uncertainty domains leads to predictions that are not sufficiently confirmed by the relevant observations; rather, the observations (measurements) reveal that when the experiments are repeated under the same conditions of observation, the possible individual values of a metrical quantity in an uncertainty domain appear as a statistical distribution. This empirical fact, obtained for metrical quantities under certain conditions that sufficiently mark the phenomena, suggests that we should mark states in uncertainty domains by means of fictitious quantities of a special kind. Since the individual values of characterizing metrical quantities (for example positions and momenta) in a large number of observations can be determined only in a statistical distribution, it here seems expedient to assign to states, or to the individual measures assigned to them, values that can be interpreted as statistical and tested by statistical methods.

Interpreting a value as statistical, that is, as a probability value, presupposes that we define a function (probability function) whose range of values includes the one in question. Now it is important that for a given set of values (measured or statistical) there are many, indeed infinitely many, functions whose value ranges include the given ones: from these infinitely many and logically equally possible probability functions for marking the individual values of states we must select the expedient ones in each case. A probability function is said to define a probability quantity. Amongst the argument values of a probability function there are always the individual values of the metrical quantity to which the probability quantity is assigned. The values of such a function are the probability values that indicate with what probability we may expect a state marked by a given individual measurement to occur. Whether a probability function is expedient depends on the predictions derivable by connecting probability propositions with observational ones. Such predictions are always probable in character and are tested by way of statistics.

What is decisive for this procedure of assigning probability functions or quantities to states or to the metrical quantities marking them is the fact that such functions can be defined only in domains at least some of whose values are fictitious. Already in the simple case where a continuous probability function is assigned to a discontinuous metrical quantity, it must be presupposed that the possible discrete individual measures are ranged into an appropriate system of fictitious values, for only in such a system can we define a continuous probability function. Thus, ascribing a probability value to a phenomenon always contains the assertion that the occurrence of that phenomenon depends on the possible (that is, more or less probable) occurrence of other phenomena. However, only one of the possible phenomena can occur, while the others are never realized and thus are purely fictitious. To put it less precisely, a probability description expresses the occurrence of a phenomenon as depending on fictitious phenomena, or ones marked by fictitious values.

4. PROBABILITY FUNCTIONS AS RELATIONS BETWEEN MEASURED AND FICTITIOUS VALUES

The choice and assignment of probability functions to states or to the metrical quantities marking them are tied to the following conditions. In the quantum physical cases concerned we can indeed obtain exact individual values by measurement, except under precisely those conditions which alone allow a sufficient characterization of the phenomena. Further, if we make no measurement on the phenomena, we can subsequently measure the final state to determine all the exact values that are needed to mark the sequence of phenomena that has occurred. In view of this possibility we can indeed speak of possible states or of the individual measures marking them.

Because we cannot obtain exact measures (of conjugate quantities as they are called) under the conditions required for a sufficient marking of states, only discontinuous changes can be ascertained in the metrical quantities concerned under the relevant conditions. This alone makes it impossible to represent regular connections between changes in state by means of differential changes in the metrical quantities: expressions for such changes become empirically empty where metrical quantities change discontinuously under the conditions required for the sufficient characterization of phenomena. This holds equally for differential laws if within the domain of quantum physical phenomena we were to give them the form of laws of proximate action in the sense of classical physics.

If now discontinuous changes in metrical quantities cannot be represented by differential expressions either, we can nevertheless complement the discrete values by fictitious ones (or range the former into a system of the latter) in such a way that in the value domain thus set up we can define functions that are continuous and differentiable. These we call probability functions, if they can be statistically interpreted, as previously mentioned. Putting it more loosely, complementing discretely measurable values by a fictitious value system amounts to setting empirical phenomena into relation with fictitiously assumed ones. Amongst these latter we must count all those that are 'possible' without occurring, in cases where only one phenomenon can occur. To such fictitious phenomena values are assigned within the value domain set up. Beyond this, the domain may contain individual values or complexes of values that denote not even a 'possible' phenomenon but purely fictitious entities, if we may talk in this way. In the probability functions that relate metrical quantities to the introduced fictitious quantities the probability for the occurrence of an observable phenomenon thus figures as depending on fictitious 'phenomena' that can in part be called 'possible' and in part are denoted by complexes of values not further interpretable.

From these methods of defining probability quantities it results that the probability functions (defining probability quantities) always relate metrical to fictitious quantities. Thus physical probability laws have the form of empirical fictitious knowledge. In the case of quantum physical uncertainty ranges we have to introduce complementary fictitious value domains before we can define differentiable quantities (namely, probability quantities) whose individual values can be assigned to phenomena as 'characterizing' them, and before we can set up differential probability laws describing these phenomena. Of course these laws now express the mutual dependence of differential changes not in metrical quantities but in probability quantities – this very clearly shows the difference in form of empirical fictitious knowledge and empirical continuous knowledge in classical physics. If, for example, we are to measure, under appropriate conditions, the metrical quantities p and q required for the sufficient characterization of phenomena, the relevant differential probability law would express the following: if the probability for obtaining a certain value for p changes differentially, then the probability for obtaining such and such a value for q changes differentially according to the function indicated in the law.

5. THE EMPIRICAL FICTITIOUS METHOD CONDITIONED BY MEASURING PROCEDURES

Between the two forms of knowledge, the empirical-continuous of classical physics and the empirical-fictitious especially as the latter appear in the probability-descriptions of quantum physics, there is a conceptual epistemological contrast which clearly influences theoretical attempts at physical description today. Here we must mention above all various attempts to interpret the same expression first as a probability function and then as a relation of differentially variable metrical quantities.[4] Thus the squared amplitude of guiding waves is interpreted by some theorists as a probability quantity marking the distribution of particles. On this view the guiding waves are fictitious 'phenomena' marked by fictitious values. Some of these values, perhaps those that mark wave interference, can be interpreted as observed values describing the distribution of particles on the intercepting screen. This corresponds to the ranging of observed values (measures) into a fictitious value system in which the functions describing the guiding waves are defined, some of the functional expressions being probability functions assigned to the metrical quantities marking corpuscular motions (that is, when the expressions marking the fictitious guiding waves have been assigned, they indicate the probabilities for the possible individual corpuscular motions). Here we clearly recognize the application of the empirical fictitious method.

In contrast, other theorists[5] try to interpret the waves 'describing' the distribution of particles as empirically real phenomena, which is possible only if amongst the quantities characterizing the waves there are energy quantities that can be proved to be genuinely metrical. Since the waves are described by differential expressions, the quantities occurring in these cases must be construed as differentially variable and metrical. Now the quantities characterizing the guiding waves are differentially variable, but the energy quantities deciding the empirically real character of the waves can here not be understood as amenable to measurement (so that they are fictitious). The importance of guiding waves for empirical description is that some of the expressions marking the waves can be interpreted as probability quantities. Thus the 'measured values' do indeed formally describe waves, but these are here fictitious and, as to content, describe phenomena of quite a different kind: namely the distribution of corpuscles under certain experimental conditions. If the guiding waves are to be construed as empirically real and differentially variable energy waves from whose equations particle motions are to be derivable, then this attempt must indicate a new theory of measurement differing from that of today and providing procedures allowing the measuring of differential changes in metrical quantities under the conditions required for sufficient characterization in each case. In other words, we must indicate measuring procedures whose use will prevent the appearance of uncertainty domains precisely under the decisive conditions.

If the guiding waves, in whatever form, are construed as differentially variable empirically real energy phenomena without indication of such a new theory of measurement, then the account is empirically empty; hanging in the air, as the phrase goes, since no procedures are known for testing the asserted relations between differential changes in metrical quantities. Without indication of an appropriate new measuring theory, our current ones at most allow us to give wave systems of the kind mentioned an empirical content in the sense that some of the expressions marking the waves are interpreted as probability functions. In that case the 'wave functions' come to signify functions defined in a fictitious value system. This interpretation becomes necessary since in the decisive cases the currently available measuring procedures lead necessarily to uncertainty domains that exclude precise measurement of changes in the conjugate quantities required for the sufficient characterization of states. This law which holds for measuring procedures involves the application of the empirical fictitious method in the form of probability functions being assigned to the metrical quantities in question.

From this epistemological point of view, attempts at construing the guiding wave as empirically real and differentially variable energy phenomena may be understood as trying to prove that empirical or grammatical circumstances by no means necessitate a probability description (and with it the application of the extreme form of the empirical-fictitious method), but that it might be replaced by a classical empirical-continuous form of description.

6. THE DEFINITION OF PROBABILITY FUNCTIONS BY REPEATED APPLICATION OF THE EMPIRICAL-FICTITIOUS METHOD

The insight that currently available measuring procedures under conditions required for sufficient characterization necessarily lead to the appearance of uncertainty (imprecision) domains is connected with the so-called procedure of 'quantisation'. Originally, this meant changes by discrete integral amounts in certain quantities characterizing states. Applying this to domains of phenomena of various kinds finally led to quantisation as an automatic application, as it were, of certain operators (for instance a gradient operator in which the quantising constant h occurs) to functional expressions, or as an assignment of quantising matrices to expressions for quantities. As with all mathematical operators, we can apply quantisation repeatedly to the same expression (just as we can with a differential operator, that is, we can differentiate repeatedly).

Physical enquiry has actually used this possibility. We speak of 'first' and 'second' quantisation of various expressions, where these steps always consist in automatically and without explanation replacing terms that, within the formulae to be quantised, have the form of characterizing quantities or functional relations between them, by other expressions that are obtained by applying a quantising operator containing the constant h to suitably chosen functions. In the theory of de Broglie and Schrödinger, which starts from unquantised classical formulae, we easily recognize these steps of quantisation, here of the 'first' kind. However, even where the individual applications of operators are not so evident, as for example when we assign quantising matrices to quantity terms in Heisenberg's theory, the procedure has exactly the same logical meaning of operators applied to corresponding quantities. This becomes especially clear if quantities obtained by first quantisation are quantised a second time.

What all quantising procedures have in common, in whatever form the operators are applied, is that the expressions thus obtained have the meaning of probability functions or probability laws. Still, we must then ask what physical meaning attaches to a repeated application of such operators. To settle that question, consider a case in which second quantisation is applied. In classical mechanics the energy of moving particles can be represented as a function of their positions and momenta. First quantisation in wave mechanics consists in replacing the position co-ordinates and expressions for energy and momenta by operators containing the constant h and applied to suitable functions (Ψ-functions), resulting in a wave equation which, failing a corresponding measuring procedure, could not be construed as an energy wave. However, some expressions marking the waves would be interpretable as probability quantities whose individual values indicate the probability that a position measurement should find a particle (with given momentum) at a place marked by determined co-ordinates.

These wave equations obtained by first quantisation did not, however, satisfy the conditions of the principle of special relativity, that is, they were not invariant under Lorentz transformation. A modification of the wave equations that would make them satisfy the requirements of relativity theory could indeed be carried out, but it turned out to be no longer possible to give a pervasive probability account of all the relevant expressions marking relativistic waves; because in the relativistic form those amongst the expressions construed as probability densities also mark a current at the point concerned. Now the current-density can be positive or negative, but if the latter, it cannot be viewed as a probability quantity which must be positive to be meaningful. Since, however, the relativistic density expressions concerned are understood also as describing a current (a current density, charge density), even though a fictitious one, and density expressions for electric charges and currents can be positive or negative, it seemed obvious to assume the existence of a charged continuum whose field points are marked by differentially variable quantities such as energy and momenta.

We must however not overlook that the expressions characterizing this continuum have been obtained by means of quantisation. By ranging the quantised values into a fictitious value system (represented by the values marking the guiding wave) it became possible to define differentiable functions in this value domain. In their original pre-relativistic form some of those functional expressions could be interpreted as probability quantities: with the relativistic modification of the formulae it turned out that the expressions previously so interpreted could in their new form take negative values as well as positive, which makes a probability interpretation difficult. Nevertheless, since these expressions were defined as functions in a fictitious value system, we may denote them, both in their pre-relativistic and in their relativistic form, as 'first-order probability functions'.

In their relativistic form these expressions marking the guiding waves may be construed as current or charge density terms. The electrically charged continuum, to which we may regard these terms as assigned, is of course itself a fictitious value system, since it was set up with the help of the fictitious value system corresponding to the guiding waves. The fictitious character of that continuum necessarily shows when we try sufficiently to mark field points by means of measurement. To start with, it is quite unclear how measurement could here be possible, that is, how we might relate the fictitious electric continuum to the possible data of observation. To mark the field sufficiently we should have to measure conjugate quantities, but if measurement is possible at all under the requisite conditions, the law holds that for the characterizing quantities only value ranges can be captured. This would involve uncertainty domains for changes in state (at field points) and once more we should have to introduce fictitious value systems and define in them probability functions to be assigned to quantities marking the states, if we are to obtain testable predictions. These probability functions, whose possibility is here at first inferred on the basis of purely epistemological considerations, we should then have to denote as 'second-order probability functions'.

Theoretical physics has indeed made use of this possibility.[6] The wave function ψ to which first quantisation was applied produced, in its relativistic form, expressions that could be construed as quantities marking the state of a fictitious electrically charged continuum. These quantities were then replaced by quantising matrices (the 'second' quantisation of ψ), which led to expressions interpretable as statistically testable value assignments, that is, as probability values. Replacing the quantities marking the fictitious continuum by matrices takes into account the insight that in measuring under the conditions required for sufficient characterisation (that is, in measuring conjugate quantities) we can only ever obtain results in the form of value ranges, while at the same time this replacement signifies the application of quantising operators to the quantities marking the fictitious states. Since this is a case of second application of operators (second quantisation) and the result consists again in obtaining probability functions, we can now indicate the general meaning of single or multiple application of quantising operators. Such operators are always applied to functional expressions whose argument values are at least in part fictitious. By this means one defines probability functions which we call first, second ... nth order, according to how often we have applied the operator. Second order probability functions are 'probability functions of probability functions'. The content of differential probability laws in which second order probability functions occur can be indicated for simple cases as follows: a differential change in the probability w_2' that the probability for obtaining a certain value by measuring the metrical quantity p is w_1', occasions a change, according to the function given in the law, in the probability w_2'' that the probability for obtaining a certain value by measuring the metrical quantity q is w_2''.

Since probability functions as used in quantum physics are only ever defined in value domains that are at least partly fictitious, it is logically impossible to reduce probability quantities to relations amongst metrical quantities only. To this corresponds the fact that the empirical-fictitious method differs from the empirical-continuous one by logical steps that preclude knowledge of the former kind from being reduced to or merely derived from knowledge of the latter kind. The decisive steps determining the logical difference between these two forms of knowledge consists in the introduction of fictitious value domains or the stipulation of ordering relations for the fictitious values, which are in principle inaccessible to measurement, and that precisely makes them fictitous. Because of the uncertainty domains we are compelled, in the sufficient marking of states, to use fictitious values along with observed ones (measured values proper). Hence the classical empirical-continuous form of knowledge, which marks phenomena only by measured values, cannot do this sufficiently when uncertainty domains arise, so that in the cases that are critical for the measuring procedures we must in its stead apply the new empirical-fictitious form.

NOTES

[1] The method that empirical enquiry uses to obtain testable predictions when uncertainty domains occur is subjected to systematic logical analysis in my as yet unpublished essay 'The Method of Fictitious Predicates'. (Published in 1959/60 and reprinted in translation in this volume, pp. 198–343.

[2] By 'predicates' we here mean one-place or many-place functions. Introducing a fictitious value system always defines relations (many-place functions) in the corresponding value domain, that connect observed values with fictitious ones.

[3] For the logical form and content of physical probability functions and statements, see my essay 'Wahrscheinlichkeitsschlüsse als syntaktische Schlussformen', *Studium Generale* 6, 4 (1953) this volume, pp. 93–104; 'Deduktion, Induktion, Wahrscheinlichkeit', *Methodos* 6/24 (1954); 'Mögliche Gesetzformen in der Quantenphysik', *Phil. Nat.* 3 (1955); 'Das Wahrscheinlichkeitsfeld', *Archiv, f, Phil.* 7, 1/2, (1957); 'Die "Metrik" als Bestandteil der physikalischen Beschreibung', ibid. 7, 3/4 (1957); 'Die "Wahrscheinlichkeit" als physikalische Bescheibungsform', *Phil. Nat.* 4, 2/3 (1957).

[4] An account of these methods of theoretical physics is given by March, A., *Die physikalische Erkenntnis und ihre Grenzen,* Brunswick, 1955, p. 86ff.

[5] Here one should mention above all the recent theories of L. de Broglie and D. Bohm. Cf. de Broglie, L., 'Une interprétation de la mécanique ondulatoire est-elle possible?', *Les Conférences du Palais de la Découverte,* series A, No. 201, Paris, 1954; *Une tentative d'interprétation causale et non linéaire de la Mécanique ondulatoire: la théorie de la double solution,* Paris 1955; and Bohm, D., *Quantum Theory,* Prentice-Hall, New York, 1951.

[6] Cf. March, A., l.c., p. 93ff.

CHAPTER X

THE METHODOLOGICAL SYMMETRY OF VERIFICATION AND FALSIFICATION*

Summary. We start from the kinds of proposition occurring in empirical science and examine how the characterizations 'completely' or 'sufficiently verified' (or 'falsified') depend on the criteria of the testing procedures used. The erroneous presuppositions of extreme generalisations like Wittgenstein's 'verification thesis' and Popper's 'asymmetric falsification theory' are exhibited. The methodological symmetry of verification and falsification is made good by reference to the equally indispensable significance that inductive steps (such as extrapolating induction, semantic interpretation and reinterpretation of formal expressions) and testing for inconsistencies have for scientific progress. Taking into account these criteria we interpet empirical universal propositions ('natural laws') as description of the real order of empirical phenomena.

1. LIKE NAMES FOR UNLIKE CONCEPTS

The multifarious enquiries as to how we decide whether a proposition is 'true' have shown that the concept of truth is used in different senses. Even in the domain of formal analytic and mathematical systems of propositions it turns out that to decide propositions different criteria are used in the various domains of forms, from which alone it follows that the 'truth' to be decided is ambiguous. Whether 'truth' is here based on some external form to which the propositions can be reduced, or whether we appeal to an intuitively obtained piece of evidence, is immaterial. Much more important for the meaning of each several concept of truth are the rules applied in the respective decision procedures, in accordance to which true propositions are used. In propositional calculus for example formulae are transformed with the help only of the substitution rule and modus ponens. In the higher calculi these rules are generalised by some additional new rules, amongst which in particular the rules for inserting or dropping universal and existential quantifiers mean something quite new. These new rules make sense only with respect to the greater diversity of forms of the higher calculi, and so the concept of 'logical truth' there gains a new and different meaning as against the corresponding concept defined in the propositional calculus.

Because the transformation and decision rules are explicitly mentioned,

* Translated from the German original 'Die methodologische Symmetrie von Verifikation und Falsifikation', first published in *Zeitschrift für allgemeine Wissenschaftstheorie* 1 (1970) 41–70.

these differences in the concepts of truth stand out particularly clearly in the calculi. The more urgent therefore becomes the question, why we use the same words 'logical truth' for the different concepts. Questions of this kind have been discussed more than once in mathematical foundation research where satisfactory answers have indeed been given. The exact definitions of natural, integral, rational, real and complex numbers in number theory have clearly shown the distinctive features of these types of numbers. If nevertheless mathematics simply calls them 'numbers' in general, this hinges on certain shared features of number concepts, often enabling us to treat the different kinds of number according to analogous rules. The common features must of course not make us overlook the differing properties of the different kinds called by the same name of 'number'. If, as in number theory, this point is clear, no dangers or difficulties arise in the use of the one name for several concepts. Error and inconsistency in ways of speaking does, however, arise where one fails to see or admit the ambiguity, and goes on to draw inferences as to the existence or non-existence of certain entities, the holding or not holding of rules, decision criteria and conceptual connections. In what follows we will examine how far the currently customary interpretation of the methods called 'verification' and 'falsification' of empirical propositions take into account the ambiguity of the concepts concerned. In particular we shall have to watch whether decision criteria valid for mathematical connections between propositions are not uncritically transferred to empirical verification anf falsification. We shall see that where this happens, distorted accounts of empirical methodology result.

2. EMPIRICAL PROPOSITIONS ABOUT FINITELY AND INFINITELY MANY CASES

As to testability, empirical propositions can be divided into two classes: one for propositions about individual cases, or, more precisely, about a finite number of such; and another for so-called universal empirical proposition, where the universal quantifier relates to an unbounded number of cases. To this second class further belong existential empirical propositions, provided existence is asserted of an unbounded domain of cases.

The difference between the two kinds of empirical proposition has often been discussed. In these discussions it turned out that one always tried to

mark the properties of the 'propositional forms'[1] as to their testability, that
is, verifiability or falsifiability.[2] The testing of a finite number of individual
cases is always in principle possible, but with statements about infinitely
many individual cases this obviously becomes doubtful. These questions
have been treated from various rather diverse prior assumptions.

   Some authors deny that even singular proposition or propositions about
finitely many cases are 'completely' verifiable, and therefore refuse to admit
any difference between individual empirical propositions and general
propositions (those about unbounded domains of cases). In this they appeal to the possibility in principle of deriving from statements about what is
empirically given (whether by measurement or experience), even in their
simplest form of basic propositions, protocols or observations, an infinity
of further propositions from which by continued use of verification further
propositions can be derived, and so on indefinitely. Here it is sought to set
aside the difference between propositions describing finitely many individual cases and the unlimited universal propositions, by urging that the verification of either kind requires infinitely many steps, so that neither can be
'completely' verified, from which it is inferred that both kinds are 'hypothetical'.[3] However, this line of argument indicates reasons that are irrelevant to the problem of verification and overlooks distinguishing marks that
persist even if all empirical propositions are regarded as hypothetical, and
it is just on these marks that the epistemological difference between the two
kinds of empirical propositions depends.

   From singular propositions, or propositions about finitely many cases,
we can certainly derive an unlimited number of further propositions, but
only if we doubt whether we have adequately denoted the empirical data (of
measurement, sense- perception, or generally experience) by means of linguistic signs and expressions. Such doubts are always possible and if we
wish to make sure whether in describing individual empirical cases we have
properly and accurately applied linguistic expressions (to which the marks
of phenomenal language belong as much as the metrical quantities of space-
time language), we can indeed derive infinitely many propositions for testing; perhaps as regards the proper working of the measuring instruments,
the absence of external disturbing causes, the reliability of our memory in
using linguistic expressions and so on. Even if such doubts and tests are always possible, it is an essential presupposition for the verification of empirical propositions, in science and in daily life, that the conditions for accu-

rately marking the given cases can be realized (whether by phenomenal expressions or by metrical quantities), provided the methods for empirical linguistic description are applicable at all in the cases concerned. If, however, the conditions for unambiguously marking individual cases are fulfilled, then propositions about a finite number of individual cases are in principle completely verifiable. Empirical enquiry in general deliberately ignores the fact that we can doubt whether empirical data have been adequately and unambiguously marked by language, a kind of doubt also described as one of logical grammar. An exception to this are the quantum physical uncertainty relations which, however, we need not discuss here. Thus in empirical science the presupposition holds, that propositions about a finite number of cases are in principle completely verifiable.

One therefore sidesteps the problem of verifiability of general empirical propositions if one appeals to the above doubt and on those grounds denies 'complete' verifiability even for individual empirical propositions, inferring that individual and universal propositions are alike hypothetical; for even if that were so, the epistemological difference between the two kinds of empirical proposition would persist: namely between statements about finitely and infinitely many cases respectively. This difference influences the conditions of verification of the two kinds of proposition whether or not they are both hypothetical. We shall see that to solve the problem of verification thus construed, different theories of verification and falsification have been sketched. Here we shall try to show that the difficulty as to verifying general empirical propositions can be met by distinguishing two concepts of verification.

3. WITTGENSTEIN'S VERIFICATION THESIS

In mathematics there are propositions about infinite domains that can be decided by indicating construction procedures. For example, the proposition that there are infinitely many primes is proved by means of Euclid's procedure. Besides, there are propositions about infinite mathematical domains which, failing appropriate constructive procedures, cannot be proved; witness Goldbach's hypothesis, Fermat's theorem, or the proposition that the decimal expansion of π contains a sequence 0, 1, 2, ...9. Amongst the latter kind, a special place is occupied by mathematical theses that are simple extrapolations to non-denumerable domains of propositions provable for denumerable ones. The 'well-ordering theorem' or the 'axion of choice' which is equivalent to it are examples of this kind, and no constructive procedures are known that would prove them in general.

These theses, provable in bounded (that is, finite or denumerable) domains, but not universally by any known constructive procedure, however, share the feature that they can be tested 'step by step' in individual cases. Thus any given even number can be tested as to its obeying Goldbach's hypothesis. For any four natural numbers we can test whether they satisfy Fermat's theorem. Up to any given place in the decimal expansion of π we can check whether the sequence from 0 to 9 does occur. For the above mentioned extrapolations to non-denumerable domains, which here interest us less, we can indicate effective well-ordering relations or choice functions for any given denumerable domain.

This circumstance, that a proposition can be tested in arbitrarily many individual cases, though universal validity for unboundedly many cases cannot be settled either way because there is no constructive decision procedure, is somewhat parallelled by the problem of verifying universal empirical propositions.

Looking at the procedures according to which propositions are tested in empirical disciplines, we note that for an individual enquirer propositions of a certain kind count as completely verifiable (either way), while propositions of some other kind count as only incompletely verifiable. This raises the epistemological problem of examining how far this traditional and inveterate distinction is theoretically justified. This distinction essentially coincides with another customary division of empirical propositions into singular ones and those that express universal regularities. The latter are often called 'natural laws', which can however easily lead to misunderstandings, since by no means all universal empirical propositions (which belong to the second class mentioned above) have the character of natural laws.

Let us provisionally leave it open whether one can justifiably speak of complete and incomplete verification. The older epistemological systems did not give critical attention to this distinction between empirical propositions. The general view was that 'empirical' propositions, by which were meant empirical universal propositions and above all laws of nature, were always 'only' probable, since our experience is finite so that empirical propositions are never completely testable. Views of this kind are found in

Hume and Kant as much as in J.S. Mill and Ernst Mach. Perhaps the first to object to the distinction between complete and incomplete verification was Ludwig Wittgenstein.[4] Starting from his thesis that the sense of a proposition is the method of its verification, he emphatically held the view that any proposition must be completely verifiable either positively or negatively. Where the conditions for complete verification are absent, as with laws of nature, we are dealing not with propositions but with linguistic formulations of some other kind. Wittgenstein[5] and following him F. Waismann[5] and Schlick have interpreted laws of nature as rules for setting up propositions of a certain form, that is, as procedural instructions, believing that thereby they had eschewed the incompatibility of incomplete verifiability with Wittgenstein's verification thesis. It has repeatedly been denied that Wittgenstein denoted the method of verification of a proposition as its sense. Amongst the countless theses in Wittgenstein's writings there may be remarks that might be construed as pointing to the independence of a proposition's sense from its verification, but in the recent book *Ludwig Wittgenstein und der Wiener Kreis*[6] there are many notes and actual pronouncements of Wittgenstein's that, unambiguously and beyond doubt, state the verification thesis in its strictest sense.

To support their view that laws of nature, being unverifiable, are not propositions but rules for applying certain forms of concepts and propositions, Wittgenstein[7] and his followers[7] have pointed to a similar view of H. Poincaré, according to which laws of nature are conventions chosen for expediency. The purposes that guide us here are the simplest possible ways of describing how phenomena are connected and the most far-reaching derivation of predictions which must of course be in suitable agreement with empirical data. As criteria for simplicity we adopt formal simplicity of the system of propositions concerned (this simplicity is given preference over simplicity of the individual laws of nature) and intervention of the minimum number of auxiliary hypotheses. If a schematic law chosen by convention turns out to be inexpedient, then it must be modified or exchanged for new conventions. This conventionalism in method is no doubt usable, but the laws of nature set up as conventions differ from those interpreted as rules and procedural instructions (by Wittgenstein and his followers) in that the latter are not supposed to be propositional, while Poincaré's conventions are analytic propositions.

Wittgenstein's thesis that empirical 'universal propositions' are not propositions, because they are unverifiable, has provoked critical reactions in epistemology, leading amongst other things to views that are incompatible partly with the logic of language and partly with the criteria used in practice for empirical enquiry. This produces distortions in methodological accounts, which, we shall try to show, are avoidable if we give up certain dogmatic theses and distinguish certain concepts that take into account the criteria of enquiry in practice.

4. PROPOSITIONS ABOUT FINITELY AND INFINITELY MANY CASES, THE METHODOLOGIES OF WITTGENSTEIN, CARNAP, AND POPPER

As already mentioned in Section 1, mathematics denotes the various kinds of numbers simply as 'numbers', although this covers different concepts treated by rules that are in part different. What determines the use of the same name 'number' for natural, integral, rational, real and complex numbers (and other types) are certain general rules applied in the various domains in more or less analogous ways. If it were only the differences that counted in concept formation, no unified theory of numbers would ever arise. If, for example, only elements of classes that can be well-ordered were to count as 'numbers', most of the arithmetic of real numbers would have to be cancelled. Subsuming different concepts, subject to rules that are partly the same, under one general concept is an essential methodological feature of theory formation.

From this point of view, we must reject as inappropriate Wittgenstein's thesis denying that laws of nature (hypotheses, universal empirical 'propositions') are not propositions because they are unverifiable and therefore neither true nor false. This takes account only of those features of natural laws or universal empirical propositions that distinguish them from statements about individual phenomena: the essential common features and rules valid for both kinds of proposition, suggesting that we subsume them under the general name of 'proposition', are left out of account. And that, in the end, is incompatible with some presuppositions and modes of expression in the methodology of empirical science and also with the procedures of empirical enquiry thus described. It was these difficulties that led some authors to mark and subdivide concepts in a way deviating from Wittgenstein's.

We must, I think, not overlook that the individual sciences denote the recognized general laws (of nature) within them as propositions that are called true, false or more or less probable, depending on the results of testing them. If now we exclude universal empirical 'propositions' from the class of propositions, declaring them to be unverifiable and thus assume these proposition-like expressions to be neither confirmable nor refutable by experience, we first of all conflict with customary modes of speaking in scientific practice. Individual lines of investigation indeed develop most precise experimental procedures for testing the laws of nature, and the result of these methods is usually decisive for marking the law being tested as true, or false, as the case may be. To say that these procedures do not verify because they never encompass all the cases whose regularity the law of nature describes, thus making it unverifiable and no true or false proposition at all, means merely that 'verification' is dogmatically declared to be just one specific kind of test, namely that of finitely many individual cases. The usual reason why, namely that in this way all the cases concerned are being decided, merely emphasizes a difference between testing procedures applied to individual propositions and laws of nature respectively. However, this overlooks what these procedures have in common, and that is precisely what customary modes of speaking in scientific methodology do rest upon, denoting both kinds of descriptive forms as 'propositions' and the tests applied to them as 'verifications'.

The situation resembles that of the above example from number theory. For each kind of numbers the rules are in part different and we may certainly speak of different entities. If nevertheless the theory of numbers and mathematics in general uses the one term 'number' for all these entities, the decisive reason lies in certain common conditions to which they are all subject. Were we to stipulate that only those entities count as 'numbers' for which we can in fact indicate well-ordering procedures, this would entail not only far-reaching changes in mathematical modes of expression but also the elimination of essential parts from number theory and therefore from higher mathematics.

Whenever a theory of knowledge is at odds with the modes of expression, methods and presuppositions of a particular science, we are led to a critical discussion of the epistemological theses concerned. This has happened in the case of Wittgenstein's verification thesis and its consequences too, although I think that these discussions have failed to give a satisfactory account of the steps actually used in empirical enquiry.

Dividing the descriptions of empirical findings into 'propositions' and 'non-propositions' contradicts normal modes of speaking in scientific methodology. This division arising from the verification thesis has not unjustly been regarded as a difficulty. For that reason R. Carnap[8] has tried to indicate criteria for an expression's being a proposition independently of the mode of its verifiability. What combinations of signs in a language have the propositional form is indeed indicated by the logical syntax of that language. In the language used by science we obtain propositions firstly by substituting individual values for the variables in the functional expressions (for example individual constants or measured values): in this way we obtain propositions about individual phenomena; and secondly, also by putting (universal or existential) operators in front of propositional functions: this is the form that general propositions have, irrespective of whether the operators relate to finite or infinite domains of arguments.

If we apply these criteria to empirical forms of proposition, then both individual empirical propositions and general empirical propositions (laws of nature) count as logically syntactically well-formed sentences. That verifiability or the form of verification has nothing to do with the character of a proposition is shown, according to Carnap,[9] by the fact that no empirical proposition is 'completely' verifiable: verification of any empirical proposition, both those about individual phenomena and natural laws, is in principle an unbounded process that can be broken off only by arbitrary decision. In empirical propositions it is thus unjustifiable to infer from the alleged difference in form of the verifications to differences in propositional character. Besides, what an empirical proposition means is not determined and conveyed by indicating how to verify it. Rather, Carnap holds[10], we understand a proposition if we know the meaning of the concepts occurring in it. This condition is just as realizable for concepts in general empirical propositions as in those about individual phenomena.

Thus Carnap detaches the propositional character or form and the meaning of empirical propositions from the criterion of verifiability. The latter is irrelevant as regards the propositional character or meaning of the descriptive expressions and serves only for deciding the truth value and probation of the propositions. Still, to obtain these results, Carnap reinterprets what empirical enquiry understands by verifying a proposition in a manner that is open to question. His view, that propositions about individual phenomena (individual data) are not completely verifiable either and that here too we can go on indefinitely deriving inferences to be verified, leads to a so-called 'consistency theory', according to which empirical propositions are verified always in terms of comparisons as to consistency between the proposition to be verified, or consequences of it, and other propositions. Which propositions are declared false and cancelled in case of inconsistency is a matter of arbitrary decision, dictated by expediency. In some cases it may be expedient to trust propositions about present sense data or observational propositions more than propositions derived from theories and hypotheses, but the reverse may also happen. On this interpretation the verification procedure serves merely to test whether contradictions might not turn up within the systems of propositions. If we meet an incompatibility, the contradiction is eliminated by modifying or dropping one of the incompatible propositions; but there is no empirical verification criterion for deciding which proposition must be given up, a question which rather rests on arbitrary decision, that is on expediency outside the process of verification.

This account does indeed avoid the difficulties arising from Wittgenstein's verification thesis, but it interprets the verification of empirical propositions in a way that not only departs from the testing procedures used in empirical enquiry but also leads to new difficulties that are possibly greater than those avoided. For if in the course of verification we obtain derived propositions incompatible with observational ones and we are free which of the two modify, then contradictions can always be eliminated by retaining as true the propositions (and system of propositions) that are to be verified. Thus, verification construed in terms of a consistency theory can always be carried out in such a way that any arbitrary proposition (or system of propositions) will stand the test of verification. This robs empirical verification of any epistemological value. We need not consider in detail objections to Carnap's verification theory and attendant difficulties, since there is an extensive literature on this.[11]

For Wittgenstein[12] every verification ends as soon as the derivations from the proposition to be tested have led to 'elementary propositions' whose truth can be decided only by observation, that is, by an extralinguistic act. Of what kind these elementary propositions might be, Wittgenstein, of course, does not indicate; indeed, in some of his theses he suggests that propositions of this form cannot be indicated at all. In any case, however, he holds that a process of verification is decided by empirical criteria (observed data). For propositions about finitely many individual phenomena there are only finitely many verification processes of this kind. In contrast, for general empirical propositions (laws of nature) which make statements about unboundedly many cases, there are correspondingly infinitely many testing processes, which are of course not all of them realizable. That, says Wittgenstein, is why laws of nature are neither verifiable nor falsifiable and therefore no propositions at all, but instructions or rules for performing certain actions. Because of this view of theirs, Wittgenstein and his followers are called by Sir Karl R. Popper 'naive empiricists'.[13] Of course, Popper also (and rightly) rejects Carnap's consistency theory. Both Wittgenstein and Carnap give erroneous though different interpretations of the 'verification procedures' used in empirical science. Carnap's view, that the meaning of a proposition is given by the meaning of the concepts occurring in it and is therefore independent of the method of verification, is incorrect. The criterion for whether an empirical proposition makes sense or not depends in definite ways on its testability (verifiability or falsifiability). Popper accordingly gives his own classification of empirical propositions.[14] Wittgenstein's view, on the other hand, is wrong in that he declares laws of nature (general empirical propositions) to be in principle 'unverifiable' (that is, neither verifiable nor falsifiable) and therefore denies that these 'kinds of proposition' are propositions at all. This consequence of his verification theory contradicts the methodology of empirical science: empirical enquiry regards laws of nature as descriptions of the real order of phenomena and therefore treats them as propositions that can be true or false. Wittgenstein's interpretation is incompatible with this way of marking universal propositions. According to Popper this arises from a misreading of certain properties of empirical universal propositions and along with it from too narrow a formulation of the verification thesis.

According to Popper it is incorrect to characterize empirical propositions generally as always meaningfully admitting either truth value; that it would make sense to think of such a proposition as true, and equally as false. Empirical universal propositions differ from propositions about (finitely many) individual phenomena precisely in that the latter can be verified or falsified, while unlimited universal propositions can indeed be falsified but are in principle unverifiable. Popper here points to certain well-known universal propositions from mathematics and asserts that laws of nature have the same epistemological character. Thus Goldbach's hypothesis could be falsified with logical necessity by a single counter-example, but however many even numbers we show as the sum of two primes we shall not have proved that hypothesis. This asymmetry of decidability is characteristic of laws of nature (empirical universal propositions): they are completely falsifiable but not completely verifiable. In this way they differ from individual empirical propositions (about finitely many individual phenomena) which are 'symmetrically' decidable, verifiable and falsifiable alike.

For Popper this gives rise to another way of marking or classifying empirical propositions and a more general form of the verification thesis. All empirical propositions that make sense can be tested, amongst them propositions about individual phenomena as well as empirical universal propositions (laws of nature). The two classes differ in that singular propositions (about finitely many individual phenomena) can be decided either as to truth or falsehood, while universal propositions can in principle be tested only as to falsehood. If a 'proposition' belongs to neither class, it is not a meaningful empirical proposition.[15]

As against Wittgenstein's criterion of sense for empirical propositions, Popper's account certainly has the advantage that on it propositions over bounded domains (about finitely many individual phenomena) and those over unbounded domains (laws of nature, empirical universal propositions) count equally as empirical. Both kinds are testable, the difference as to verifiability or falsifaibility is merely methodological and does not touch their empirical propositional character.

Over Carnap's account of empirical propositions, Popper's criteria have the advantage that, in agreement with the procedures of empirical enquiry, they anchor sense, form and empirical character of a proposition in its testability, while with Carnap these properties of a proposition are determined only by the descriptive concepts occurring in it and 'verifiability' and 'falsifiability' are merely procedures of practical trial. As we have seen, this interpretation of verification leads to a consistency theory of what is empirical truth, in that only formal compatibility and incompatibility of propositions are recognized as criteria of truth and falsehood respectively.

Although Popper's theory avoids these difficulties in Wittgenstein's theses and Carnap's account of science, his own statements and presuppositions likewise give rise to objections, namely as regards the fact that his account of laws of nature and the criteria for verifying them departs so strongly from the methods, concepts and forms of laws actually used in empirical enquiry. Carrying over the form of certain mathematical propositions to laws of nature is effected in Popper on the basis of analogies that are, however, of limited validity and, if assumed otherwise, lead to erroneous characterisations of laws of nature and testing procedures in individual sciences. This shows up especially clearly in the theses of those who start from Popper's falsification theory and draw the consequences with extreme rigour.[16] In the next section we shall consider the difficulties arising from this approach and then try to obtain, by means of new conceptual distinctions, a verification theory that continues to agree with the scientific methods, concepts and forms of propositions actually used.

5. 'ASYMMETRICAL' FALSIFICATION THEORIES

Investigations concerning the epistemological character of laws of nature very often consider especially simple forms of empirical universal propositions, which, in my opinion, however, lack some essential structural marks of laws of nature. The often used proposition 'all ravens are black' is perhaps the best known example of a simple empirical universal proposition, on which one then goes on to study the properties of laws of nature. It is certainly tempting to compare such a proposition with mathematical propositions having the form of Goldbach's hypothesis. If, with empirical enquiry, we assume that we can ascertain the presence of individual states (data, objects, phenomena) with adequate unambiguity and precision — an assumption that is surely justified leaving aside certain microphysical conditions –, then a proposition of the above form can be construed as definitively refutable by indication of a counter-example. By exhibiting a white raven, the proposition 'all ravens are black' would be completely falsified. In contrast, the 'truth', that is, general validity, of this empirical universal proposition would not be proved, however many black ravens we exhibit. Thus, the proposition, which says something about infinitely many cases (objects), can in principle not be 'completely' justified, and therefore not at all. Here we seem to have complete analogy with Goldbach's hypothesis and Popper's falsification theory is nothing but a general extension of this analogy to all empirical universal propositions, laws of nature amongst them.

However, it is a mistake to assume that this analogy is pervasive and that the two kinds of proposition have the same form. The error of extending the analogy becomes obvious if we reflect how (genuine) laws of nature are tested. From the fact that empirical universal propositions can never be verified by any number of positive instances, that is, agreement with actual data, Popper infers firstly that agreement with reality is irrelevant for verification and secondly that empirical induction (which infers from a 'sufficient' number of confirmations to the truth of a universal proposition) is invalid.[17] On the other hand he holds that a negative instance, that is, disagreement with actual data, necessarily falsifies an empirical universal proposition. As to that, we must point out that all empirical testing procedures, especially by experiment, presuppose the validity of the principle of induction. If we were to declare that principle invalid any kind of experimental testing of laws of nature would become senseless. Of any experimental result we presuppose that it would be repeated if we repeat the experiment under the same or similar conditions. Likewise we presuppose the principle of induction as valid wherever we infer, from observed repetitions of sequences of events when the same conditions recur, to the regularity of these repetitions.

The real error in assuming asymmetry of verification and falsification in testing of empirical universal propositions consists in transferring the form of certain kinds of mathematical propositions to the empirical propositions mentioned. The view that laws of nature are necessarily falsified by a single counter-example does not hold, not even for simple universal propositions if interpreted in the ordinary or empirical scientific sense. We continue to hold on to the truth of the proposition 'all ravens are black' even though we know of a few cases of non-black (white) ones. Empirical universal propositions are declared false only when 'disagreements with reality', that is, counter-examples, themselves occur with regularity. Thus a law of nature (empirical universal proposition) can be falsified only by another empirical law, as can easily be illustrated in the simple examples above. The proposition 'all swans are white' was given up as false upon discovery of the Australian black swans which in their habitat occur as regularly as white ones in the Old World.

The example from classical physics that Popper himself mentions in order to establish his asymmetry thesis also shows clearly that falsification of empirical universal propositions (laws of nature) requires the regular appearance of disagreements with reality. Kepler's law states that all planets move round the sun in ellipses. Popper holds that the single 'counter-example' of Mercury suffices of necessity to falsify the law. However, this would not be so if the deviation of Mercury's orbit had been observed only once or a few times, while for the rest Mercury, too, moved according to Kepler's law. Mercury's orbit counts as a sufficient (though by no means logically necessary) falsification of Kepler's law only because its deviation from Kepler's law can be ascertained as an anomaly occurring regularly. From the fact that a law of nature can be falsified only by another law[18] it follows not only that we presuppose the empirical principle of induction for falsifying and verifying procedures alike but also that falsification and verification are symmetrical in yet another way.

Popper gives as his reason why empirical universal propositions are not verifiable the fact that one cannot test infinitely many cases as would be required to verify these propositions. Since we saw that to falsify such propositions (for example laws of nature) we need another law that states the regular occurrence of deviations from the proposition concerned, that the falsifying law does hold can in turn be 'verified' only by observing infinitely many cases, or else a further law of nature is required to falsify it and so on indefinitely. Thus, falsifying such a proposition, like verifying it, would require the testing of infinitely many cases: on this point too there is complete symmetry between verification and falsification. The 'truth' and likewise 'falsity' of general universal propositions can be decided only by arbitrarily breaking off the testing procedure because the latter is unbounded in either case. If for this reason Popper declares agreement with reality to be irrelevant for the verification of such propositions, to be consistent we should have to say for the same reason that disagreement with empirical data is irrelevant for falsification of these same propositions.

A further argument, habitually advanced by holders of the one-sided theory of falsification[19] to prove that verification is irrelevant for establishing general empirical propositions as true, points to the possibility that for any law of nature, hypothesis and theory one can indicate arbitrarily many 'truth-establishing' propositions, i.e. propositions about empirical data that agree with the universal propositions and theories to be tested. In this way, by using the principle of induction, any arbitrary law, or system of laws, of nature could be verified to an arbitrary extent, from which it follows that verification, and agreement with reality, are worthless and irrelevant for the holding and establishing of the propositions (or systems of these) to be tested. However, this argument overlooks that the possibility of thus indicating arbitrarily many verifying propositions in order to establish any general empirical proposition or empirical theory exists in exactly the same way for falsifying propositions. For every law of nature one can indicate arbitrarily many propositions about empirical data that disagree with the proposition to be tested. For consistency, we should have to infer from this that falsifying propositions and disagreement with reality are likewise worthless and irrelevant for the non-validity or refutation of general empirical propositions and theories. The thesis of irrelevance of agreement or disagreement with reality for the holding or not holding respectively of empirical universal propositions is diametrically opposed to the usual empirical testing procedures which always decide this in terms of agreement or disagreement with reality.

This discrepancy between their theses and the testing procedures (methods of verification) actually used in the real sciences the one-sided ('asymmetrical') falsification theories try to meet by distinguishing, without legitimate epistemological justification and really in contradiction with the propositions and basic theses of their own methodology, between relevant and irrelevant agreements or disagreements with reality on the part of derivations from the universal propositions or theories to be tested. In fact, this completely annuls the asymmetry between verification and falsification. For to decide what is thus relevant or irrelevant for the holding or otherwise of these derivations, we use criteria which, according to falsificationists, completely agree with the verification and falsification criteria of inductive empiricism, that is, empirical enquiry. There is only this difference, that asymmetric falsification theories base their selection criteria on metaphysical sham concepts or simply forego any justification, while inductive empiricism does not need such dubious means. To see that point we must of course clarify the verification and falsification concepts used in empirical enquiry as well as the connected concepts of empirical truth and falsity, above all as to their differences from the 'analogous' concepts in asymmetric falsification theories.

6. CRITERIA OF SCIENTIFIC PROGRESS

Empirical enquiry ought to lead to knowledge of objective reality. Science gradually approaches this goal and in this light the sequence of theories that replace each other can be viewed as a progressive development. If, however, empirical universal propositions are in principle only falsifiable but not verifiable, as asserted by asymmetric falsification theories, we must show how determining and looking only for disagreements of the propositions to be tested with observed data can lead to a progress of scientific development. Popper himself[20], at many stages of his enquiries, points to the fact that looking for disagreements and contradictions in a theory and demanding that these be eliminated is by no means sufficient to guarantee scientific progress. Contradictions can be removed by modifying the propositions under test or by introducing auxiliary hypotheses, but such removal constitutes scientific progress only if by means of the new propositions we achieve agreement of the system of propositions not only with the empirical data concerned but also with further data not used in setting up the new propositions. Such relevant agreements are, however, simply genuine verification of the propositions or theories newly set up. Analogously, disagreements of propositions or systems of propositions with reality become relevant for falsification when from the auxiliary hypotheses assumed for the removal of imcompatibilities there constantly arise further derivations that in their turn contradict new observed data not used for setting up the hypotheses. Thus in choosing new propositions and theories one uses agreements and disagreements of derivations with reality in the sense above explained. Only by thus 'symmetrically' considering verification and falsification of propositions under test does the setting up of the new propositions, or of systems of these, lead us to laws of nature or theories that constitute scientific progress. Amongst the criteria of progress there is the additional requirement that newly set up propositions or systems of propositions should give the widest possible scope for predictions whose coming true or otherwise is ascertained as agreement or disagreement respectively of the derivations with observed (experimental) data.

We are therefore not denying that observing contradictions, incompatibilities and disagreements constitutes a very important methodological device for empirical enquiry. Rather, we merely wish to show the incorrectness of the view that a necessarily final falsification of general empirical propositions can be obtained, while finding out agreement of propositions with reality (that is, verifications) is irrelevant for the holding of the propositions to be tested. As we have shown, as regards completeness or incompleteness of the testing procedures, there is complete symmetry between verification and falsification. Moreover, in deciding the validity of empirical universal propositions, we use 'verifying' agreements and 'falsifying' disagreements in symmetrical manner. Symmetrical application of the criteria is amongst the necessary methodological presuppositions for progress in scientific knowledge.

This fact is taken into account by Popper[21] through one of his complementary theses. He is of the opinion that there is an 'absolute truth' which is indeed in principle unknowable; yet scientific enquiry comes steadily nearer to it. For this 'coming nearer', which is precisely scientific progress, we must of course indicate criteria. One might expect Popper to look for them only in falsification methods, which on his theory of science are alone relevant to deciding the validity of general empirical propositions and theories. If now we have to discover propositions and theories that contradict our traditional hypotheses and systems, this can be achieved (as Popper[22] himself admits) in an unlimited number of ways, without this providing any criteria for scientific progress in the setting up of new propositions and theories. Popper is therefore compelled to seek such criteria outside his falsification theory. As stated, he assumes that there is an 'absolute truth' that, though in principle unknowable, influences in some inexplicable way the choice of new hypotheses and theories so as to produce a 'progressive approximation to the absolute truth'. As to that, we must first of all remark that any 'absolute truth' such as this one, is a metaphysical concept explaining nothing. If next we look more closely at the criteria Popper says we must obey in order that the propositions (hypotheses, theories) set up may be viewed as 'approximations to absolute truth', it turns out that they merely prescribe, within the new propositions and theories, a growing and more accurate measure of agreement with experimental data as compared with that in the earlier systems. Thus verification (agreement of derivations with observed data), having first been declared 'irrelevant', is now reinstated as 'relevant' in terms of the rules that govern scientific progess. From this procedure of mutually correcting steps of falsification and verification it follows that these last are completely symmetrical in the method of approximation to 'absolute truth' that complements Popper's theory.

The incompatibility of Popper's initial 'asymmetric' falsification theory and the symmetry of verification and falsification in this complementary method has been noticed or at least felt[23] by some of his adherents and followers. In extreme consequence, the one-sided falsification theory has been carried further by Paul K. Feyerabend. Basing himself on the insight that the contradiction in Popper's theory of science (between the initial assumption that one-sided falsification is valid and the symmetry of verification and falsification that is after all required for progress in scientific knowledge) arises from the view that empirical enquiry aims at describing the real order of phenomena, Feyerabend denies that the so-called empirical sciences make statements about real states and their order. He thinks it is in principle impossible to verify so-called empirical propositions and theories by observation. Whatever agreement between derivations and observed data there might be is quite irrelevant, since general propositions cannot be verified by individual cases. The only 'progress' of scientific enquiry consists in seeking, for each system of propositions, hypotheses or new systems that contradict the old systems. From the empirical systems of propositions we derive all possible consequences, amongst which there will always be some that contradict propositions already accepted, or if one prefers, some observed data or other. Accordingly, any theory, will sooner or later be falsified, from which it follows at once that so-called empirical scientific systems say nothing about reality, or as Feyerabend[24] puts it, 'there are only false theories'.

Now for any proposition and any theory there are always infinitely many new theses that contradict the traditional propositions (that are 'to be tested'). This, Feyerabend thinks, must be taken into account by confronting any proposition 'to be tested' (or system of these) with as many contradicting assumptions as possible, and then examining by which new hypotheses or theories the incompatibilites can be eliminated. This interpetation of the procedure of scientific testing, which requires us to seek and construct the greatest possible number of hypotheses and theories contradicting traditionally accepted theories, Feyerabend calls 'methodological pluralism', holding that this procedure aims at falsifying the system of scientific propositions under consideration as extensively and multilaterally as possible.[25] Sooner or later this must lead to a revolutionary demise of any empirical theory and to the invention of new theories to which the revolutionary pluralist method must at once be applied in turn, and so on indefinitely. Accordingly a 'permanent revolution' in the above sense would have to count as the ideal method of empirical enquiry.[26] Since on this view any theory will necessarily be falsified sooner or later, it would follow that all scientific theories are false, both the refuted and the as yet unrefuted ones and above all the refuting theories themselves; from which it further follows that systems of empirical scientific propositions state nothing about reality or about phenomena and their real order.

Rigorously carrying the asymmetric falsification theory further we are led to a methodology that might be called 'contradiction for contradiction's sake'. We must indeed ask ourselves what purpose is served by this aimless game of constantly constructing falsifying propositions or systems of these for any given theory, if in any case we know in advance that this only ever leads us to false theories. Here the difficulty of explaining development and progress in science while cleaving to asymmetric falsification theory stands out even more clearly than in Popper. Indeed, for every proposition or system of propositions we can indicate arbitrarily many and quite different falsifying theses and theories, but which of these 'counter-instances' is now to count as the relevant and authoritative falsification? If one declares that agreement of derivations from hypotheses and theories with observed data are irrelevant, then as regards falsification all incompatibilities are equally valid and there is no criterion for what contradictions are to be viewed as relevant and which of the infinitely many possible hypotheses and theories, set up for the purpose of removing the incompatibilities, are to count as progress in knowledge. This clearly illustrates that Feyerabend's pluralist falsification theory is one-sided in taking into account only one kind of testing criterion amongst many that are actually used in empirical enquiry. This makes 'methodological pluralism' a distortion of the methods applied in scientific enquiry; a picture that is not in the least suitable to lead to the goals which empirical enquiry sets itself and which it approaches in progressive development through the use of its methods.

The discrepancy between his pluralist permanent revolution and the method and testing criteria of empirical scientific enquiry is noticed by Feyerabend himself. Therefore he too distinguishes between relevant and irrelevant incompatibilities and gives restrictive rules for choosing, from the infinitely many hypotheses and theories capable of removing contradictions and disagreements, those that must then count as scientifically relevant systems.[27] As to the restrictive rules, Feyerabend, like Popper, introduces them as directive principles for enquiry, without being able to justify this step and in contradiction to the basic theses of his asymmetric falsification theory ('methodological pluralism'). No very deep logical analysis is required to see that the criteria prescribed by these rules are inductive verification criteria[28], as with Popper. According to these prescriptions, values derived from the new hypotheses or systems of propositions must agree with the observed data with equal or, if possible, greater accuracy than already existed for them as regards deductions from the earlier theories. Moreover, the observed data must confirm other propositions (values, predictions) derivable from the new theoretical system and incompatible with deductions from the old, these latter appearing as falsified by the measured values in the corresponding cases. Experimental confirmation of laws derivable from the new theories, but devoid of sense under the presuppositions of the theory to be refuted and thus not derivable from it, counts as criterion for verification and the scientific relevance of the new theories. We need not consider in detail what standards apply to empirical enquiry in setting up or choosing a new theory from many other theoretically possible ones. Without argument, Feyerabend simply takes over these criteria from practical empirical methodology as rules that restrict the choice of theories relevant to empirical enquiry to just a few theoretical formulations amongst the infinitely many possible falsifying systems. In this he does, however, overlook the fact that these restrictive rules without exception indicate certain agreements of values derived from the new theoretical formulations with observed data as decisive verifying criterion, that is, as criterion for the holding and scientific importance of the new hypotheses and theories concerned. This flatly contradicts the basic thesis of asymmetric falsification theory. Feyerabend does not try to justify his now suddenly introducing what he had at first declared to be in principle irrelevant, namely 'agreements with reality' ad decisive criteria for setting up and choosing scientifically relevant hypotheses and systems of propositions. In any case he thereby revokes the asymmetry and along with it the 'permanently revolutionay' character of his falsification theory. By his rules for choosing scientifically relevant theories he has let back in all the verification criteria (agreements with reality) that he had at first outlawed as insignificant and misleading. If we obey these rules, then verifying and falsifying steps in method enjoy completely symmetrical roles in empirical enquiry.

7. 'COMPLETE' AND 'PROGRESSIVE PARTIAL' TESTABILITY

That empirical universal propositions and theories can be tested only by one-sided falsification, the adherents of asymmetric testing procedures also often base on the argument that all new revolutionary theories with their new forms of concepts and cognition arose by way of refutation (falsification) of previous theories and never by inductive formation of hypotheses on the basis of observed results. However, this is not in general correct. In particular, the founders of modern natural science discovered the first mechanical laws by extrapolating induction applied to series of measured values. One of the most revolutionary innovations in the forms of concepts, laws and theories in empirical enquiry, namely Newton's system of mechanics, did not arise by refutation of preceding theories, for there were none, but by inductive assumptions on the basis of regularities already known or observed (Kepler's laws, centrifugal phenomena, free fall). Of course, incompatibilities are also of great methodological importance for the progress of enquiry, whether the enquirer meets them in experiment or seeks them by systematic analysis. However, in every formation of hypotheses or theories we must always take into account the relevant agreements with observed data, alongside disagreements (if any) with earlier propositions or systems of these, if the new theory is to mean not only a merely formal removal of contradictions but also progress in our knowledge of reality.

Alongside a procedure for ascertaining and removing incompatibilities, a methodology of empirical science must therefore always take account as well of inductive criteria, such as ascertaining agreements with reality (i.e. verifying data), generalisations and theoretical formulations based on them, and the like. From the symmetrical roles that verification and falsification play in empirical methods, guidelines arise for conceptually marking the valid forms of empirical propositions.

From his thesis that only propositions about finitely many cases can be verified or falsified while general empirical propositions cannot because this would require infinitely many steps, Wittgenstein[29] inferred that empirical universal propositions (for example laws of nature) are by no means propositions that could be true or false, but rather have the character of instructions for setting up propositions of certain forms in individual cases. Wittgenstein bases this interpretation of laws of nature (generally: of empirical universal propositions) on his verification thesis. By the verification procedure belonging to a proposition we indicate under what conditions the latter is to count as 'true' or 'false', which is said to indicate its sense. If now an empirical universal proposition pronounces on infinitely many cases, its truth and falsity conditions can never be completely enumerated and tested so that the propositions are not testable and it is in principle impossible to indicate what is their sense. Thus the sense of a proposition is here linked with its verifiability and all later verification and falsification theories come to grips with this formulation, as we have seen.

Here we must indeed ask ourselves what right individual empirical sciences do have to test their universal propositions (laws of nature) with the help of highly complicated and exact methods, when these 'propositions' can be neither true nor false, since they are neither verifiable nor falsifiable, nor therefore testable. That the answers attempted in the various theories of science are unsatisfactory was discussed above. To interpret the testing of laws of nature as aiming not at truth or falsity but at their probation in practice is incompatible with the surely well-founded epistemological view that laws of nature describe the real order to phenomena. Just as unsatisfying is the interpretation provided by asymmetric falsification theories, which assumes all empirical universal propositions to be one-sidedly falsifiable but then, in order to explain scientific progress, has to introduce the actually used verification procedures as methods, not further justifiable, for 'approximation to absolute truth' (or even without attempt at justification). Moreover, these theories are compelled to declare as devoid of sense propositions that on their view can neither be verified nor falsified (as for example statistical laws).[30]

We must not overlook here that empirical enquiry, in applying its verification and falsification procedures, does not bother with the question whether only individual and empirical propositions have the character of true or false propositions, while laws of nature do not, nor what character the latter do have. Empirical enquiry presupposes that the laws describing empirical phenomena are in principle testable and only where this is not so do they deny empirical content to the universal proposition concerned.

What natural science regards as verifiable and falsifiable are thus not only propositions describing finitely many cases but also empirical universal propositions and laws of nature. A theory of science that is to give an adequate rather than a distorted picture of the methods and criteria used in the practice of enquiry ought, I think, to take this into account. Verification and falsification must be marked in such a way that we can meaningfully speak of verifiability and falsifiability or truth and falsity of individual as well as of universal propositions.

In testing procedures it certainly makes an epistemological difference whether after finitely many steps the test is complete and need not be repeated or whether it can be continued without limit or has to be undertaken in infinitely many cases. However, does it follow from this that the concepts 'verification' and 'falsification' are admissible only for the one kind of test, namely the procedures that can be in principle decided after finitely many steps? Empirical enquiry says no. In the individual empirical sciences we do indeed distinguish between propositions whose truth or falsity is completely decidable and those where this is in principle possible only 'partially', 'to a higher or lower degree' or 'sufficiently but not completely'. However in scientific methodology both kinds of propositions are denoted as verifiable or falsifiable. The difference between finitely and infinitely many methodological steps is thus not viewed by practical enquiry as a disjunctive distinguishing mark for the verifiability or falsifiability of propositions. Logically speaking we must here object that the two descriptions ('finitely' and 'infinitely many steps or cases' respectively) necessarily give rise to two different concepts, even if we denote them both as 'verification' or 'falsification'; but then to verify or falsify propositions about finitely many cases means something different from doing the same for empirical universal propositions (laws of nature) and we must ask for the distinguishing marks of the two kinds of verifying or falsifying.

What is the basis for thus using the same terms for logically and methodologically different testing procedures is hinted at by a comparison with an analogous conceptual situation in mathematics. The differences between the different kinds of numbers were early noticed: we know that the Pythagoreans were astonished by the properties of irrationals of being in principle not representable by constant bisection of a line, rational numbers having been assigned to its points. The thinkers of this school were therefore inclined to deny that irrationals were numbers. In the sequel the differences were recognized but so were the common rules by which the different kinds of numbers are to be treated. The logical mathematical enquiries of recent times have largely clarified how far each kind of number can be defined with the help of the others. The result was that each kind of number involves a new form of concept and a new entity corresponding to it so that for each kind new rules of operation must be defined, but that the one term 'number' is nevertheless expedient and justified for denoting all the various kinds.

The analogy with individual and universal empirical propositions goes quite far. Here too, epistemological analysts, insofar as they follow Wittgenstein's view, wish to recognize as 'propositions' that can be true or false only statements about finitely many verifying series or steps. They deny that empirical universal propositions are propositions, because their 'complete' verification would require infititely many verifying series and steps. However, for empirical universal propositions (laws of nature) we always know of procedures by means of which we can in principle push the testing of propositions as far as we wish, even if we cannot effectively carry out all infinitely many possible tests. We might for instance carry out verifications of growing accuracy, which in the case of confirmation or deviation is taken as approximation to the positive or negative result of testing respectively. Moreover, quite new and hitherto untested values derivable from an empirical universal proposition can be tested by new experimental procedures, where again we take the positive or negative result in the sense of increasing verification or falsification respectively. This progressive though never completed testability of laws of nature is somewhat analogous to the calculability of irrational numbers which can be carried arbitrarily far without ever yielding a completed result. Nevertheless we call irrationals or real numbers in general 'numbers'. In my view, therefore, epistemological analysts who declare empirical propositions about finitely many cases to be alone verifiable and deny that empirical universal propositions are propositions because they are not terminably verifiable, resemble the Pythagoreans who wished to reserve the status of 'numbers' to those numbers that could be completely calculated or captured by rational division.

If in characterizing the kinds of empirical proposition and their testability we proceed on the model of mathematical concept formation, then we must distinguish between two forms of verification or falsification. It is expedient to speak of 'complete' verifiability or falsifiability of propositions about finitely many cases. In contrast the testability of empirical universal propositions, and in particular of laws of nature, can be described as 'progessively partial'. If in testing a universal proposition we dispose of procedures that enable us to obtain far-reaching results (perhaps of various kinds) with increasing accuracy, then it is expedient to speak of the propositions concerned as 'sufficiently' verifiable or falsifiable in the case of positive or negative results respectively.

We are surely not departing from the methodological criteria and modes of expression in individual empirical enquiry if we collectively denote the two forms of empirical testability (namely 'complete' and 'progressively partial') as 'empirical testability'. Correspondingly it will be appropriate to call 'completely' verifiable or falsifiable individual propositions and 'progressively partial' or perhaps 'sufficiently' verifiable or falsifiable universal propositions collectively 'verifiable' or 'falsifiable'.[31] This would eliminate the distortions of the theory of science discussed above, which contradicts the methodology of empirical practice. For we can then denote both individual and universal propositions as testable, that is, verifiable or falsifiable. What now disappears as superfluous are the unnaturally cramped and artificial distinctions between verifiable individual empirical propositions which alone are to have sense, and empirical universal propositions that are to be in principle unverifiable and to be denied the status of proposition, being treated merely as ways of acting and their 'verification' merely a matter of proving themselves in practice.

'Progressive partial' or 'sufficient' verification and falsification presuppose that the empirical principle of induction holds. This agrees with the experimental testing procedures of empirical enquiry. As we saw, experimental results are taken as verifying or falsifying only under the (necessary though not sufficient) condition that on repeating the experiments under the same or similar circumstances they will likewise recur. Since this holds in general for progressive partial testing and thus for verification and falsification alike, agreements and disagreements of observed data with derived values are in principle symmetrically weighted as regards evaluation of experimental results. This is in harmony with the application of verifying and falsifying criteria in empirical scientific enquiry. The artificial assumption of assymmetric falsification theories – that for empirical universal propositions there are no verifying but only falsifying criteria – thereby becomes quite superfluous, along with all the inconsequent theses about absolute truth and the equally inconsequent injection of verification criteria in the theoretically unjustifiable restrictive rules.

One might here object that the marks of the concepts of progressive partial testing, and of sufficient verification and falsification to be applied to empirical universal propositions (laws of nature), are incompatible with the criteria for sense in Wittgenstein's verification thesis. He does not allow that the 'inductively' inferred holding of empirical universal propositions

(laws of nature) on the basis of repeated observations and experimental results amounts to verification[32]. However, the view that only those propositions can count as true or false, and therefore as meaningful 'propositions'
whose truth value is completely decidable, is just as dogmatic as the notion,
several times mentioned above, that only completely calculable numbers
can really count as 'numbers'. If we widen the concepts 'verification' and
'falsification' in the sense explained above by including 'progressive partial'
testing amongst verification and falsification procedures, the verification
thesis can likewise be extended as to sense. The 'sense' of empirical propositions is given by the possible complete or progressive partial tests. Certainly
we need to distinghuish different forms of the concept of sense, as in the
case of the different number concepts, but this quite corresponds to the
usual distinction in empirical enquiry between individual propositions
and hypotheses. Amongst the latter are empirical universal propositions
(laws of nature), to which after sufficient verification we assign 'empirical
hypothetical truth'. This form of 'truth' is rightly distinguished from the
truth of completely verified individual propositions.

    The distinction between 'complete' and 'progressive partial' testability or
between 'complete' and 'sufficient' verifiability and falsifiability further enables us to retain the usual interpretation that laws of nature describe the
objective order of the sequence of phenomena. As we saw, Wittgenstein
and his adherents, by cleaving to the original verification thesis, were
forced to interpret laws of nature as instructions concerning what form of
proposition to use for describing individual cases. On this view laws of nature do not describe the real empirical order of phenomena. Even in the
consistency theories following Carnap, laws of nature do not describe the
objective order of phenomena, but are rather set up with the aim of enabling us to obtain consistent systems of propositions of the amplest possible scope. Since to this end observational propositions and hypotheses may
alike be modified, it follows that the objective content of laws of nature
(namely the description of the objective order of phenomena) is lost. In
asymmetric falsification theories inspired by Popper, agreement with reality is declared irrelevant for the holding (verification) of the laws of nature.
In these theories there are thus no criteria for the objective content of empirical universal propositions. In order nevertheless to be able to interpret
laws of nature as describing the objective order of phenomena, Popper
indeed assumes an 'absolute truth', which description by natural laws is

gradually to approach, while, agreements with observed data, previously excluded, are introduced after all as criteria of approximation, as we have mentioned several times above. Extreme adherents of Popper's initial theses condemn the (mystical) approximation to absolute truth and rightly, from their point of view, deny that empirical theories have objective content concerning empirical reality. That in order to account for scientific progress they finish by illogically setting up rules according to which enquiry must respect agreement of derived propositions (predictions) with reality, clearly shows that the presuppositions of these theories are inadequate.

All of these difficulties drop away if we decide to widen or subdivide the concepts 'testability' and 'verifiability' or 'falsifiability' in the sense mentioned above. If we recognize the possibility of progressive partial testability of hypotheses (empirical universal propositions and laws of nature) which can always lead to sufficient verification or falsification of laws of nature, then we thereby have criteria of scientific progress without the need for artificial assumptions and interpretations. Moreover, comparison of the formal order represented in the laws of nature with the real order of the course of phenomena, by testing whether and how the values derived from laws agree with observed ones, is quite in tune with the view that laws of nature describe the empirical objective order of phenomena.

NOTES

[1] Some authors are inclined to deny that universal propositions are propositions at all. Thus Wittgenstein, L., *Philosophische Bemerkungen*, Frankfurt a, M. 1964, p. 66f, 174, 200f, 282, 287, 289. Waismann, F., *Ludwig Wittgenstein und der Wiener Kreis*, ed. McGuinness, B.F., Oxford 1967, p. 47, 48, 53ff, 70f, 158, 159. Schlick, M., 'Sind Naturgesetze Konventionen?', *Actes du Congrès international de Philosophie scientifique*, Paris, 1934, 8-17.

[2] Cf. Wittgenstein, L., l.c. and *Tractatus logico-philosophicus*, 5.512. Schlick, M., l.c. and 'Meaning and Verification', *Philosophical Review* **45** (1930). Carnap, R., 'Die physikalische Sprache als Universalsprache der Wissenschaft', *Erkenntnis* **2** (1931), 432–465; 'Über Protokollsätze', *Erkenntnis* **3** (1933), 215–228; 'Testability and Meaning', *Philosophy of Science* **3** (1936), 419–471 and **3** (1937), 1–40; 'The Methodological Character of Theoretical Concepts', *Minnesota Studies in the Philosophy of Science* **1** (1956), 38–76. Juhos, B., 'Über die empirische Induktion', *Studium Generale* **19** (1966), 259–272. Popper, K.R., *The Logic of Discovery*, 4th ed. 1961; *Conjectures and Refutations*, London 1963, 3–292, 385–410. Stegmüller, W., *Das Wahrheitsproblem und die Idee der Semantik*, Vienna 1957, 262–268. Weingartner, P., 'Probleme, die sich bei der Darstellung von wissenschaftlichen Lehrsystemen ergeben', *Salzburger Jahrbuch für Philosophie* **7** (1963), 199ff.

[3] Cf. Stegmüller, W., *Einheit und Problematik der wissenschaftlichen Welterkenntnis*, Munich, 1967, pp. 3–22 (Münchner Universitätsreden, New Series, 41).

[4] Cf. Wittgenstein, L., *Philosophische Bemerkungen,* Frankfurt a. M. 1964, p. 64, 174, 287, Waismann, F., *Ludwig Wittgenstein und der Wiener Kreis,* ed. McGuinness, Oxford 1967, p. 47, 71, 79, 97f, 126, 158ff, 211, 221, 226, 227, 232, 243ff, 247.

[5] Cf. ibid, p. 97, 99ff, 159ff, 162f, 188, 210f, 214, 229, 250, 255ff, 282ff. Schlick, M., 'Philosophie und Naturwissenschaft', *Erkenntnis* 4 (1934), 379–396.

[6] Cf. Waismann, F., l.c., pp. 46–48, 70f, 79, 97f, 126, 158ff, 186, 211, 221, 226, 232, 243, 245f, 258f.

[7] Cf. Wittgenstein, L., *Philosophische Bemerkungen,* p. 216. Waismann, F., l.c. p. 38, 61ff, 71, 162f. Schlick, M.,l.c. and 'Über das Fundament der Erkenntnis', *Erkenntnis* 4 (1934), 71–99. A view similar to Poincaré's 'methodological conventionalism' is held by Kraft, V., *Erkenntnislehre,* Vienna 1960. He calls his view 'normative epistemology', in which the norms that determine the form of knowledge play a role similar to Poincaré's conventions.

[8] Cf. Carnap, R., *Logical Syntax of Language,* London and New York 1937; 'Remarks on Induction and Truth', *Philosophical and Phenomenological Research* 6 (1950), 590–602.

[9] Carnap, R., 'Über Protokollsätze', *Erkenntnis* 3 (1933), 215–228; 'Wahrheit und Bewährung', *Actes du Congrès international de Philosophie scientifique,* Paris 1936 cf. also *Logical Syntax of Language.*

[10] Cf. Carnap, R., l.c. and 'Empiricism, Semantics and Ontology', *Meaning and Necessity,* Chicago 1956; 'Die physikalische Sprache als Universalsprache der Wissenschaft', *Erkenntnis* 2 (1932), 432–465.

[11] Cf. Juhos, B., 'Kritische Bemerkungen zur Wissenschaftstheorie des Physikalismus', *Erkenntnis* 4 (1934), 397–418, this volume, pp. 16–35; *Die Erkenntnis und ihre Leistung,* Vienna 1950; 'Empiricism and Physicalism', *Analysis* 2 (1945), 81–92, this volume, pp. 36–46, 'Deduktion, Induktion, Wahrscheinlichkeit', *Methodos* 6 (1954), 259–278.

[12] Cf. Wittgenstein, L., *Philosophische Bemerkungen,* p. 106ff. Waismann, F., l.c. p. 42f, 73ff, 93, 248ff. Antiseri Dario, *Dal neopositivismo alla filosofia analitica,* Rome, 1966, pp. 106–146.

[13] See Popper, K.R., *Conjectures and Refutations,* London 1963, l.c. and p. 28ff; *Logic of Scientific Discovery,* 1959, p. 38ff.

[14] Cf. Popper, K.R., *Conjectures and Refutations,* p. 28ff, 41. Note 8, 115, 193, 256, 257, 287f, 391. Weingartner, P., 'Probleme die sich bei der Darstellung von wissenschaftlichen Lehrsystemen ergeben', *Salzburger Jahrbuch für Philosophie* 7 (1963), 199ff. Juhos, B., 'Über empirische Induktion', *Studium Generale* 19 (1966), 259–272; *Die Erkenntnis und ihre Leistung,* Vienna 1950.

[15] To be consistent one would have to complement the above division by means of a third class of propositions. On Popper's account of empirical testing, empirical existential propositions over unbounded domains would be verifiable only, but in principle not falsifiable. A proposition of this kind would be for example 'There is a fluid whose compressibility is smaller than Lamé's constant'. Cf. Juhos, B., 'Über die empirische Induktion', *Studium Generale* 19 (1966), 259–272.

[16] Cf. Agassi, Joseph, *Towards an Historiography of Science,* The Hague 1963 (History and Theory, 2); *Science in Flux,* Dordrecht 1975. Feyerabend, Paul, 'Explanation, Reduction and Empiricism', *Minnesota Studies in the Philosophy of Science* 7, ed. Feigl, H., and Maxwell, G., Minneapolis 1962, pp. 28–95; *Knowledge without Foundations,* Oberlin, Ohio 1961; 'How to be a good Empiricist', *Philosophy of Science, The Delaware Seminar, Vol.2,* ed. Baumrin, Bernard, New York, London and Sydney 1963; 'Realism and Instrumentalism, Comments on the Logic of Factual Support', *Popper Festschrift,* Glencoe and London 1964. Spinner, Helmut, *Theoretischer Pluralismus – eine Theorie des Erkenntnisfortschritts,* Philosophical Seminar of Mannheim University, 'Falsifizierarbeit und Falsifikation' (address given at the University of Erlangen-Nürnberg 1967).

[17] Cf. Popper, K.R., *Conjectures and Refutations*, 1965, p. 15, 193, 256. Also Lakatos, I., 'Proofs and Refutations', *British Journal of Philosophy of Science* **14** 1963/64, 1ff., 120ff, 221ff, 296ff. For criticism of Popper's theses see also Juhos, B., 'Über die empirische Induktion', *Studium Generale* **19** (1966), 259–272.

[18] Cf. Juhos, B., l.c. p. 267ff.

[19] Cf. Popper, K.R., *Logic of Scientific Discovery*, 1959, p. 38ff; *Conjectures and Refutations*, 1963, p. 28ff, 39ff, 41 Note 8, 115, 193, 256, 257, 287f. Also Agassi, J., l.c.; Spinner, H., l.c. Feyerabend, P., 'Explanation, Reduction and Empiricism', *Minnesota Studies in the Philosophy of Science* **3**, ed. Feigl, H., and Maxwell, G., 1962; *Knowledge without Foundation*, Oberlin, Ohio 1961; 'Problems of Empiricism', *Beyond the Edge of Certainty*, Englewood Cliffs 1963, pp. 145–260 (University of Pittsburgh series in the Philosophy of Science, ed. Colodny, Robert G., vol. 2).

[20] Cf. Popper, K.R. *Conjectures* p. 217, 232, 243, 256, 384, 391.

[21] Ibid., p. 157, 174, 216ff, 226, 229, 246, 384.

[22] Ibid., p. 58, 217, 256, 391.

[23] Cf. Agassi, J., l.c.; Feyerabend, P.K., l.c.; Spinner, H., l.c. Also Lakatos, I., 'Proofs and Refutations', *British Journal of Philosophy of Science* **14** (1963), 1ff, 120ff, 221ff, 296ff.

[24] Cf. Feyerabend, P.K., *Knowledge without Foundation*, Oberlin, Ohio 1961; 'Realism and Instrumentalism, Comments on the Logic of Factual Support', *Popper Festschrift*, Glencoe and London 1964; 'Problems of Empiricism', *Beyond the Edge of Certainty*, Englewood Cliffs 1965, pp. 145–260 (Univ. of Pittsburgh series in the Philosophy of Science, ed. Colodny, Robert G., vol. 2).

[25] Ibid.

[26] Ibid.

[27] Ibid.

[28] Cf. Feyerabend, P.K., 'How to be a good Empiricist', *Philosophy of Science, The Delaware Seminar*, vol. 2, ed. Baumrin, Bernard, New York, London and Sydney 1963. Also in references in Note 24.

[29] Cf. Wittgenstein, L., *Philosophische Bemerkungen*, p. 62, 66ff, 174, 200f, 282, 287, 289. Also Waismann, F., *Wittgenstein und der Wiener Kreis*, p. 71, 97f, 158ff, 211, 232. Cf. Schlick, M., 'Sind die Naturgesetze Konventionen?', *Actes du Congrès international de Philosophie scientifique*, Paris 1935; *Actualités scientifiques et industrielles No. 391, Paris 1936*, pp. 8–17; 'Philosophie und Naturwissenschaft', *Erkenntnis* **4** (1934), 379–396; 'Meaning and Verification', *Philosophical Review* **45** (1936).

[30] Cf. Juhos, B., 'Über die empirische Induktion', *Studium Generale* **19** (1966), 259–272; to appear shortly, 'Logische und empirische Induktion', *La philosophie contemporaine. Chronique des années 1956-1966*.

[31] A similar characterization of the 'truth' of empirical universal propositions is given by Törnebohm, H., 'On the Confirmation of Hypotheses about Regions of Existence', *Synthese* **18** (1968), 28–45. He there tries to show that by means of verification we ascertain the partial truth of hypotheses and that this degree of validity depends numerically on the degree of confirmation.

[32] Cf. Wittgenstein, L., *Philosophische Bemerkungen*, p. 66f, 174, 200f, 282, 289. Also Waismann, F., *Ludwig Wittgenstein und der Wiener Kreis*, p. 158ff, 204ff, 211, 232.

CHAPTER XI

THREE SOURCES OF KNOWLEDGE*

1. THE TRIADIC METHOD

Epistemological analysis, besides investigating the methods, presuppositions and results of scientific enquiry, examines in particular the elements of the logic of language with the help of which knowledge in its theoretical connections is represented. The development of modern epistemological analysis began in the first half of the 19th century, with the examination of certain elementary concepts used in daily life and in various individual sciences where they are so-called basic concepts. In this it turned out that these fundamental concepts, of which it was thought that their unambiguous meaning was known and that certain objects belonged to their extension, are used in essentially different senses in the different scientific domains, so that in the cases concerned one denotes by the same name concepts of different meanings, covering objects of very different kinds. These epistemological investigations finally showed that science uses some of its basic concepts in three different characteristic senses.[1] It thus became possible, when examining other elementary concepts, to ask methodically whether in the expressions concerned we ought not to distinguish three different senses characterized in determinate ways. We then reach a 'triadic method' which, in analysing concepts, expressions and propositions systematically, examines three questions. Much of recent analytic investigations in epistemology suggest these questions even if they have not yet been explicitly discussed.

In what follows we shall show by means of examples how these analytic steps of epistemology were determined, how its methods are to be systematically applied and what results it has led to up till now.

2. DATA OF CONSCIOUSNESS – LOGICO-MATHEMATICAL CONSTRUCTIONS – METHODS OF EMPIRICAL CONTENT

When a scientific discipline has reached a high degree of development, criti-

---

* Translated from the German original 'Drei Quellen der Erkenntnis', first published in *Zeitschrift für philosophische Forschung* **26** (1970) 335–347.

cal questions are bound to arise (either in that discipline or in epistomology) as to the methods by which knowledge in that field is obtained and on what its validity is based. At the end of the 18th and beginning of the 19th century geometry developed so fast that in that field of mathematics certain limits were reached beyond which it was impossible to go with the usual methods of enquiry and proof.[2] The propositions of geometry (then only Euclidean) counted as indubtitable, like all mathematical propositions. That some of them made statements about infinite domains was nothing mathematically extraordinary, since even in number theory and classical analysis the basic propositions rest on presuppositions about infinite domains. However, geometry had a special position amongst the other mathematical disciplines. Although it was counted amongst the branches of pure mathematics its propositions were at the same time construed as statements about reality, namely about a definite real object 'space'.

Amongst the axioms of Euclidean geometry there are statements in which infinite extent is either ascribed to space or presupposed for it. Where statements about infinity in arithmetic and analysis were not then regarded as problematic, because the propositions of those branches of mathematics were not taken as pronouncing on real objects, the universally assumed realistic interpretation of geometry (namely, that it describes the regularities of the real space in which phenomena occur) compels us to ask the critical question: how can we attain to absolutely certain knowledge about the nature of really existing infinite regions of space? This is one of the well-known critical questions of Kant's transcendental theory.

Kant answered both questions, namely how we obtain the propositions of geometry and on what their validity is based, by assuming a special cognitive faculty of pure or transcendental intuition. The data conveyed to us by pure intuition are supposed to be given as intuitive data of consciousness. It is surely correct that we can intuitively grasp certain spatial geometric situations, which is of course not to say that we dispose of a special (namely transcendental) faculty of intuition. Nor is it established what domain of geometrical relations we can grasp intuitively. If then we denote intuitive data of experience, whether sense data, ideas, feelings and the like, generally as 'data of consciousness', Kant's foundation for geometry at least points to one, if as yet questionably characterized, source of our knowledge about space.

Mathematicians have objected to Kant's transcendental theory that whatever the character of our data of consciousness our intuitive experience contains no data that would give us infinite regions of space. We know that the mathematicians' constant critical discussions of Kant's transcendental foundation for geometry have led to the detachment of mathematical geometries from any kind of intuition and intuitive data of consciousness. Starting with Bolyai's and Lobatchevsky's indirect proofs for the independence of Euclid's axiom of parallels from the other axioms right down to the latest possible generalisations of the mathematical concept of space by Riemann, Felix Klein and Hilbert, this consistent line of proofs led to the insight that 'space', for mathematics, is always a formal or analytic conceptual system of relations that is independent of all intuition. Certain narrowly restricted parts of the infinitely many logical mathematical systems of geometric relations may be intuitively presentable, but that is of no account whatever for the validity of logical mathematical geometries. Mathematical spaces are formal analytic systems of concepts that are established quite independently of any intuition and in the great majority of their theses transcend all bounds of possible intuition. From this it follows to begin with that we must distinguish several meanings for the concept 'space'. The formal conceptual systems of space in mathematics are something quite different from the intuitive spatial relations given in experience, that is, different from experiential spaces. In logico-mathematical formal concept-formations we therefore recognize a second source for obtaining knowledge or a second constitutive element of knowledge in general.

However, in the further course of this debate on Kant's transcendental theory, mathematicians as well as physicists and philosophers have pointed to the question: if there are so many types of logico-mathematical forms of space that are admittedly for the most part inaccessible to intuition, what then is the space in which we live and in which all phenomena occur? This 'real' space is indeed partly given to us by intuitive data such as sensations and perceptions and so on. Might not Kant's transcendental concept of space relate to this real space? In favour of this view is his thesis that by our forms of consciousness, those of transcendental intuition amongst them, we have shaped the formless 'thing in itself' into what we then call real world. Maybe, then, the real space of Kant does have Euclidean properties and we do recognize this by means of a special faculty. From this transcendental real space we should then have to distinguish the many surely possible logico-mathematical constructions as analytic non-real 'spaces'.

Gauss had already objected to the transcendentality of real physical space, for the space in which all empirical processes occur, and physical ones in particular, is given to us by empirical observational data just as any other empirical phenomena are. Therefore, Gauss thinks, the properties and regularities of real space must be ascertained by the same method by which we come to characterize empirical phenomena and their connections. This is the method of measurement. Gauss himself undertook the first such measurements of space, although they were of course not sufficiently accurate. Carrying these reflections further, Riemann and particularly Poincaré gave a very precise account of the methods by which we come to know the metric of empirical physical space, as well as considering the possibility (again following Gauss) of representing the metric of real space by non-intuitive mathematical constructions. This yielded not only the new concept of 'physical' space to be distinguished from mathematical and experiential spaces, but also showed up and examined in a very clear example yet a third source of knowledge. This is the method of attaining to a knowledge of reality by starting from and basing ourselves on empirical data, whether of sense or generally of perception and observation, and using mathematical constructions.

Before discussing these three sources of knowledge in more detail, let us in passing point to an epistemologically remarkable development of the critical discussion on Kant's transcendental concept of space. From the arguments of mathematicians and physicists it emerged that the propositional systems of mathematical geometry are analytic and those of physical geometry empirical. (We can ignore the possibility of a conventionalist and therefore analytic propositional system of geometry in physics, because it can be ranged into the epistemological methods already mentioned.) To this, the neo-Kantians have objected that while admittedly all bounds of intuition are transcended in either kind of system, the question remains whether our intuitive spatial experience, which does indeed exist, might not give us a transcendental geometric space in the sense of Kant. This question has been investigated by psychologists. Intuitive experiential data of whatever kind are always the object of psychology which now enquired whether our experience of space did in fact give us a Euclidean space.

We know that these investigations taught us to distinguish experiental spaces of various kinds: visual, tactile, aural, equilibrium, and perceptual. These 'spaces' differ not only qualitatively but also not a little in geometrical metric. Accurate measurements showed that the visual space of the one-eyed comes nearest to being Euclidean, although even here we note deviations from the Euclidean metric, especially at the edges. For bifocal vision, only limited portions of space are approximately Euclidean. In this experiential space there are discontinuously bordered domains in different areas, and partly we note an interpenetration of different non-Euclidean metrics. At all events we cannot here speak of a uniform Euclidean metric. Likewise for tactile space, where in some parts of the human skin parallel lines intersect at a finite distance. In the remaining sensory spaces we can find only imprecise attempts at establishing metric relations. Highly complicated properties exist in perceptual space, where again Euclidean conditions can be demonstrated only sporadically. Investigations on the psychology of space thus led to the result that amongst intuitive experiential spaces none is uniformly Euclidean. Therefore the concept 'space' contains not only the two meanings of logico-mathematical' and 'physical' but also a third, namely 'space of psychological experience'. Description of spaces of this last kind rests on the first mentioned source of knowledge, namely data of consciousness, amongst them sense data and ideas.

Those who adhere to Kant's transcendental theory, especially the neo-Kantians, tried to maintain yet a fourth concept of space, as against the three that result from the above results of epistemological analysis, which resolved the original uniform and unambiguous concept of space into three different concepts as if by logical microscope. This is precisely supposed to be 'transcendental space' recognizable by a special non-empirical faculty of intuition that, free from the criteria of logic, mathematics or experience, shows transcendental space to be uniformly Euclidean. However, it is one of the insights of modern epistemological analysis that concepts and propositions not subject to the testing criteria (of logic, mathematics and experience) in the individual sciences, are empty and say nothing. Such items can be interpreted as one will, but for scientific enquiry they have no bearing and can be dropped. Therefore the concept 'transcendental space' and statements about it in speculative philosophy are now no longer discussed in the epistemological analysis of the theory of science.

The critical epistemological analysis of the concept of space showed up three sources of knowledge. This suggests the triadic method of testing further basic concepts in scientific enquiry, to see whether they might not have different meanings in relation to the three sources of knowledge.

3. THREE CONCEPTS OF 'TRUTH' AND 'PROBABILITY'

Relativistic time relations and the criteria indicated for them clearly underlined the difference between the subjective and often very variable time of
experience and the physical order of time. The concept of possible logical
mathematical time orders has been investigated above all by H. Reichenbach.[3] The three time concepts 'experiential time', 'logico-mathematical
times' and 'physical time' were conceived of in close analogy with the corresponding concepts of space.

Of greater interest to epistemological analysis is the application of this
critical method (namely, examining the meaning of basic concepts in relation to the three sources of knowledge) to new concepts and propositions
not connected with the concepts of 'space' and 'time'. Thus we can remove
certain problematic questions within epistemology itself, if we apply the
method to the concept 'truth'.[4]

In the epistemological analysis of the logic of language the question has
often been discussed whether in a linguistic system in which there are so-
called 'elementary' or 'atomic' propositions, the truth value of a proposition put together out of elementary ones depends only on the truth or falsity of these latter. The thesis asserting this view is known as the 'extensionality thesis' and we readily see that it has different meanings depending on
what we understand by 'truth': an experienced taking to be true ('experiential truth'), or a truth amenable to logico-mathematical decision ('formal truth'), or a truth decidable with the help of extra-linguistic, perhaps
empirical, criteria ('truth of content'). If we fail to observe these distinctions, the discussion of whether the extensional thesis holds may lead to 'in-
soluble' problems. A well known example is the controversy whether the
truth value of the proposition 'A believes that p' depends only on the truth
or falsity of the proposition p or whether the extensional thesis does not
hold here. To save that thesis, propositions of this kind have often been
rather artificially reinterpreted: for example, by stating that A's belief concerns only the symbol 'p' but not the proposition represented by 'p', so that
'p' would not be a partial proposition within 'A believes that "p"'. However,
such reinterpretations are untenable. In the present example, the inappro-
riateness of querying the extensionality thesis becomes clear as soon as A's
'taking to be true' as experiential truth is contrasted with the 'formal' or
'content' truth of p. Thus the extensional thesis loses its general and uniform significance and can be discussed at best in a rather restricted sense

in relation to whichever of the three concepts of truth is involved.

Perhaps the first to recognize that experiential truth is independent of formal and content (empirical) truth were the sophists. They praised their method for enabling them to make the true false and the false true at will. An especially abrupt contrast between experiential truth and the two other kinds is recorded in the mediæval saying 'credo quia absurdum'. This proposition is by no means nonsense. The 'credo' means the experiential (emotional) 'taking to be true' which is possible even where the proposition believed is false as to form or content, or even lacks any sense. A striking example for this view is provided by the famous scholastic controversy about the proposition 'there is an omnipotent being'. A logician here objected that the proposition could not be true because it was formally false, that is, self-contradictory. For an omnipotent being must be able to create a stone so heavy that he himself cannot lift it. In reply it was rightly urged that experience of truth could occur even where the formal criteria of truth, that is, criteria of consistency, are not satisfied. This independence of experiential from formal and content truth is likewise enlisted by the modern method of so-called 'brain-washing'. The methods of astrology and often even of speculative philosophy similarly rest on this independence being presupposed.

The use of the two concepts of 'formal' and 'content' truth is of some interest in Gödel's undecidability theorem.[5] He shows that in all systems satisfying certain criteria of the logic of language we can form propositions that assert their own undecidability which latter can then not be proved with the means available in that system. However, this enables us to recognize that those propositions are true as to content. If we made no distinction between formal and content truth, then Gödel's theorem would yield a contradiction. For undecidability concerns the truth value. If a proposition can be formally proved to be undecidable, while at the same time we can decide its truth as to content, the proposition would be decidable and undecidable at once. The contradiction does not appear if we distinguish formal from content truth.

In inductive logic too many of the controversies and paradoxes currently rife can be resolved in part or must be reformulated in quite a different sense, if we apply the triadic method to the concept of 'probability' and examine its different meanings when related to each of the three sources of knowledge.[6] Here too we observe a difference of principle between the experience of taking to be probable ('experiential probability') and the numerically calculable or expressible probability values in mathematics and physics. 'Experiential probability' (also called 'probability of belief') is a purely subjective mental attitude to which different people can be quite differently disposed. 'Credulity' denotes a special form of this disposition. We can indeed note different intensities of 'taking to be probable', for example greater or smaller credulity, but because of the subjective character of the probability of belief it is in principle impossible to capture it with numerical precision; for experiential probability is independent of any logical relations between propositions and of the objective empirical order of phenomena. That the probability of belief often differs abruptly from numerically describable ones can readily be shown from everyday examples. Somebody might hold a lottery ticket and think it very likely that he will win the main prize, clinging to this view even if he knows that the rules of probability calculus or statistical counts allow him to assign only a very small probability.

A specially important form of the probability of belief is the taking as probable the repetition of hitherto observed regularities when similar conditions recur. That this assumption, in the past loosely called 'inductive inference', was a belief that cannot be theoretically founded, had been pointed out already by Hume. The ordinary probable belief that an event A will always be followed by an event B if the same circumstances recur lies at the foundation of all empirical enquiry and its methods. Experiments in natural science are significant for enquiry (signifying, say, confirmation or refutation) only if we presuppose that on repetition of the experiments the same experimental results will come true. The probable belief at the basis of this presupposition finds expression in the saying that a hypothesis is the more probable the more often it is confirmed by individual events: a thesis for which there can be no theoretical foundation of any kind, which is why the probability of hypotheses can in principle not be adequately marked by numerical values however great the number of observational data.

That this can be done after all is an error at the basis of the inductive logic founded by Carnap. In his account two concepts of probability are mixed up, namely the 'experiential' probability just mentioned and the 'logico-mathematical' one to be discussed presently, from which in part result the paradoxical inadequacies within his 'inductive logic'. (These two concepts are by no means congruent with Carnap's 'probability$_1$, and 'probability$_2$', about which more later.) Logico-mathematical probability expresses logically possible relations between propositions or between classes of propositions and their elements and sub-classes. It is therefore purely analytic and any system of such relations is likewise purely deductive and analytic. This ·is best illustrated in probable inferences. These are inferences from disjunctive classes of propositions. From the disjunction $p \vee q$ we can only probably infer that one of the elements holds. From the form of disjunctive classes of propositions that always figure amongst the premises of probable inferences we can see that we here infer to several possible mutually exclusive conclusions. A precise numerical probability for such a conclusion becomes possible only when we stipulate a probability metric. The best known metric of this kind is the stipulation of 'equiprobability' for the individual elements of the disjunction. Another often used probability metric stipulates that for the possible individual cases or the propositions describing them, limiting values of distributions ascertained by statistical counts are to count as probability values. Carnap calls values derived with the help of this second probability metric 'probability$_2$'. Today we know that alongside these two systems infinitely many other probability metrics can be stipulated. Any chosen consistent metric enables us to calculate determinate probability values for the individual derivable propositions ('possible cases') with analytical exactness. Whether in real cases these values do, or even can, adequately convey either the intensity of probable belief in the coming true of these cases or the probability of a hypothesis on empirical data, is a matter about which it is in principle impossible to infer anything from the analytic stipulations of logical mathematical probability theories. Carnap was and is of the opinion that we can do this, but I think that view is based on a confused mixing up of the two probability concepts so far explained. That is also why Carnap's inductive logic is so inadequate, for whatever the amount of confirmatory empirical data ('evidence'), a law of nature (that is, a hypothesis about infinitely many cases) will always have the probability zero.

The meaning of the concepts 'experiental' and 'logical mathematical' probability becomes clearer still if we apply the triadic method and ask what is the meaning of numerically exact probability statements in the natural sciences, above all in physics. That is, what does the concept of probability mean in relation to the third source of knowledge. The analogy to the three concepts of space becomes clearly visible: there too we had to distinguish experiental spaces from logico-mathematical systems of geometrical relations, and these two in turn form the concept of 'physical space'.

Statements of physical probability say nothing about subjective experiential probabilities. They do indeed use mathematical systems of probability metric, perhaps in the form of reinterpreted wave expressions or correspondingly interpreted matrix schemata, but given this content the logico-mathematical relations no longer denote merely logically possible relations between classes of propositions, their elements and sub-classes, but empirically real orders of phenomena in determinate form. Whether such orders exist in the real cases to be described can be accurately tested empirically. This we may make clear as follows.

From the causal laws of classical-relativistic physics, especially from laws of proximate action, which are indeed regarded by general relativity as the universally valid form of physical description, we can derive one-one relations between initial and final states. Given an initial state A sufficiently characterized, we can derive by means of the relevant laws of proximate action an unambiguously characterized final state B, excluding the possibility of other final states B_1 or B_2 or ... B_n. Conversely, given a final state B sufficiently characterized, we can derive by means of that law which unambiguously characterized initial state A must have preceded the given final state and exclude the possibility of other initial states A_1 or A_2 or ... A_m. Thus we may say that the continuous causal laws of classical-relativistic physics (above all, laws of proximate action) state one-one relations between states and changes in them.

In contrast, physical probability laws are characterized precisely by the fact that they express relations other than one-one relations between states and changes of state. Given an initial state A, a probability law enables us to derive the occurrence of several possible and mutually exclusive final states B_1 or B_2 or ... B_u. Likewise, given a final state B, the probability law allows us to infer back to several possible initial states. The fact that initial and final states, or in general the states and changes of state within a sequence, are related in a way that is not one-one, is by no means tantamount to total lawlessness. It is an insight of epistemological analysis that physical probability laws, amongst them the statistical laws used in everyday life, do not describe one-one orders but one-many, many-one and many-many orders of states and changes in them, and that these individual orders are empirically testable with no less accuracy than continuous causal laws. The interpretation of wave expressions (e.g. squared amplitudes) or matrices as probability quantities means nothing other than choosing and stipulating a probability metric that allows exact characterization of the order of phenomena when it is not one-one. By this metric we assign probability values to the individual propositions or cases that can be derived, which values can be tested in the light of the relevant empirical facts of each case. The simplest way of testing such orders between phenomena or the probability values representing these orders is statistical counting. Amongst more complicated testing procedures are the measurement of the number, width and distribution of spectral lines construed as statistical frequencies of changes in state.

From the recognition that by 'empirical' or 'physical' probability we must understand real orders of phenomena of determinate form, it follows that we must give new definitions for the concepts of law of nature, causality, determinacy, indeterminacy, which we cannot pursue here. What we did wish to do is to explain the application of the triadic method, that is, the analysis of the meanings of fundamental scientific concepts in relation to the three sources of knowledge: firstly, data of consciousness; secondly, logico-mathematical constructions or stipulations; and thirdly, empirical methods for recognizing reality and the attendant formation of concepts and hypotheses. As we have said, this method of epistemological analysis acts like a logical microscope that reveals a multiplicity of concepts where previously one used a concept that was wrongly felt to be uniform and unambiguous. Therefore I believe that the triadic method of epistemological analysis promotes not only epistemology and the theory of science, but is also suited for influencing enquiry in individual fields. In particular, the establishment of new individual disciplines, such as geometrical measurements in the psychology of spatial perception, mathematical systems of inductive logic, the determination and application of stochastic forms of law in biology, psychology and in the social and political sciences, these in my opinion are the effects of analytic epistemological investigations that have used the triadic method mentioned.

NOTES

[1] Cf. Juhos, B., 'Drei Begriffe der "Wahrscheinlichkeit"', *Studium Generale* **21** (1968), 1153-1173.

[2] Cf. what follows with Reichenbach, H., *Die Philosophie der Raum-Zeitlehre,* Berlin 1926.

[3] Cf. ibid.

[4] Cf. Juhos, B., 'Die "intensionale" Wahrheit und die zwei Arten des Aussagengebrauchs', *Kant Studien* **2** (1967), 173–186.

[5] Cf. ibid.

[6] Cf. Note 1 above.

CHAPTER XII

THE TRIADIC METHOD*

I. ELEMENTS OF LINGUISTIC REPRESENTATION AND THE TRIADIC METHOD

Scientific enquiry not infrequently starts from presuppositions that count as commonly intelligible and indeed as necessary for thought. Often, such presuppositions are long used with great success, but finally enquiry tends to reach problems that compel us to examine whether our initial presuppositions were justified. In recent times it is precisely the presuppositions of the exact, indeed the most exact, individual disciplines that have been subjected to analysis as to their validity or realisability. Here it turns out that by applying the formal syntactic rules of the linguistic system used we sometimes reach expressions whose content can be semantically interpreted only by breaking or abolishing the original presuppositions. This leads to ever sharper distinctions and oppositions between the syntactic formal and semantic content elements of the representing or descriptive linguistic expressions. The provisional end point of this analytic development is the independent application of formal constructions on the one hand and the interpretation of content on the other, for the purpose of obtaining adequate linguistic representations. It is not at all as though correctly applying the formal syntactic rules of the linguistic system used must always give us expressions that either have or can be found a definite content. For in using a mathematical formalism well suited to describing the content, one may easily reach expressions for which no sensible interpretation of content can be given. In such a case we speak of the formal system as 'idling'. This possibility alone shows that principles of syntactic construction and of interpretation of content are mutually independent.

The two elements of logical grammar are dissociated even more clearly when we use a linguistic system whose content has already been interpreted and, by syntactic transformation, we obtain expressions for which the prior interpretation loses its sense although they can be sensibly construed by applying a new principle of interpretation incompatible with the old one.

* Translated from the German original 'Die triadische Methode', first published in *Studium Generale* **24** (1971) 924–945.

Such cases often occur in modern theoretical physics, about which more later.

The distinction between the two types of elements of linguistic representation or description suggests a new method of epistemological analysis. For we can relate any linguistic expression, whether concepts or propositions, to domains of elements of form or content respectively, and then ask what is the significance of the expression with respect to the various element domains. By explicitly applying this procedure we place the theory of science on new foundations. In the past, expressions, concepts, propositions and relations between propositions of the theory of science were judged from the prior supposition that the basic concepts and basic propositions of the system of propositions have definite meanings, from which definite meanings are obtainable for derived expressions also. The fact that in the derivations there appeared the occasional 'idling' in the sense explained could be ignored as long as the uninterpretable empty expressions remained unimportant for the application and further development of the theories. However, it was precisely in well advanced and highly developed theories that continued extension and modification of their formal (logical mathematical) starting points led to results that contradicted the initial presupposed interpretation of the systems and their expressions. Such cases then always occasion independent modification of the syntactic and semantic elements of the theories concerned, with examination of the resultant consequences; and this can lead to quite new formations of systems which nobody had even thought possible before.

For modifying the two kinds of elements of a theory and for the analysis of resultant consequences there are certain guidelines that appear in each case to be determined by the conditions governing the theories concerned. The classical example in which this new epistemological method has been applied is the analysis of the concept of space as initiated by mathematicians in their discussions of Kant's transcendental theory. In Kant, the concept 'space' had a uniform meaning, as with his contemporaries and predecessors. The view was that philosophers, mathematicians, physicists and phychologists all mean the same object when they talk about 'space'. The regularities of this one object 'space' was generally assumed to describe the only geometric system then known, namely that of Euclid. The first step towards formalising his systems was taken by Euclid himself by means of axiomatisation. This provided a point of leverage from which to test Kant's thesis that the propositions of Euclidean geometry are necessary for thought, or synthetic *a priori*. One merely had to test Euclid's axioms to see whether Kant's way of characterizing or justifying geometric propositions is correct.

We need not here discuss in detail the critical examinations that occurred, since this epistemological analysis and its results are largely familiar. However, what we must stress is the step-like examination as to the relation of 'space' to three definite element domains: namely that of contents of consciousness (sense data, ideas, feelings and so on), of logical mathematical relations, and finally of (for example empirical) realities or entities independent of consciousness. Because of the three domains with respect to which we are to conduct a critical analysis (here of the concept 'space'), we call this epistemological procedure the 'triadic method'.

It is well known that applied to 'space' this method produced a multiplicity of space concepts that are in part quite different as to their formal properties and content. As regards the data of consciousness we must distinguish the spaces of the different senses (sight, taste, hearing and so on) both as to quality and geometric metric structure. To this domain belongs the space of intuitive perception which arises from rather complicated empirical mental processes. None of these experimental spaces exhibits entirely Euclidean properties, as the exact empirical investigations have shown. The insight that experiential spaces have preponderantly non-Euclidean metric properties could of course be gained only after the epistemological method had previously investigated the meaning of the term 'space' as regards the domain of logical mathematical relations. Historically this step was an incipient application of the triadic method. The doubts as to an intuitive justifiability of Euclid's axiom of parallels finally led mathematicians to use the indirect form of proof. From new geometrical axiom systems which were incompatible with Euclid's system and with each other one could thus deduce theorems without meeting contradictions, so far as one could see. From Riemann's recognition that one could assume an arbitrary number of dimensions for the types of spaces to be constructed and a continuous order (represented by the functional values of the curvature tensor) could be indicated for the possible metric systems, there resulted a new concept of a logical mathematical space. Attempts based on Kant's transcendental theory to save for Euclid's system of space a privileged position over the other possible systems, failed both formally and as to content. The argument that the consistency of Euclid's geometry is transcendentally guaranteed while for that of the other geometrical systems no proof exists was refuted by the well-known proof by Felix Klein, according to which concepts and propositions of Euclidean and non-Euclidean systems can be so correlated that unambiguous correspondences likewise result for derivations in the various systems. That is, if one geometric system is consistent, so necessarily is any other; likewise, it follows from Klein's proof that all geometric systems must be inconsistent if in one of them a contradiction can be derived.

The thesis as to 'content' in Kant's theory, namely that it is necessity of thought for the laws of Euclidean geometry to be those of real physical space, was shown to be untenable by the third step of the triadic method. Gauss was the first to raise the question what is the meaning of the expression 'space' with regard to the domain of empirical phenomena. The question is obviously based on the view that physical space in which all empirical processes take place is given to us as an empirical phenomenon like any other empirical state. Thus the expression 'space' is here being related to the domain of empirical reality independent of consciousness. Just as we seek to characterize any other empirical phenomenon by metrical values, so, Gauss thinks, we likewise must determine by measurement what are the geometrical metric properties of physical space that is given to us as an empirical phenomenon. Later physical enquiry has adopted just this empiricist conception with regard to the description of physical space, as against the conventionalist possibilities noticed by Poincaré. On the basis of empirical data we nowadays ascribe approximately Euclidean metric properties to 'physical space' in restricted domains, but non-Euclidean ones in extensive domains.

By application of the triadic method, the initially uniform concept 'space' was resolved, as under a microscope, into the three groups of concepts: 'experiential spaces', 'logical mathematical spaces' and 'physical empirical space'. Before we use further examples to explain the steps and functioning of this method of epistemological analysis, more clearly, let us in this connection recall another method used by many philosophers. After it was shown that statements about 'space' in the individual sciences of mathematics, physics and psychology are either analytic or empirical, the adherents of Kant thought they could interpret his transcendental theory as asserting that there is a special transcendental 'space' whose properties we can recognize not by the methods of the individual sciences (not, that is, by logical mathematical constructions nor by empirical observations and measurements); but that this requires philosophical transcendental intuition given to us as a special cognitive faculty that allows us to recognize 'transcendental space' (a fourth concept of space) to have Euclidean properties, as Kant had maintained.

However, even if there were such a space inaccessible to the methods of science, this would not topple Kant's doctrine of a 'pure' space of intuition, for he held that the synthetic a priori space was precisely the one of which the individual sciences of mathematics, physics and psychology speak. On his theory it is inconceivable that, alongside transcendental space, individual sciences could represent spaces of various kinds with different geometrical metric properties, or that such spaces exist. The objection of Kantians that Kant's theory of transcendental intuition concerns only transcendental space which cannot be captured by the methods of individual sciences and therefore has nothing to do with the spaces set up in mathematics, physics and psychology, thus does not amount to saving his transcendental theory. Indeed, this view has consequences that show the assumption of a space unknowable by means of any method in individual sciences to be empty. For if we grasp the properties of 'transcendental space' by a special cognitive faculty that cannot be tested by any of the criteria of the individual sciences, then everybody is free to assert for himself that this faculty has shown him that transcendental space has any arbitrary properties whatsoever. Nobody else can test 'knowledge' about space gained by this transcendental faculty. The concept of 'transcendental space' and statements describing it thereby become irrelevant to scientific enquiry. Accordingly this empty concept is now no longer discussed in the theory of science. This result too was obtained by applying the triadic method.

2. THE TRIADIC ANALYSIS OF 'TRUTH'

The analysis of the concept of space has clearly exhibited the three steps of the triadic method. This suggested that one might apply them analogously to the concept of time. H. Reichenbach has distinguished the subjective times of experience from the possible structures of mathematical time and the empirical objectivity of physical time. The epistemological analogy with the distinction between three concepts of space is obvious. This is not to deny that triadic analysis raises new problems of a special kind, particularly as regards the characterization of physical time. However, for the theory of science it is to begin with of greater interest to use the triadic method to bring out new conceptual connections that are independent of the concepts of 'space' and 'time'.

Already in antiquity there had been frequent discussion of the question whether there was only one kind of 'truth' or whether in view of secure and less secure knowledge several concepts of truth had to be distinguished. Indeed, a decisive turn in the development of ancient philosophy, namely the conquest of sophism by Socrates and his followers, was brought about by a critical polemic as to the concept of truth. The sophists regarded 'truth' as relative only to the domain of data of consciousness, that is, experiential truth. Their definition of truth expresses this very clearly: 'truth is what one takes to be true'. They interpreted the concept 'truth' by means of the first step of the triadic method and from the view that truth was experiential they drew the well-known sophistic consequences. The insight that experience of truth can occur quite independently of objective and scientific criteria of truth is stated by the sophists in the thesis that by their method one can arbitrarily turn truth into falsehood and conversely. This thesis clearly operates with several concepts of truth and falsity, although the sophists recognize only the experiential ones. It is by no means as though the truth of experience were to be regarded merely as a sophistically degenerate concept, but the use the sophists made of it in their 'rhetoric' (the method of producing experiential truths at will) was a very degenerate move as compared with the scientific enquiry into truth.

The 'truth of faith' ('divine truth') of the Middle Ages likewise belongs to the domain of experiential truth. Here there arose a clear distinction of two truths. The application of logic to theological dogmata in scholasticism often led to contradictions. Thus, logicians argued that the proposition 'there is an omnipotent being' could not be true because the concept involved was contradictory.[1] Such objections were met by the doctrine of two 'truths' (the 'divine' and the 'secular'), according to which propositions that are false for human understanding, that is, for science and all its criteria, can be believed as 'true'. The truth of faith, like the sophists' experiential truth, is characterized by experiential criteria that are independent of scientific ones. This independence is for example expressed in the mediæval statement 'credo quia absurdum' (I believe because it is absurd). When Socrates or the opponents of scholasticism insist on the 'objective scientific' (that is, logical mathematical and empirical) criteria as the only valid ones for truth, they did not thereby prove that the concept of 'experiential truth' is logically impossible, but were merely relating the expression 'truth' to other element domains than the sophists and scholastics.

Concepts of truth defined with the help of 'self-evidence' differ from 'experiential truths' in that the experience of evidence cannot appear independently of scientific criteria of truth that are objective as to form and content, or that its appearance is tied to certain kinds of propositions, for example experiential ones.

Recent epistemology has closely examined the meanings of concepts of truth related to the domain of logical mathematical relations, and of 'objective' reality independent of consciousness. These two steps of the triadic method applied to the expression 'truth' led to many new problems and investigations both in the foundations of logic and mathematics and in the methodology of the empirical sciences.

Today we count as 'formal truth values' all those that can be decided with the help of grammatical means within a linguistic system, that is, syntactic stipulations and semantic meaning rules. It is not as though all analytic propositions representable in formal systems could be decided by means of the logical grammar of those systems, that is, that all analytic propositions are to count as 'formally true'. Gödel[2] has shown that in systems sufficiently rich in formal rules we can form propositions of a determinate form using only analytic elements of the system, but that the validity of these propositions cannot be decided in terms of the domain of logical mathematical relations of that system. If now, as Gödel has likewise shown, the 'truth' of such analytic and formally undecidable propositions can be intuited by 'considerations about content', this means nothing else than interpreting the expression 'truth' in terms of the domain of entities independent of consciousness, which in this special case are 'mental' or 'ideal'. A proposition that asserts its own undecidability and whose undecidability can be proved, can thus be seen to be true 'as to content'. By applying the triadic method we here recognize firstly that even in formal systems we can, under certain conditions, sensibly speak of the 'truth as to content' of analytic propositions, and secondly that alongside empirical reality there are 'mental' or 'ideal' entities to which we can relate logical mathematical expressions and propositions. The view sometimes uttered that mathematics investigates the domain of ideal existences (entities) and could thus be characterized as 'mental experience' thus appears in a new light, given the third step of the triadic method applied to analytic mathematical propositions as just explained.

If, in the theory of science, sciences 'of reality' are set over against 'analytic', 'ideal' or 'purely conceptual' ones, 'reality' is understood as objective and empirical. The methods of empirical science can be used and its propositions tested by anybody, that is, intersubjectively. It is therefore natural that the triadic method should investigate the significance of expressions generally used in science as regards the domain of empirical reality independent of consciousness, too, that is, objective intersubjective reality. Applying this to the expression 'truth', we reach the much discussed problems of verification and falsification. That general empirical propositions (laws of nature, hypotheses) are not completely verifiable hardly anybody will deny today; the view that they can be completely falsified[3] rests on a mistaken identification of empirical with logical mathematical falsity.[4] Closer analysis of how we determine the truth of empirical propositions leads to a distinction between two concepts of empirical truth: the non-hypothetical and the hypothetical. The former belongs to experimental propositions (such as 'I feel pain'). Propositions of this kind, which I call 'constatations', are used according to rules that exclude the possibility of empirical error. They are marked by the fact that their truth or falsity is decided by the way they are obtained, that is, the process for obtaining them is also that for verifying them; which suggests that we may regard the truth of experiential propositions (contatations) as immediately obvious 'evidence' and therefore as 'experiential truth'.[5] However, empirical non-hypothetical truth differs from the experiential truth' described above, firstly by belonging only to propositions of a certain form and secondly by not being arbitrarily creatable or modifiable.

The second form of empirical truth is the 'empirical hypothetical': the truth of a proposition in that sense depends on the truth of other propositions in such a way that the hypothetical proposition concerned may on occasion be denoted as an error. Some epistemologists deny that this kind of truth is justified.[6] However, to construe empirical truth in such a way that we cannot sensibly ascribe truth to empirical hypotheses (laws of nature) involves a most artificial reinteretation of hypothetical propositions and ways of talking about them. By applying the third step of the triadic method we obtain a natural classification of the forms of empirical propositions into non-hypothetical and hypothetical, where the latter can be subdivided into singular and general hypothetical propositions. This enables us to speak of the truth or falsity of any empirical proposition. Empirical non-hypothetical truth is introduced as a self-evident undefined basic concept, and empirical hypothetical truth defined with the help of it and the concept of 'sufficient agreement'; and analogously for falsity.[7]

3. THE TRIADIC ANALYSIS OF 'PROBABILITY'[8]

The application of the triadic method to the expression 'probability' has led to much discussed problems of epistemological analysis in modern theory of science. Here, too, conceptual difficulties arose from mixing up probability concepts, which are related to different element domains, and differ in meaning.

If we relate the expression 'probability' to the domain of experience, it takes on the meaning of 'considering to be probable', which can appear with greater or smaller intensity, though we cannot give adequate numerical marks to the probabilities assigned to propositions or phenomena on such a basis of 'probable belief'. 'Experiential probability' is subjectively variable and can deviate widely from calculated or statistically observed probability values. This is expressed by the often heard everyday sentiment that the credulous and those given to wishful thinking will often regard the most improbable things as very likely. We see at once that here at least two concepts of probability are being contrasted. To experiential probabilities belong the probability assumptions that observed regularities in the occurrence of phenomena will be repeated in the same way if the same circumstances recur; likewise the assumption that the (experiential) probability of a law of nature grows with the number of individual confirmations. Such inferences from individual propositions to general laws (hypotheses) continue to be infelicitously called 'inductive'. Already Hume pointed out that such inferences cannot be justified on theoretical logical grounds and that probabilities ascribed to regularities by means of them merely express subjective probable belief.

It would however be a mistake to think that such probabilities (assigned on the belief that observed regularities will recur if similar conditions return, and thus subjectively variable and inadequately capturable by numerical measure) do not have their role to play both in daily life and in science.

Here too it was Hume who pointed out that our practical everyday behaviour rests on the probable belief mentioned above. This presupposition, which is not justifiable theoretically, stands out with particular clarity in the experimental methods of natural science: experimental results have verifying or falsifying force only if we assume that repeating the experiments will always lead to the same results.

The belief that the probability of a general empirical proposition grows with the number of individual confirmations suggests a comparison of the experiental probability based on it (which cannot be captured with adequate precision) with those relations between propositions that allow calculation of numerically exact probability values for individual propositions and their combinations. By systematic application of the triadic method this comparison leads us to distinguish two new concepts of probability. Examples of relations between propositions that can be represented with numerical precision are given even in classical probability calculus. Modern logical analysis has shown that numerically exact probability values can be ascribed to propositions only if the latter bear relations of quite definite form to other propositions. Where this condition is formally satisfied, that is, where from the form of premises with numerically exact probability values we can infer conclusions without having to watch the content of propositions and expressions, there we speak of 'logical mathematical probability' of the 'derived' propositions. Where on the other hand these relations between premises and conclusions (or between classes of propositions and their elements) exist (and are calculable) only under certain empirical conditions, we speak of 'empirical' or 'empirical physical' probability of the propositions concerned.

The conditions under which relations between propositions can be represented as numerical values of logical mathematical probabilities are investigated by 'inductive logic'. The name is misleading insofar as the propositional connections represented in this system are purely deductive. The purely formal and logical mathematical character of analytic probability is best understood if we compare probable inferences and propositional connections appearing in them with other mathematical inferences and propositional connections that have the same formal properties as the former. Here it is noticeable that traditionally used extrapolating and intrapolating inferences in mathematics have the same elementary form as formal probability relations. In every extrapolation or intrapolation there are several orders (representable by functions) towards wich we can work. If for example from a given finite and bounded segment of a curve we are to extrapolate to the shape of the total curve, there are many, indeed infinitely many different functions each satisfied by the values marking the given segment but describing a total curve of different shape. Another mathematical example where from given premises we can 'infer' several possible but mutually exclusive conclusions is the frequently occurring case of interpolation; for instance if a given number of points are joined by a curve, or a given number of numerical values are to be subsumed under a function to be determined, then there are in general any number of curves or functions all of which satisfy all the conditions mentioned and from which a curve or function can be chosen. These extrapolations and intrapolations have the same form as probability inferences. In both cases the inference to a definite conclusion is not logically necessary, or, as we say in logic, the relation between premises and conclusion is not a tautological implication.

The formal sameness of extrapolations and intrapolations on one hand and probability inferences on the other goes further still. In an endeavour to obtain exact deductions even where the relations between premises and conclusions are not necessary and unambiguous, we introduce into extrapolations and intrapolations certain restrictive conditions by which the choice of a conclusion must be guided. For example, in extrapolating to the shape of a curve we stipulate that the curve shall have as small a homogeneous curvature as possible; or, we demand that the curve that must be satisfied by given individual values is to be differentiable and of as low a degree as possible. By these and similar restrictions one may finish with a single definite form or order to which one can extrapolate or intrapolate. For probable inferences, the same restrictive role is played by the stipulation of a probability metric. The best-known such metric systems are the stipulation of equiprobability for all possible cases and the rule that the probability of a case is the limiting value of the average frequency with which that case appears in statistical series. Once a probability metric for (ambiguous) probability relations between premises and conclusions has been chosen, it becomes possible to infer with numerically exact probabilities to the individual conclusions.

Thus with regard to the domain of logical mathematical relations the expression 'probability', just like 'space', reveals itself as a certain form of analytic conceptual connections. This is made especially clear by the fact that we can indicate a continuous order for the possible metrics both of probability and of geometry, represented in the latter case by the curvature tensor and in the former by the inductive-logical G-function.[9]

In view of the explained character of logical mathematical probability we are justified in denoting 'inductive logic' as the 'logic of extrapolations and intrapolations of a certain form'. However, Carnap wished to go further and construe the (logical mathematical) probability defined in his inductive logic as identical with the one that empirical enquiry usually ascribes to general empirical propositions (hypotheses, laws of nature) on the basis of confirming evidence (data of observation, experimental results). This erroneous interpretation rests on a conflation of the two concepts of 'experiential' and 'logical mathematical' probability. The assumption that a law of nature is the more probable the more often it has been confirmed by evidence is a probable belief that is subjectively based and unjustifiable, as mentioned before, and that is why it can never be adequately captured by exact numerical measures. Carnap's attempt to achieve this in his system of inductive logic is known to have led to the paradox that on his approach the probability of general empirical propositions not only fails to grow with increase of confirmatory evidence but always receives the value zero.[10]

Leaving aside the already mentioned interpretation of the content of c-functions by which Carnap defines logical mathematical probability, his system of inductive logic is consistent. The formal system even then retains the peculiarity that 'general' propositions within it, that is, propositions about infinitely many cases, receive the probability value zero, however many positive individual propositions there are; but this is not a formal contradiction. Only if we interpret the c-functions semantically as representing that experiential probability which we usually ascribe to empirical hypotheses on the basis of confirmatory individual cases, are we led to the paradoxical inadequacy that the probability values for general propositions derived in the system of inductive logic contradict the values usually assumed for them in the practice of enquiry, for these latter values do grow with the number of confirmations. Many attempts have been made, by special semantic interpretations and by formal changes in the system of inductive logic, to remove this discrepancy, but since the conflation of the two concepts of probability was retained, new inadequacies always appeared, even if not always in the same place.[11]

The question how one can describe empirical connections of phenomena with the help of inductive logical (that is, logical mathematical) probability systems can be aswered by using the third step of the triadic method. If we relate the expression 'probability' to the domain of empirical reality, logical analysis shows that (empirical) probability description is used where we can distinguish or delimit states and their changes by ranges of measures. In such a case we speak of a discontinuity in the course of phenomena. Physically such ranges signify that there is no ascertainable continuous causal law in those ranges. Even though discontinuity in the sequence of states does not exclude one-one relations between initial and final states, still, in the case of discrete changes in state such relations cannot be derived from testable continuous causal laws (since such laws are here impossible because of the ranges that occur). Thus, in discontinuous sequences of states any regularities that might exist can be determined only by ascertaining average frequencies, that is, by statistical counting. It thus becomes clear what is the meaning of 'probability' as relating to empirical reality: by the 'probability' of empirical phenomena we understand a testable order of phenomena that is a natural law of a certain form. Amongst the presuppositions of this kind of order is the discreteness of sequences of states or changes in them and the fact that relations between initial and final states are not one-one.

That physical probability laws are based on these presuppositions could be observed even in probability descriptions in classical physics, except that there one was disinclined to recognize discreteness of states and the lack of one-one relations between initial and final states as a real order of the phenomena concerned. Rather it was felt that discontinuity and the attendant statistical counting of states or changes in them was merely the result of imprecision and imperfection of our instruments and procedures of observation. On this view, probability laws have only an 'as if' character: as our methods of measuring become more refined and precise, we shall be able to eliminate physical probability laws more and more, that is, progressively replace them by continuous causal laws. In reality, it was assumed, all phenomena are subject to strictly continuous causal laws, which can however be recognized only in those domains where measuring instruments and observational procedures are available.

This view can be maintained as long as we seem to be justified in presupposing that there are no limits to progessively improving measuring procedures and making them more precise. It therefore amounted to a revolution in the domains of descriptive forms in physics when Heisenberg proved in his uncertainty relations that sufficient characterization of states by measures is subject to limits below which we cannot go. Still to assume that in reality all phenomena are subject to continuous laws, only in principle we cannot know or test them in domains of the (finite) least possible magnitude (that is, in micro-domains), would amount to assuming metaphysical other-worldly realities with absolutely valid but unknowable laws, about which anybody can speculate at will, since they cannot be tested. However, such assumptions are not tolerated forever in empirical enquiry. Accordingly, quantum physics has abandoned the assumptions that phenomena can pervasively be described by continuous causal laws, recognizing the irreducible autonomy of the statistical type of description. In this way the expression 'probability' gains a new meaning in relation to the domain of empirical reality independent of consciousness: it now marks real orders of phenomena in which the relations between initial and final states are not one-one. Let us call this concept 'empirical (physical) probability'.

There is a large (indeed unbounded) number of forms of order that are not one-one. Mathematically, such orders, if satisfying certain conditions, may be represented by functions defining probability metrics. Which of the possible orders that are not one-one and used for describing real sequences of phenomena is to be chosen will be decided by empirical considerations in each case. What is epistemologically important here is that physical probability laws, and therefore the assumption of certain orders that are not one-one in the course of phenomena, can be tested by empirical methods (statistical counts, measurements of distribution and width of spectral lines and the like) to arbitrarily close approximations. As regards precise testability there is thus no difference of principle between testable continuous causal laws (laws of proximate action) in macro-domains and probability laws holding in micro-domains. Indeed, the same accuracy of testing is possible also for probability laws that describe macro-phenomena such as the occurrence of states in games of chance. This must be stressed, because it refutes the basic misconception that, as against the continuous causal laws, probability laws are merely a kind of imprecise description.

Connected with the concept of physical probability are the expressions 'causality', 'determinacy',[12] to which the triadic method can be likewise applied. Let us merely point out that Hume construed 'causality' as belief, ex-

pectation, in short, experience of a certain kind. Maxwell thought that he
could find the criteria of causality in certain formal thematic properties of
the formulæ representing laws of nature. According to him, causal determinacy in the course of phenomena exists where there is no explicit occurrence of space and time co-ordinates in the descriptive laws of nature. This
criterion also involves continuity of the describing function as a condition
for causality.[13] The emptiness of such formal criteria shows itself in this example in that for every sequence of phenomena one can set up a (continuous) function in which space and time co-ordinates appear only implicitly.
That this might at times be done only in the form of ad hoc hypotheses is
formally indifferent. The same holds of all the other purely formal logical
mathematical properties of the functions, as soon as one wants to declare
them to be causal criteria. It follows that the expression 'causality' is meaningless with regard to the domain of logical mathematical relations. Only
when we relate the expression to empirically confirmed regularities of real
sequences of phenomena, that is, when we take the third step of the triadic
method, do we obtain concepts of causality that are adequate to the concepts used in daily life and in empirical enquiry. When logical analysis is applied, the 'classical' concept of causality, like the space concept of classical
physics, is resolved into several (two) concepts. The classification into first
and second order causality corresponds to the two kinds of description in
modern physics: by means of (absolute) probability laws and by (relative)
laws of proximate action.[14] Here the causal criteria are two real sequences
of phenomena differing in form and noted through experience. The effect
of these conceptual distinctions on the expressions 'determinacy' and 'indeterminacy' has repeatedly been investigated as regards the domain of empirically real regularities.[15] In contrast the relations of 'experiential causality' ('experienced causality') to the experience of complete determination
('predestination') or of indetermination ('free will') has been given less consideration, precisely because the three steps of the triadic method have not
been separately applied to the expression 'causality'. The distinction between 'causality of faith' ('experiential causality') and the orders of 'empirical physical causality' leads to a resolution of the much discussed problems
of determinism and indeterminism into several questions.

4. THE TRIADIC ANALYSIS OF 'NUMBER'

Amongst the attempt to indicate the meaning of the expressions 'number', 'kind of number' and of the operations to be applied to them, we find in enquiries on the foundations of mathematics interpretations that we can mark more precisely with the help of the criteria of the triadic method. Intuitionists and operationalists relate the expressions 'number' and 'elementary operation' to data of consciousness. The number 1, or the natural numbers, or some other symbols, are interpreted either as intuitive data of a special kind of 'mathematical intuition', or exhibited as normal intuitive data to which intuitive operations of ordering and transformation are applicable. What these orientations have in common is that they do not recognize as legitimate proof procedures (for establishing the non-denumerable) those forms of mathematical proof which, like the diagonal procedure, presuppose the existence of infinite classes. Connected with this is the restriction of indirect proof to domains (classes and orders) that can be captured constructively (that is, with the help of intuitive elements and operations). This restriction makes consistency proofs trivial and superfluous. However, under these presuppositions a dilemma arises: when may one add extensions to the rules of an intuitionist operational statement? Such rules cannot be eliminated. If, then, they are to be admissible, we should have to be able to indicate appropriate criteria for them. These criteria, too, are sought by the intuitionist operational orientation in intuitive elements ('atoms') and operations. To assume the existence of infinite classes or infinitely many steps in a proof or construction counts as non-intuitive. If we can prove indirectly that an extension rule when applied will not lead to contradiction, this is not enough for intuitionist-operational logic to admit that rule (because the indirect proof presupposes the existence of infinite classes). On carrying out this programme we note that with the help of intuitive elements ('atoms' and 'operations') we can construct only a fragment of classical arithmetic.[16] The second step of the triadic method does not relate the expressions 'number' and 'operation' to the domain of conscious data as given to us in intuition or in the sense perception of symbols and their constructible orders. With regard to the domain of logical mathematical relations 'numbers' are defined as higher order classes, for example as 'classes of classes of the same number'. Formally this manifests itself in that formation and ordering rules are indicated for variables and logical constants, by means of which we then combine symbols into 'propositional form'. We next test the formulæ as to when they may or may not be written down in combinations of propositions. From the logical (syntactic and semantic) rules there will then emerge two extreme forms of propositions: those that may be added to all or none of the combinations of propositions (in a calculus) respectively. Thus we have found the regulative criteria for setting up and extending arithmetic, if it is construed as a purely logical mathematical system of relations. The choice of the basic propositions (axioms) represented with the help of the basic logical mathematical signs (individuals, predicates, classes, class membership, logical constants and so on) is largely dictated by the scope of classical arithmetic, analysis included. For the axiom systems and for all their extensions, for example by continued formation of predicates and classes or by introduction of new relations and rules, there is now only the regulative stipulation that in the system, of which classical arithmetic (of real numbers) must be a proper or improper part, no contradictions shall arise. It did indeed turn out that consistency proofs even in fairly limited calculi met with difficulties of principle. However, it is a remarkable fact in foundation enquiries, that those who relate the expressions 'number' and 'operation' only to the domain of logical mathematical relations, by no means infer from the impossibility of general consistency proofs to the inadmissibility (illegitimacy) of parts of 'number theories' that can be axiomatically set up with the help of logical mathematical elements and formation rules, but not decided as to consistency. Rather, those who take this view (not the least amongst them all non-philosophical mathematicians) go on extending arithmetic and its fictitious parts, merely setting up regulative prohibition rules for the formation of certain complexes of signs or application of rules under the relevant conditions, where contradictions actually occur in derived combinations of propositions. Such prohibition rules are called 'regulative ad hoc principles' because they cannot be justified from theory. Even if this be regarded as a blemish, applying this procedure to the domain of purely logical mathematical relations nevertheless makes it possible to go on consistently extending 'number theoretic' systems and thus to continue with enquiries into the foundations of logic and mathematics.

The third step of the triadic method relates the expressions 'number' and 'operation' to the domain of entities independent of consciousness. Here, the question how numbers can be used to describe empirical phenomena has often been discussed. It was recognized that this requires 'spatio-temporal delimitations' of states, objects and phenomena in general. Such 'delimitations' can be given in nature or created by artificial (experimental) means. We then mark phenomena by numbers according to two methods: either by measurement or by statistical counting.[17] This shows what is to be understood by the numerical values marking empirical reality: they are either measured values marking states or limiting values of average frequencies. These meanings of 'empirical numbers' stand out especially clearly where we have to seek an empirically testable meaning for the expressions derived from the physical formulæ. From the mathematical formalism used in physics we can not infrequently derive expressions whose content is meaningless so that their possible numerical values merely signify formal logical mathematical relations or classes. In other cases we obtain expressions whose individual values do have a definite empirical meaning, that is, they are either measured values marking states or statistical distributional values. In yet other cases, however, we obtain expressions for whose individual values we must seek an empirically testable meaning. The practice in such cases is to try out two interpretative principles on the formal expressions. The two principles are conditioned by the empirically performable methods of assigning numbers to real phenomena, namely measurement and statistical counting. Accordingly, we try to interpret derived expressions whose empirical content is not clear as marking some phenomena or other by measured or by statistical distribution values, while presupposing that these last are always empirically testable; which makes them 'empirical numbers'.

Empirical numerical expressions, that is, expressions for quantities whose individual values are measured or statistical, figure in physical formulae as continuous and differentiable. As regards the domain of empirical reality, however, there is a difference of meaning for 'empirical numbers' describing measured values of continuous changes in state on one hand and statistical distributions of discontinuous phenomena on the other. In the first case, what corresponds to differential changes in the metrical quantities to be marked are differential changes in the states concerned (say, of energy fields). In the second case, what corresponds to differential changes in probability quantities (which in the functions might formally appear as 'metrical quantities', for example as wave amplitudes) are differential changes in the probabilities assigned to the discretely variable states (for example corpuscular states). In principle, the differential changes in either kind of quantity can be tested equally accurately (by measurement and statistical counting respectively).

By the third step of the triadic method we thus obtain two concepts of 'empirical number' that differ in meaning, because of the different form of the two real orders of phenomena that are marked by means of them respectively. With the help of continuously variable metrical quantities we mark initial and final states between which we can ascertain one-one relations; where this last condition is not fulfilled, we cannot give a testable description of the connections between phenomena by means of continuously variable metrical quantities. In that case, however, it is possible to represent the relations (now not one-one) between discretely variable states with the help of differentially variable probability quantities. The derived numerical values can then be tested with arbitrary close approximation by statistical counts. The two kinds of empirical number thus receive their meaning by the two forms of order of real connections between phenomena described by means of corresponding functional expressions of the two kinds of number.

As with numbers, so with the mathematical operations applied to them we must examine what is their meaning with respect to the domain of empirical reality. Here logical analysis shows that to some operations there correspond real and sometimes even experimentally realizable processes. For example, the mixing of two amounts v_1, v_2 of the same kind of fluid can be described by simple addition. As description of an empirical process addition here becomes an 'empirical operation'. However, such assignments of operations on empirical numerical expressions in formulæ to definite physical processes is impossible in the overwhelming majority of cases.[18] Formulæ are often subjected to operations of various kinds while an interpretation will be sought only for the content of the resultant expressions. This procedure is especially clearly in evidence in the transformation of given formulæ by application or insertion of quantising operators. This shows what is the meaning of mathematical operations (and mathematical expressions in general) when thus applied in empirical formulæ. Here the syntactic and semantic elements of empirical description part company and are handled independently. The mathematical expressions and operations are either logical mathematical or empirical content elements, according to whether we use them as pure formalisms or can find interpretations for them. The third step of the triadic method applied to numbers and mathematical operations thus shows firstly that, as to meaning, there are two different kinds of expressions and operations used in physical formulæ (as logical mathematical formal and as empirical content elements respectively); secondly, this step of epistemological analysis gives insight into the mechanism of the two kinds of elements in physical description, that is, into the ways in which syntactic (formal logical mathematical) and semantic (interpretational) elements are handled in the obtaining of general empirical propositions (laws of nature) under the extreme conditions of modern physics.

5. THE TRIADIC ANALYSIS OF 'PHILOSOPHIC EXPRESSIONS'

The triadic method is an analytic procedure of epistemology. We could show in other examples, how, by applying it, expressions in general use, when viewed in relation to the three domains, are resolved into a multiplicity of concepts quite different in meaning. Failure to distinguish, or conflation of, such concepts that go by the same name but differ in meaning, leads, both in individual scientific enquiries and in philosophy, to difficulties (paradoxes) that can be removed by analytically revealing the differences in meaning. Once we have recognized these differences, we are led to think of new concepts, which previously never even had, nor could have, occurred to us. Examining the new conceptual connections can lead to the foundations of new methods and branches of science. The rise and development of non-Euclidean geometric systems, the representation of geometry as parts of arithmetic, the psychology of space, the setting up of 'inductive logic', the contrasting of the presuppositions for the two kinds of physical description (by continuous causal laws and probability laws respectively, with all the resultant insights as to epistemological analysis and physics), these are only some examples for the way in which the triadic method can influence enquiry both in individual sciences and in philosophic epistemology. Amongst these achievements we must count also the investigation as to the sense in which one relates basic philosophical concepts in various philosophical orientations to the three element domains in question. Looking back on the history of philosophy we clearly see for example that in certain orientations the concept of 'reality' is constantly confined to the domain of conscious data (data 'dependent on consciousness'). Thus arise sensationalist and solipsist systems, and some narrowly limited forms of empiricist and positivist ones. In other directions the expression 'reality' is viewed with regard to logical mathematical relations, which leads to the foundation of so-called 'platonist' systems, where conceptual abstractions and relations are said to have 'absolute' existence. If we relate the expression 'reality' to the domain of entities independent of consciousness, we come to philosophic lines that seek reality criteria in proved or empirically valid propositions and systems of propositions (theories) of individual sciences. This happens in 'logical empiricism' and 'logical positivism', where the concept 'reality' comprises alongside conscious data ('sense data') the regularities that can be tested by just those data. Moreover, in these epistemology-based orientations one investigates the question of how far one must assume entities independent of consciousness in logic and mathematics.

In the many other types of philosophic system, these three forms of the concept of reality are either confused or set into mutual 'relation' in speculative form, often by the introduction of metaphysical transcendental 'layers of being'. At all events, this example shows that the triadic method is suitable for making evident the meanings of basic philosophic concepts and thereby rendering intelligible the relations and differences between philosophic systems as well, which opens up prospects for possible further developments.

NOTES

[1] One argument in point runs thus: 'an omnipotent being must be able to create a stone so heavy that he cannot lift it.'

[2] Cf. Gödel, K., 'Über formal unentscheidbare Sätze der Principia Mathematica und verwandter Systeme', *Monatshefte f. Math. u. Phys.* **38** (1931), 173–198.

[3] This view is held by K.R. Popper and not a few of his followers. Most recently, however, arguments have been put forward within Popper's school, which remove the originally asserted asymmetry between verification and falsification. Cf. Lakatos, I., *Criticism and the Methodology of Scientific Research Programmes,* Meeting of the Aristotelian Society, London 1968.

[4] The main argument of this orientation on theory of science is the (erroneous) interpretation that laws of nature have the same character as Goldbach's hypothesis. Cf. Juhos, B., 'Über die empirische Induktion', *Studium Generale* **19** (1963), 259–272.

[5] Cf. Juhos, B., 'Negationsformen empirischer Sätze', *Erkenntnis* **4** (1934), 41–55, this volume, pp. 47–59; 'The Truth of Empirical Statements', *Analysis* **4** (1937), 65–70; *Die Erkenntnis und ihre Leistung,* Vienna 1950. 'Die empirische Wahrheit und ihre Überprüfung', *Kant Studien* **59** (1968), 434–447.

[6] This view is held, though from quite different starting points, by Wittgenstein, L., *Tractatus logico-philosophicus* 1922; Schlick, M., 'Gesetz und Wahrscheinlichkeit', *Gesammelte Aufsätze,* Vienna 1938, pp. 323–396; Popper, K.R., *Logic of Scientific Discovery* 1959 and 'Theories, Experience and Probabilistic Induction' in *The Problem of Inductive Logic,* Amsterdam 1968.

7 Cf. Juhos, B., l.c. and 'Die "intensionale" Wahrheit und die zwei Ärten des Aussagenge-brauchs', *Kant Studien* **2** (1967), 173–186.

8 Juhos, B., 'Drei Begriffe der "Wahrscheinlichkeit"', *Studium Generale* **21** (1968), 1153–1173.

9 Cf. Johnson, W.E., 'Probability', *Mind* **41** (1932), 1–16, 281–296, 408–423. Carnap, R., *The Continuum of Inductive Methods*, Chicago 1952. Kemeny, J.G., 'A Contribution to Induc-tive Logic', *Philosophy and Phenomenological Research* **12** (1953), 371–374; 'A Logical Mea-sure Function', *Journal of Symbolic Logic* **18** (1937), 289–308.

10 Cf. Popper, K.R., 'A Set of Independent Axioms of Probability', *Mind* **47** (1938), 275–277. *Logik der Forschung*, 1966, pp. 198ff, 77, 80, 214, 352. Kraft, V., *Erkenntnislehre*, Vienna 1960, pp. 228f, 365. Carnap, R. and Stegmüller, W., *Induktive Logik und Wahrscheinlichkeit*, Vienna 1959, p. 251f.

11 Cf. Lakatos, I., 'Changes in the Problem of Inductive Logic', in *The Problem of Inductive Logic*, Amsterdam 1968.

12 Cf. Juhos, B., 1931/2, 'Stufen der Kausalität', *Jahresber. d. Philos. Ges. zu Wien* (1931/2), 1–19, this volume, pp. 1–15; 1934, 'Praktische und physikalische Kausalität', *Kant Studien* **39** (1934), 188–204; 'Die zwei logischen Ordnungsformen der naturwissenschaftlichen Beschreibung', *Studium Generale* **18** (1965), 582–601; *Die erkenntnislogische Grundlagen der modernen Physik*, Berlin 1967. Juhos, B. and Schleichert, H., *Die erkenntnislogische Grundlagen der klassischen Physik*, Berlin 1963.

13 Cf. Juhos, B., l.c. and 'Die Systemidee in der Physik', in *System und Klassifikation in Wis-senschaft und Dokumentation*, Studien zur Wissenschaftstheorie Vol. 2, Meisenheim/Glan 1968, pp. 65–78.

14 Cf. Juhos, B., 'Schlüsselbegriffe physikalischer Theorien', *Studium Generale* **20** (1967), 785–795.

15 Cf. Juhos, B., l.c. and 'Wie gewinnen wir Naturgesetze?', *Zeitschr. f. philos. Forschung* **22** (1968), 534–548.

16 Cf. Lorenzen, P., 'Methodisches Denken', *Theorie* **2**, Frankfurt a. M. (1968; *Differential und Integral*, Frankfurt/M. 1965.

17 Simple counting to ascertain the number of 'objects' is used as a preparatory step both in measuring and in statistical procedures.

18 The possibilities for interpreting formal physical expressions and operations are examined in Schleichert, H., *Elemente der physikalischen Semantik*, Vienna and Munich 1966.

CHAPTER XIII

THE METHOD OF FICTITIOUS PREDICATES*

INTRODUCTION

1. SCIENTIFIC AND SPECULATIVE PHILOSOPHY

The exploration of the foundations of knowledge, as pursued in the scientific disciplines, has led within philosophy to a division of methods and orientations, which must be regarded as the outstanding feature of the current state in philosophic development. One direction examines the procedures by which the sciences aim to attain their findings and beyond this, what further procedures might be possible alongside those already used. Moreover, this kind of philosophy is not all concerned to reach new specific findings in those disciplines whose methods are being examined. Starting from previously obtained theses in a specific science, clarification of assumptions and steps in the cognitive procedures may of course lead to new specific items of knowledge, or it may guide research in the special sciences into new pathways; but the goal of philosophic insight consists in the laying bare of the mechanism of the methods of enquiry and of how these are anchored in elements that can count as foundation of the relevant systems of findings. This philosophic orientation is sometimes called 'foundational research'. More generally, let us call it 'scientific philosophy'. Not only because its object of study is scientific knowledge in general so that it undertakes to determine the aim, achievements and limits of the cognitive methods in the special sciences as well as their possible extensions, but also because of the peculiarity of the method itself here used to analyse knowledge. Indeed, this method shares with the procedures in the special sciences the characteristic of subjecting its propositions to criteria of testing (criteria of truth, falsehood, refutation, or generally: decision). As with the findings of research in the special sciences, therein lies the guarantee that the results of this philosophic orientation are objectively scientific.

Against the claim of scientific philosophy, that it differs from the second, so-called 'speculative' orientation precisely by its character of an exact

* Translated from the German original 'Die Methode der fiktiven Prädikate', first published in *Archiv für Philosophie* 9 (1959) 140–156; 9 (1959) 314–347; 10 (1960) 114–161; 10 (1960) 228–289.

science, it has been objected that this difference is non-existent, because it is impossible to indicate precise distinguishing marks for scientific and speculative metaphysical methods.[1] However it is incorrect to think that a distinction is justified only if in all instances we can indicate exact and usable decision criteria. There are well known fallacious arguments, by which one tries for example to prove that there is no difference between whites and negroes, because there are hybrids of which one cannot decide whether to count them as one or the other. Such borderline cases certainly exist, but the distinction mentioned is nevertheless epistemologically justified and has a precise scientific sense. We cannot deny that there are, in the special sciences and in scientific philosophy, propositions and expressions, whose sense and validity cannot be marked by exact formal criteria, but only grasped intuitively.[2] Such expressions occur in logic and mathematics as much as in the empirical disciplines. Besides, we can point to affinities between intuitive interpretations and explanations of certain scientific concepts and propositions and metaphysical interpretations. However, it is obviously fallacious to go on to infer that the methods and theses of speculative philosophy are as exact as those of the exact sciences, on the grounds that the latter too contain imprecise intuitive elements. These specifiable elements defying exact formulation that are partially shared by scientific methods and speculative methaphysical procedures, might at best allow us to infer that science and metaphysics had the same imprecise and unscientific character.

Yet this inference would rest on inadequate premisses. For what is decisive for the scientific character of a method is its aptitude, through constant application, to make possible a progress in knowledge. Exact formulation of concepts and decision criteria are certainly helpful for attaining this goal, although the conceptual and methodological resources of the various sciences are of rather different degrees of precision: at the same time, in strict contrast with the speculative methods of metaphysics, they have in common the characteristic of allowing us to gain new knowledge in such a way within the disciplines we can identify as undeniable objective progress of knowledge. There is such a thing as a universally observable progress of knowledge in the special sciences. There is no such progress in speculative metaphysical philosophy.

From this point of view, and thus not merely as regards formal logical criteria of the form of concepts and propositions, we must examine the difference between scientific and speculative or metaphysical expressions and therefore also between scientific and speculative or metaphysical philosophy. From this standpoint it is indeed worth noting that the sciences along with scientific philosophy recognize, for the results obtained by applying their methods, test-criteria, whether logical or empirical – even if in borderline cases opinions differ at times as to whether the available criteria are adequate or applicable –, while metaphysical methods reject precisely the scientific criteria (such as 'consistency', 'agreement with experience') as irrelevant and invalid. In order to justify its theses, the speculative method can therefore rely only on subjective 'intuition' or a subjective position, for which there are no objective grounds of validity (an objective 'compulsion to validity').

2. 'OPERATIVE' AND 'FICTITIOUS' EXISTENCE

For the following enquiry it is important to ask, in what sense empirical descriptive science speaks of the existence of qualities and relations, or in the language of modern logic: of the existence of one-place and many-place predicates or their corresponding extensions (classes and relations).

In examining this question we are helped by considering certain modes of procedure in mathematics: operations like adding, dividing, raising to a power, differentiating, and many others, are applied to mathematical expressions, and, being performed, result in new expressions. Amongst mathematical operations there are those which can again be applied to the result, in general even arbitrarily often. For example, raising to a power, taking a root, differentiating are of this kind. If a function is differentiated, the result is an expression which in principle can again be differentiated. In theory, the procedure can be repeated arbitrarily often. With certain functions repeated differentiation leads in the end to a constant for which the procedure, if further applied, yields zero. We can now go on differentiating as often as we like and always obtain zero as result.

In such cases the question does arise, what could be the sense of such continued repetition of an operation. To represent operations, mathematics uses so-called 'incomplete' symbols. By this are meant symbols which written on their own have no meaning. Not until operation signs are combined with other kinds of signs, according to fixed rules, do significant expressions (complexes of signs) result. To represent operations mathematics uses symbols that are called 'operators', whose external appearance shows that they must be completed if we are to obtain meaningful expressions. As an example of a mathematical operator let us cite the symbol d/dx, for the formation of differential coefficients. From our point of view it is worth noting that the application of operators has been taken over into theoretical physics where it is used in a special sense. For example, tensors for rotation and other physical magnitudes are formed by automatic application of the corresponding operators. Recently, physics has developed operators for 'quantisation', which in principle are just as capable of repeated application to physical magnitudes as is differentiation. Accordingly one speaks of first and second quantisation (logically an nth quantisation would be possible too), which is formally analogous to the first, second (nth) application of differentiation (or any other operator).

There is a certain analogy between the possibility of constantly re-applying mathematical operations to expressions and the possibility of constantly (indefinitely) forming extensions (of classes and relations). This becomes especially clear if we identify the concept of class with that of 'set' in set theory.

Starting from any arbitrary elements (for example an individual), we can form the class of these elements, then the class of this class and thus proceed without limit to classes of higher order. Likewise for relations. Since classes and relations are defined by predicates, what corresponds to forming repeated extensions is a steplike repetition of predicate formation. We speak of 'predicates of predicates', 'properties of properties', 'properties of relations', 'relations between properties', 'relations between relations' and so on, thus forming a manifold and unlimited hierarchy of orders. We must not overlook that to the repeated (hierarchical) formation of extensions there corresponds the repeated application in a certain form of the existential operator. By asserting the existence of an individual, for example, we also assert the existence of a class containing it. With the existence of this class, we assert the existence of the next higher class containing the former as element, and so on.

The general question as to the kind of 'existence' of entities formed by the application of operators does not interest us here. We are concerned with the question what is meant when empirical science, and here above all theoretical physics, uses the expressions obtained by applying operators, or in what sense they customarily speak of the existence of objects designated by such expressions.

In mathematics we presuppose the existence of predicates of various orders (for example of numbers of different logical form) when setting up definitions or carrying out demonstrations of certain kinds. Except for set theory, mathematical disciplines speak of the existence of predicates (numbers, functions) only so far as this is required for the propositions and proofs in question. It is logico-mathematical foundational research that has first shown, by analysing the forms of concept used, that asserting existence here is always linked with a certain form of concept formation (formation of predicates). Set theory systematically applies the various kinds of concept formation in question, using consistency as its only criterion for the existence of the extensions so formed and the predicates defining them.

In mathematics, foundational research is concerned not least with establishing from several points of view the existential assertions made for predicates (numbers, functions) and extensions (classes, sets, relations), or else it subjects their justification to critical review. It is here worth noting that the performance of the operations (that is, the application of the operators) in special cases on given elements (for example forming classes of specifiable numbers) is generally sufficient ground for the holding of the existential assertions. In such cases we shall speak of the 'operative existence' of predicates and extensions.

Existential assertions become problematic only when we speak of the existence of numbers and functions, that is of predicates, functors and functions, in mere analogy to the operative existence of 'objects' of a similar kind. Since there are procedures that allow us to select from sets (for example of numbers) those elements (certain numbers) that possess a certain predicate, it is occasionally inferred that there are elements that have a predicate P even when there are no given performable procedures for obtaining elements of the extension 'defined' by P. In such cases it is expedient to speak of the 'fictitious existence' of the elements in question.

Now it would be a mistake to suppose that the concepts of 'operative' and 'fictitious' existence are semantically independent. We said earlier that operators in mathematics are schematically applied to expressions. It may happen that in some cases applying the operator can be interpreted for instance as a constructional procedure and that the results turn out to be interpretable as designations for existing objects. (For instance, the solution of two linear simultaneous equations may be taken as designating an existing point). However, there may be cases in which applying the same operator yields expressions not thus interpretable as designating existing objects. (There are for instance pairs of linear equations which solved for the variables yield the value ∞, which expression cannot be taken as designating an existing object). In such cases it is at times expedient to speak of the existence of 'fictitious objects', in analogy with the operative existence of 'constructible' objects (where the existential assertion is based on the possibility of interpreting the application of the operator as a constructive procedure). Here the existential assertion is based solely on the schematic application of the operator, not now interpretable as a construction. (In this sense geometry speaks of the 'fictitious' existence of points at infinity, in the above example).

However, there are also cases in which application of an operator leads to expressions that are not used as designations of fictitiously existing objects, but count as empty complexes of signs. Such cases occur quite often when functions are interpreted in a geometric or physical way. The application of operations to functions, which can be done by schematic application of operators, often leads to several solutions of which in certain cases only one or at most some are appropriately interpretable as designations of fictitiously existing objects. Such a case obtains when applying the operator leads to some real and some imaginary solutions, where amongst the real ones there are for example extreme values usable as symbols for fictitiously existing objects whilst the imaginary ones are of no use for further representation at all and must therefore be eliminated as empty complexes of signs. This shows clearly that assertions of fictitious existence are used exclusively for the purpose of linguistic convenience and find a use only insofar as this is required for simplicity and clarity of expression. What confirms this further is that 'fictitious' existence of objects (entities) is only ever asserted on the basis of an operation (application of an operator) such that its application to other cases enables us to assert the 'operative' existence of objects (entities). Fictitious existence is thus only ever asserted by appeal to a logico-linguistic operation whose application in other cases leads to expressions interpretable as designating operatively existing 'objects'.

Here it might be objected that the examples mentioned relate to mathematical operations or their application in physics. For it is not unknown that the application of mathematical operators is confined to special systems of objects (number systems) and leads to empty combinations of signs if these limits are exceeded. In some cases it is then expedient to designate such empty complexes of symbols as signs for 'fictitiously' existing objects. Similarly with the application of mathematical operations to domains of physical phenomena. Here it may happen that the results of empirical measurement (say, the distribution of corpuscular impacts on a screen) have applied to them operators of a kind such that the resulting expressions are regarded partly as designating fictitious phenomena (for example fictitious waves), partly as physically devoid of physical meaning. However, even if we grant holders of this nominalist view that in mathematical language or its application there occur ways of speaking about 'fictitious' existence of objects in the sense explained above, the opponents of nominalism maintain that the logical operations of forming predicates and extensions are after all different in kind from the application of purely mathematical operators: the assertion that the predicate 'redness' exists, whether or not there are particular red things, cannot be interpreted as though this was purely a matter of applying some logic-linguistic operation or of its result; rather, the 'entities' formed in the formation of predicates and extensions enjoys an 'absolute' existence that cannot be further described.

However, this objection is mistaken. For the existential assertion of universals (predicates, extensions) necessarily presupposes the continued hierarchical applicability of a logic-linguistic operation. Assertions about the existence of predicates, classes and so on can be set up only if we admit concept formations such as 'predicates of .predicates', 'classes of classes' and the like in an unlimited series of steps. Admitting such a purely logicolinguistic operation along with the criterion of consistency is a necessary and sufficient condition for establishing statements about the existence of universals.

If, overlooking this, one attributes to universals, (predicates, extensions, propositions) an existence independent of all linguistic operations and systems, beyond the 'fictitious' existence based only on the logical linguistic operations mentioned, we obtain the most senseless 'problems'. Thus one might ask from the 'platonist' point of view whether the predicate (class) 'horseness' existed even millions of years ago before there were any horses, or whether this universal arose only together with individual horses. Or one might ask whether the proposition 'the moon is spherical' existed as a true proposition when there were as yet no men or linguistic systems, and further, in what way this proposition did exist when there was as yet no moon, so that this proposition, true today, was false then.

In our opinion the so-called problem of the existence of universals concerns exclusively the question of applying logico-linguistic operations, and that therefore there can be no scientific statements about the 'existence' of universals (that is predicates, extensions, propositions) outside the domain of logical linguistic operations. Thus the possibility of a real science of universals seems to be excluded.

PART I: PREDICATES AS DESCRIPTIVE CONCEPT-FORMS

1. THE EXISTENCE OF LOGICO-LINGUISTIC FORMS

The central position of predicates in all scientific languages suggests the question how we recognize by what predicates the objects of a domain or their interconnections are to be described.

Mathematical logic by its methods seeks to ascertain the forms and systematic connection of predicates in the most varied ways possible. It is sometimes argued in epistemology that logico-mathematical forms are there ('exist') and the task of logic and mathematics is to find or discover these forms and their connections. The above-mentioned comments on 'existence' of universals alone show that this position fails to do justice to the methods of logico-mathematical research. The decisive mark of the various systems of forms is not their greater or lesser richness in proved formulae but, if we may so put it, the intuitive and qualitative difference between the methods used by the various systems. Every logico-mathematical orientation that attempts to give the foundation of the system of mathematical forms, in the end chooses elements (forms and rules of concepts, statements, inference) that it declares to be intuitively given or valid. What elements are chosen then decides what domain of forms counts as provable according to the method in question. For example, that in one system the inferences in an indirect proof count as binding but in another not, does indeed mean that in the former certain formal connections can be proved that cannot occur (are not admitted) in the latter; but this difference can by no means be explained by trying to establish the 'existence' or 'non-existence' of those connections.

On the contrary, we gain a much better grasp of the systems of forms or the differences between them, if we see the task of the various orientations

as being the examination of how much in the way of forms and their connections can be represented by the chosen apparatus of method. If, for example, one orientation admits as the only evident distinction that between
arguments and predicates (without restrictive conditions as to either) while
another in addition regards the difference between individual and predicate
as logically secure, then this alone will result in differences as to formal connections provable in one system or in the other. Differences will enter even
more clearly if a system chooses, as the ultimate and only intuitively
grounded elements of its method, a certain intuitive 'object' (or objects of a
certain kind), for example the number 1, or the signs '0' and '+', and an intuitively performable operation (such as +1 or writing the symbols alongside each other in a certain sequence); while other systems admit infinite inference schemata, according to which classes of operational steps may be
written down even if in principle the steps themselves severally cannot.

So construed, epistemological questions as to the foundations of logic and
mathematics always concern those elements that are indicated in the various foundational approaches as ultimately and intuitively given forms of
expression, forms of inference, and rules, which can only be intuitively
grounded. The working out of each system then aims at ascertaining all the
formal connections describable in terms of the chosen elements. The question, which of the methods used is the 'correct' one thus seems unjustified.
This not infrequently debated question presupposes that there exists a domain of logico-mathematical connections, the object being merely to convey them as fully as possible. However, as explained in the preceding chapters, we cannot justify the assertion that there exist logico-linguistic forms
independent of logico-linguistic operations (to which logico-mathematical operations too belong). For this reason we further hold it to be incorrect
to ask whether this or that foundational attempt is right or wrong. As soon
as the systems represented by means of the chosen elements are consistently
established, one can compare them only as to their richness in forms, or one
can critically examine the elements that they employ as intuitively obvious
to see whether they are usable and appropriate. If however one admits
them, – and if the rules and forms of expression and inference are applied
consistently there is no reason for rejecting the methodical apparatus of
any of these orientations – the question as to correctness or otherwise of
any of the systems becomes superfluous.

2. DISPOSITIONAL CONCEPTS AS FICTITIOUS PREDICATES

In verbal and symbolic languages science has at its disposal systems of forms that it can use to describe domains of empirical objects. That attempts at description may lead us to extend the domain of forms in the (for example mathematical) language used by setting up new forms of expression and perhaps also of new rules, we need not discuss, since such cases do not directly affect either the foundations of the analytic.formation of forms or the conditions of empirical description.

When one examines the relations between language and the empirical phenomena to be described the question is often discussed how far the concepts used in science or even in everyday life can be defined by means of such expressions, or more generally: how can these concepts be 'reduced' to expressions denoting only observational data. Recent epistemology here adduces certain concepts used in scientific and ordinary language, whose reduction to observational expressions yet raises difficulties. These are above all the so-called 'dispositional concepts'[3] The expression 'disposition', which is not usual in natural science, has been adopted by epistemology from psychology or physiology (or medicine). Psychology speaks of dispositions in several senses. Thus it distinguishes between momentary and lasting dispositions. Amongst the former belong for example 'irritable moods' or the 'momentary inclination to fulfil requests even if disagreeable'. Amongst the latter are all so-called character traits, such as willpower, inclination to compromise, good nature, irascibility, a tendency to melancholica or self-absorption. Physiological dispositions are for instance allergies and liability to certain diseases.

Reducing dispositional concepts of the physiological kind to observational expressions meets with relatively the least difficulties: here one tries to characterize 'dispositions' by means of the special nature of parts of organs, nerves and the like. Recent attempts at reducing psychological disposition words to observational expressions aim at indicating for these concepts too, purely physiological or physiological-behaviourist characteristics. How far this succeeds in sufficient measure we need not examine specially here, since the properly epistemological problems concerning dispositional concepts stand out most clearly when applied in inorganic natural science.

Terms like 'liquefiable', 'soluble' and the like are to mean the 'dispositions' of material bodies to react in certain ways under certain conditions. Ascribing a disposition thus states that the body in question is subject to a certain natural law. Amongst the criteria for deciding whether a body has such a disposition (for example of being soluble in water) there always also belongs the observation whether under the appropriate conditions the object in fact reacts correspondingly. But then we can no longer ascribe this disposition to a body whose reaction under those conditions we have not yet observed. If amongst the necessary criteria for the disposition 'soluble in water' there is an object's reaction of dissolving in water, then a lump of sugar that is never put into water could not be denoted as soluble in water. Since moreover a lump of sugar does dissolve in water and vanishes, it would be impossible to denote any lump of sugar as soluble in water.

In spite of numerous attempts, it has till now remained impossible to find a complete and formally satisfying reduction of dispositional concepts to observational expressions. Nor, in our view, can this goal be reached, because dispositional concepts are always set up on the basis of natural laws, and it is part of these latter that they indicate for observational data an order in which these data are linked with elements specified as fictitious and in principle untestable by observation.

Natural laws always indicate an order for the quantities that characterize phenomena. However, not all of these quantities are susceptible to observation (measurement) to ascertain whether they actually belong to the phenomena to which they have been ascribed. The empirical content of statements of natural laws presupposes that some phenomena have quantities assigned to them by convention. Here we do not mean stipulations required for unambiguous definition of the metrical quantities concerned. Rather, we are referring to ascriptions of 'measured values' to phenomena by stipulation in those cases where the measuring procedure associated with these quantities is in principle incapable of being applied in order to ascertain the values characterizing the phenomena. However, this makes the order indicated by a natural law a relation between quantities that are partly observable and partly stipulated according to expediency. It is one of the chief tasks of the present essay to lay bare this logical mechanism of empirical description as most clearly evidenced in exact laws of nature.

From this interpretation of natural laws it follows that dispositional concepts, which are always based on a natural law, are set up by means of stipu lations that have to show themselves appropriate in terms of the further use of those concepts. Thus if a body is designated as 'soluble in water', it is precisely characteristic for the ascription of this disposition, that it is not made on the basis of that body's reaction of dissolving in water. Rather, the ascription is made on the basis of observable features of the body that are in principle different from the observed dissolution in water. There are observable facts that we have noticed with other bodies whose dissolution in water was also observed. The ascription of the disposition of being soluble to all similar bodies is then a stipulation that may turn out to be appropriate. The property ('disposition') of solubility is thus a fictitious predicate ascribed to bodies by stipulation and possibly connectable to other observationally given predicates by means of numerical functional relations. Such fictitious predicates thus ascribed by convention to an object and thus functionally linked with others given by observation allow us to derive predictions. Whether or not the ascription is appropriate is then decided by whether or not what is predicted comes true. Ascribing the disposition of being soluble to objects that share some properties quite different from 'solubility', allows us to derive the prediction that these objects will dissolve if put in water. In some cases we shall feel obliged to check the predictions and if they come true preponderantly, it counts as expedient to ascribe the fictitious property of solubility to objects of that kind. All that can be understood by a disposition is a fictitiously introduced predicate expression that allows us to derive predictions when it is functionally connected with other expressions denoting other observational data. What further favours this interpretation of dispositional concepts is the fact that names for dispositions are simply dropped or declared unusable if the predictions derived from observational data by means of them do not come true. The 'dispositions' that astrology ascribes to men according to the constellation under which they were born must be regarded as such inappropriate unusable fictitious predicates. Here a connection between empirically observable data (the constellation of the stars at a certain time) and a fictitious predicate (say, the disposition of being irritable) is based not on a natural law but on an arbitrary stipulation. That the ascription is inappropriate is shown by the fact that the derived predictions largely fail to materialize, so that we speak of a senseless ascription of a disposition.

Thus we can characterize dispositions as being predicates for which by convention we assert fictitious existence. Such stipulations have sense only if the predicates introduced as fictitiously existing are set in functional relation with observed data or realisable empirical conditions, so that we are enabled to derive predictions.

In this way it turns out that significantly introducing dispositional concepts presupposes ascertaining a natural law. Thus to ascribe a disposition does not mean new knowledge beyond the natural law that provides for asserting that the disposition 'exists'. Rather, the ascription is merely a logico-linguistic operation, which may prove appropriate for empirical description. If in this way we have recognized that the assertion 'disposition D belongs to body B, because bodies of this kind usually react in such a way' does not amount to an increase of knowledge as against the assertion of the lawlike course of reaction itself, but rather merely to a more convenient way of talking, then we can no longer count queries like 'What are dispositions?', 'What is stated by ascriptions of dispositions?' amongst those epistemological questions concerned with obtaining empirical knowledge. The knowledge expressed by ascription of a disposition is only the natural law on whose basis the ascription is made; that law can equally be stated without the use of dispositional expressions.

One may indeed ask the further question about dispositional concepts, whether they might not be used only in connection with natural laws of a certain form. This question is certainly of epistemological interest, for it equally concerns form and content of natural laws and the conditions under which the laws are obtained.[4]

3. CHARACTERIZATION OF PHENOMENA BY FICTITIOUS QUANTITIES

In epistemology it is often said that natural laws are obtained and tested by observation. Of course natural laws are also characterized as stating more than has been observed, more than merely the sequence of those observational data that have been used for establishing the laws. Is it then perhaps the case that in natural laws we connect expressions denoting possible observational data with others that denote nothing observable?

To be able to answer this question we must give a more precise account of the concepts 'observational data' and 'observable'. For these are used both in physics and in psychology and the humanities and doubtless not always quite in the same sense.

The question, whether symbols for observational data are constituted only by expressions denoting conscious mental contents, like the experience of the senses, of configuration, feeling, will and so on, or whether what denotes perceptions and spatio-temporal objects may also count as observational data in a wider sense, may well be important for certain epistemological distinctions. However, we shall for the moment confine ourselves to those scientific statements in which metrical quantities are assigned to phenomena or relations are stated between such quantities. Let us begin by saying of values obtained by measurement that they denote observational data or observables in a wider sense. In the further course of our inquiry we shall on occasion discuss the question whether ascribing metrical values doe not go beyond the boundaries of the observable or whether beside observational data we are not compelled to assume the 'existence' of unobservable elements, in order that we may ascribe metrical values to phenomena or objects. In descriptions within natural science metrical values generally function as argument values in functions that express natural laws. Thus the functions may be regarded as relations between metrical quantities. As mentioned earlier, relations may also be called 'many-place predicates'. Thus natural laws state the existence of certain phenomena (states, changes of state). To know natural laws consequently means to ascertain many-place predicates that hold as relations between the measured values. Knowledge of nature is thus knowledge of predicates.

However, it is characteristic of natural laws that the relations between metrical quantities stated in them concern changes in these quantities. We can generally represent the content of a natural law as follows. If a state is specified by the quantities $(a_2, a_2, \ldots a_n)$ and $(b_1, b_2, \ldots b_n)$ then changes in the b's are functionally related to the changes of the a's. Now when a natural law states a relation between the changes in different metrical quantities, then these changes must be at least in principle ascertainable by measurement. What in natural science we call events or phenomena, are changes in state. The states are characterized by means of metrical quantities so that every change in state is represented by changes in those quantities. However, in a mathematical representation it may happen that changes in the values of argument and function are readily calculable, while no measurable change of the quantities corresponds to the calculated changes in value. In such cases we call the formulae or the calculated values physically empty, they no longer state anything further about empirical reality. In these by no means infrequent cases there are several possibilities for empirical research to give interpretable content to the formal functions or predicates, by producing a correlation with observed connections. If some such attempt is sufficiently successful, then we speak of the 'existence' of the quantities and predicates characterizing the connections between phenomena. If nothing observable corresponds to the values derived from a law, then the predicate ascriptions stated in it must either be abandoned as inappropriate and unusable, in which case one must look for new descriptive predicates (relations between the metrical quantities); or one can look for a new interpretation as to the empirical content of the calculated or calculable changes in the quantities. By means of such interpretations we correlate these changes with observational data or complexes of such data occuring under certain circumstances, so that the calculable changes of the 'metrical quantities' can be checked by observation.

However, this procedure is applicable only on occasion and after the event. In general, the description of connections between phenomena by means of predicates relates objects, or states characterized by measurable quantities, with 'states' to which we can formally ascribe 'metrical quantities' though there are no corresponding measuring procedures for them. Nevertheless, as we shall make plain by empirical examples later, if we assume such predicates exceeding the domain of the measuring procedure and combine them with statements about observable quantities, we can derive new propositions that can be checked by measurement. If such predicate ascriptions are sufficiently confirmed by experience, we are wont to denote the phenomena (states) and the predicates (relations) characterizing them as 'existing'. In this the linguistic practice of natural science usually draws no (or at least no precise) distinction between the 'existence' of quantities, predicates, phenomena, states given by measurement and the states characterized by quantities that are in principle unmeasurable (and perhaps only calculable).

Thus classical mechanics denotes transmissions of effects propagated at finite speed as existing ('real') in the same sense as the propagation of gravity which, it assumes, spreads at infinite speed. Here gravitational phenomena are characterized by a 'metrical quantity' (namely a speed), which can be derived (from them as a prerequisite for the laws of gravity), that is calculated, but not measured because its value is infinitely great.

As regards the development of modern physics it has occasionally been said that epistemological progress in the formation of physical theories consists not least in gradually eliminating from our statement-systems, all propositions and expressions about states, phenomena or objects designated as existing but characterized by values unmeasurable in principle. However, this view is only partly correct. Certainly it is a step forward to recognize that the 'existence' (whatever we are to understand by it) of states or phenomena which cannot be ascertained and checked by measurement, is at all events not the same as the existence of phenomena whose characterizing quantities can in principle be so ascertained. If however we were to infer from this that the description of phenomena is justified only in terms of relations between quantities for which there are practicable measuring procedures, this would be wrong. In the formulae of classical physics one does indeed relate quantities such that the values used in the description can almost without exception count as possible values of measurement. However, in what follows we shall try to show that in general every description of phenomena by functions whose arguments are 'metrical quantities' – and these are the predicates used in physical description – necessarily relates expressions of which some appear to be defined by 'quantities' or 'values of quantities' to which measuring procedures do not apply.

This circumstance, which is brought out especially clearly by the crucial cases dealt with in modern physics suggests that in the empirical sciences too we should – in a certain analogy with the difference between 'operative' and 'fictitious' existence of logico-mathematical entities – distinguish between the measurement-based existence of phenomena or of quantities and predicates characterizing them, and the 'fictitious' existence of phenomena (states) characterized by non-measurable quantities.

PART II: PREDICATES AS QUANTITIES CHARACTERIZING STATES

4. CHARACTERIZING BY INFINITE VALUES OF QUANTITIES

Characterizing phenomena by metrical quantities, and representing the lawlike order of their sequence by functions whose arguments are metrical quantities, leads to the question which extreme values are physically admissible for arguments and functions. Epistemologically it turns out that infinitely large (and under special conditions also infinitely small) values, if a physical theory admits them for metrical quantities, are always problematic. Firstly, because such values in principle elude measurement. (Such values are sometimes denoted as 'transcending experience', not too felicitously). Secondly, however, infinite values of metrical quantities within a theory not infrequently produce consequences that must count as physically impossible, that is as inconsistent with the continued existence of the empirical world or with the lawlike course of phenomena. Thus the assumption of Newtonian physics that space is infinite (that is, the assumption of an infinite length) has led to difficulties that could not be satisfactorily resolved within classical mechanics.

Epistemologically perhaps even more interesting are cases where certain metrical quantities take infinite values. This happens in relativity theory, where with a set upper limit c for translational velocities (of transmission of effects) the metrical quantity of 'mass' grows to infinity as the velocity reaches the limit. A physical interpretation of 'infinite mass' can hardly be given. Even if observation nicely confirms the value for growth of mass with velocity derived from relativistic laws, it is always a case of masses moving at velocities below the limit. Infinite mass can only be calculated and in principle defies measurement. Thus we here have a 'value of measurement' obtainable only by calculation and entitled only to fictitious existence with regard to the measuring procedure. In relativistic laws this value appears connected with values that can be obtained by measurement.

Because of the problematic consequences that may arise from infinite values of metrical quantities, physical research aims to look for lawful bounds to the growth of quantities ascertainable by measurement. Here we must distinguish two possibilities. The conditions under which the three elementary metrical quantities 'time', 'length' and 'mass' take on infinite values are epistemologically different from those in which only relations between these quantities grow to infinity.

From the stipulation of the relativistic space-time metric, that the greatest possible velocity for transmitting effects shall be $c = 300,000$ km/sec, it follows that mass grows to infinity as the limiting velocity is reached. Here we fix a finite upper limit for the metrical quantity 'velocity'. In 'velocity', we relate 'length' with 'time'. For 'mass' the relativistic metric sets no bounds.

If here one were to ask, under what conditions the values derived for the growth of moving masses might be kept within finite limits – which, since infinite quantities cannot be measured and admitting them raises difficult-ies, is a logically justifiable question –, the most obvious answer emerges if we recall how physics usually deals with such cases. Every restriction of possible changes in empirical quantities by finite limiting values is given by introducing constants.

The constant c of the relativistic space-time metric has the dimension of a velocity and thus concerns only a relation between 'length' and 'time'. Since c is to count as the limiting value of empirically possible translational veloci-ties (for transmission of effects), this appears to impose a limit only on pos-sible changes in lengths and times. While thus velocities are to be able to change only within a finite domain of values, the relativistic metric allows infinite changes in mass depending on finite changes of velocity. Infinite values of metrical quantities may be called 'discontinuities' in the lawful or-der of phenomena. To eliminate them one must introduce a new constant, which must give a finite value for a relation between mass, length and time such that infinite changes of mass can no longer be considered as empiri-cally possible 'values of measurement'. Relations connecting mass, length and time are for example the metrical quantities impulse, energy, action, and so on. However we might also introduce a physical constant for a rela-tion expressing a connection between the three elementary metrical quanti-ties mass, length and time but not in the form of an already defined quantity (such as 'impulse' or 'energy'). In that case we have constants whose physi-cal dimensions are not those of the usual metrical quantities (as for exam-ple with constants that we might introduce for a particular value of the rela-tion m/v or m/v^2).

Logically, we can say that introducing limiting constants for only one of the three elementary metrical quantities or for relations of only two of these quantities has the almost unavoidable consequence that the other usual quantities not limited by the constant will under certain empirically possi-ble conditions take on infinite values. In that event we have the problematic cases where one speaks of the empirically possible existence of phenomena characterized by values that are in principle inaccessible to measurement.

As against this logically unsatisfactory mode of physical description, epistemology would admit infinite values of metrical quantities only as characterizing 'fictitious states'. The empirical laws would then have to be such that if we insert finite values of measurement (namely those that characterize empirically possible states) we can derive only finite values for all metrical quantities concerned. Expressions with infinite values of measurement would then not denote empirically possible phenomena, but would, in a certain functional connection with possible observational values, enable us to derive predictions. In this sense we may call the conditions characterized by infinite values 'fictitious states' ('phenomena').

It is with this meaning that infinite values arise when Heisenberg's uncertainty relation is employed. The constant h which limits the accuracy of measurement has the physical dimension of action and thus relates the three elementary metrical quantities mass, length and time in a certain way. The imprecision of measurement of quantities is expressed by a finite range of values. In precise measurement of a quantity, when that range practically equals zero, the imprecision of a certain other quantity becomes infinite, according to the uncertainty relation. To this corresponds an 'infinite' range in value characterizing a fictitious state, which can be interpreted as the practical impossibility of carrying out the measurement of that quantity.

Thus there is no objection to using infinite values for empirical description, so long as such ascriptions characterize fictitious 'states' (in principle not ascertainable by measurement), which have no other meaning except that when they are connected with empirically possible states we can derive new predictions. On the other hand, infinite values are always epistemologically dubious if they are ascribed to phenomena (objects) that are to be possible (can exist) under empirically given or realizable circumstances, so that they are supposed to exist not just fictitiously but on the basis of empirically performable procedures of observation.

5. CHARACTERIZING PHENOMENA IN UNCERTAINTY RANGES BY MEANS OF FICTITIOUS VALUES ('VALUES OF MEASUREMENT')

Using fictitiously existing predicates along with the ascription of empirically existing properties and relations (namely those, whose existence is asserted with regard to empirically performable observation procedures) in empirical description is one of the most important methodological instruments in the pursuit of empirical knowledge. The logical significance of this procedure stands out even more clearly within finite domains of values than in characterizations by infinite 'values of measurement'.

Recalling that in defining a metrical quantity one always must also define the procedure by which in actual cases we can determine the values of measurement that characterize the phenomena, we are led to the insight that the definitions of the quantities 'time' and 'length' satisfy this condition only if these metrical quantities are defined as relativistic relations. Signals procedure is the only method by which time and length can be measured under all conditions. Measuring by sginals is based on the prior assumption that measurements of a phenomenon from a reference system are possible only if we can establish transmission of effects between the phenomenon and the system. Signals are precisely an instance of such transmission.

Since in experience we have at our disposal only transmissions (signals) of finite speed, it follows necessarily that there are phenomena between which there can be no such transmissions (causal chains). In such cases it is in principle impossible to carry out measurements of time and length by way of observation. The spatio-temporal order of these phenomena remains necessarily indefinite relatively to the system from which one is measuring. The schema for events for which we cannot ascertain a temporal order by observation (that is with the help of signals) is best illustrated by the sending, reflection and return of light-signals (or, in general, electromagnetic waves) at two widely separated locations. Denoting the latter by A and B, the sending of the light-signal by E_1, its arrival and reflection at B by E_2 and its return to A by E_3, then it is in principle impossible to ascertain anything by observation (measurement) as regards the temporal relations between E_2 and the events occuring at A in the interval $E_1 - E_3$: they remain relatively undefined as to temporal sequence so long as we confine ourselves to using observational data of measurement – in the case of signals, the data of sending and return.

Such ranges of phenomena, in which the correlation of values cannot be performed by means of the measuring procedure itself, frequently occur when physical methods of measurement are applied. The resulting uncertainty of the phenomena can sometimes be corrected by refining the methods of measurement. With signals, if we use transmissions of effects at the greatest empirically possible speed, that possibility of refinement lapses, for it could consist only in raising the speed of the signals used, which would indeed reduce the intervals of uncertainty.

If the possibility of refining measurement appears excluded, there is only one way to remove uncertainty in the correlation of 'values of measurement' with the phenomena concerned: namely by stipulating a metric. As is readily seen from the schema mentioned above, it is possible to choose from infinitely many metrical systems a metric for obtaining an unambiguous correlation of characterizing quantities with the phenomena that lie within the range of uncertainty. Choosing a metric means stipulating a convention that always follows considerations of expediency. What physical description aims at is a representational system of propositions that is as simple and perspicuous as possible, and as empirically complete. The predictions derivable from the propositions of the system are to agree as fully as possible with observation; and if they disagree, as can happen in any theory, the gaps are to be removable by simple and readily testable hypotheses. It is indeed with this in view that one chooses conventions in physics.

When linked with observational statements, those introduced by convention allow us to derive predictions that may be put to the test of observation. If, for example, some phenomena in a domain are characterized by measured values and others by conventionally assigned ones then a state thus characterized may well lead to the derivation of testable predictions. Assigning 'values of measurement' by convention in such cases counts as appropriate, if the derived predictions very largely come true.

In using the procedure of signals it turns out that we must conventionally fix a temporal order or time metric for the phenomena that are undefined as to temporal sequence in relation to the observational reference frame. Einstein's relativistic metric has here proved itself particularly appropriate as regards the simplicity desirable in a representational system. However, this metric leads to theses and concept-forms that underline the urgency of inquiring into the epistemological character of the predicates (spatio-temporal values of measurement) conventionally ascribed to the phenomena.

The relativistic time metric fixes a temporal order for phenomena that are undefined as to temporal sequence in relation to an observational reference frame S, by means of two conventions. Firstly, in all systems electromagnetic waves are to move (in vacuo) in all directions at a constant speed c, independently of the motion of the systems or of the field sources. Secondly, mechanical and electromagnetic phenomena in systems that are in relative translational motion to each other shall obey laws invariant in form. If we accept this metric, then simultaneity of phenomena is no longer defined as a two-place relation but as a three-place one. Classical physics – or, in view of the above, the classical time metric – defines simultaneity (and therefore all other temporal relations as well) for two events e_1 and e_2 as a transitive, symmetrical two-place predicate sm (e_1, e_2). According to the relativistic time metric indicating the simultaneity of two events is unambiguously significant only in the reference frame in which the 'measuring' signals are sent and received. The predicate 'simultaneous with' must therefore here be represented as a three-place relation sm (e_1, e_2, S). Since S is variable, we cannot speak of the relativistic simultaneity relation (that is, a three-place predicate) as transitive and symmetrical.

Analogously for the predicates 'prior to', 'later than', 'running synchronously with'. In the classical metric these relations are defined as two-place transitive predicates holding only between the phenomena (that is indepently of any system), their arguments being just the two event variables e_1, e_2 or clock variables c_1, c_2. In symbols we can represent them thus: $P(e_1, e_2)$, $L(e_1, e_2)$, synchr (c_1, c_2). In contrast, the relativistic time metric defines these time relations as three-place predicates $P(e_1, e_2, S)$, $L(e_1, e_2, S)$, synchr (c_1, c_2, S) where we can no longer speak of transitivity.

Now the way a metric is chosen and fixed is arbitrary. That in this we let ourselves be guided by a certain expediency does not alter the logical character of the chosen metric. Under certain conditions we can indeed observe (measure) the temporal order of phenomena. Thus we can ascribe simultaneity or being 'earlier or later' to the sending or receiving of signals at the same place, by means of observation. However, where such observed or measured determination of a temporal order is in principle impossible we shall have to ascribe temporal values to phenomena and thus stipulate their temporal order by a fixed metric. If we attribute empirical existence to temporal predicates ascertained by observation and ascribed to phenomena, then in the sense of the above account of 'fictitious' existence we must denote as fictitiously existing those temporal predicates that are conventionally ascribed to phenomena. What supports the fictitious character of such predicates is not only that we can arbitrarily change the metric on the basis of which the time predicates are ascribed but above all that the values so ascribed in principle elude measurement in every case.

To ascribe fictitious predicates to phenomena while relating the fictitious values to the predicates obtained by observation (measurement) and to their individual values within the framework of the chosen metric, constitutes an application of empirical epistemological method whose decisive importance stands out especially clearly in modern physical theories. The crucial cases that have occasioned the establishing of the new theories appear marked by the fact that the empirically possible measuring procedures leave open ranges of phenomena, for which it is in principle impossible to ascertain a spatio-temporal order by means of measurement. If for these phenomena we are to reach an empirical order of events, empirical knowledge logically compels us to take the step of choosing and fixing a metric whose suitability is tested by linking it with known laws of nature and observational propositions (that is, those in which predicates are ascribed to phenomena by means of observation (measurement)) and with new predications derived from them. Introducing such a metric amounts to ascribing fictitious predicates to phenomena, as explained above.

One might think that empirical research had always ascribed predicates to phenomena not by measurement alone but also by hypotheses. The crition for the hypothesis being valid is taken to be the confirmation of predictions derived by connecting the hypothesis with known propositions. Nevertheless the predicates introduced in the hypothesis and ascribed to phenomena are not usually denoted as 'existing fictitiously'. Rather, if the hypotheses are successful, characterizing predicates introduced by them count as empirically existing in the same sense as the descriptions ascribed to phenomena on the basis of observation and measurement.

This objection however overlooks the epistemological difference of meaning between hypotheses and metric introduced by stipulation to remove ranges of uncertainty. Characterizing values ascribed in hypotheses count as in principle measurable, whether individual values are ascertained by measurement or obtained on the basis of other values of measurement. This feature of hypothetically introduced values thus correlated, namely that of being measurable in principle, shows itself further in that setting up a hypothesis is not a matter of selecting from an infinite number of possibilities with equal logical title. Rather, in a hypothesis one ascribes to a phenomenon predicates in principle asserted to be ascertainable by observation and measurement. In this hypotheses are distinguished from predicates ascribed, in the sense explained, by stipulated metrics. For such predicates ascertainment by measurement is a logical impossibility. Thus they can be ascribed only by convention and it makes no sense to speak of their being testable by observation and measurement. Where phenomena in ranges of uncertainty have ascribed to them predicates and their individual values by means of a chosen metric, there we cannot have observations (measurements) by which to ascertain the (empirical) existence of those predicates. Here we can speak only of the fictitious existence of such predicates, in contrast with hypothetical predicates which appear to be in principle ascertainable by observed measurement and therefore have empirical existence ascribed to them when they are confirmed by measurement.

Ranges of uncertainty unavoidably occur when values are ascribed to phenomena by way of measurement. As is clear from the above, we are here referring not just to 'uncertainty relations' in quantum physics; rather, by ranges of uncertainty in general we understand domains of phenomena in which transmission of effects ('causal chains') between phenomena is impossible. We therefore hold that the 'method of fictitious predicates', that is the ascription of fictitious predicates to phenomena in ranges that are undefined as to measurement, is an indispensable device for attaining knowledge and unambiguously describing the empirical order of events. The hope that this method could be eliminated or cancelled by changing or refining the measuring procedures so that phenomena might once again be characterized exclusively by means of measurable quantities, is a vain hope, justified neither by empirical knowledge nor by epistemological analysis. Rather, it would be advisable to draw the proper conclusion from the fact that modern empirical inquiry must use the method of fictitious predicates if it is to indicate a testable order of events. In what follows we shall try to show how the direction of empirical inquiry in natural sciences seems to be determined by the deliberate application of this method: by means of it, recent natural science has gone beyond the procedures of method in classical physics.

6. LOGICAL ANALYSIS OF THE CONCEPTS 'MOTION' AND 'REST'

The progress of human thought can well be illustrated by changes in traditional conceptual forms and the working out of new ones. For thought and knowledge to progress it is decisive that we ascertain those concepts that constitute our predicates. If we recognize that, when the concepts are exactly defined or interpreted on the basis of empirical results, a one or two-place predicate is to be understood as a two or three-place one, this constitutes a progress in thought that may result not only in an intellectual, but also in an imaginative, shift of perspective.

We have stated already that we must consider the definition of times and lengths as three-place predicates as one of the essentially new epistemological steps of relativity theory. Classical physics understood the relations 'simultaneous with', 'as long as', 'earlier than', 'shorter than' and so on as two-place predicates, while relativity theory, in view of the signalling procedure used for empirically ascertaining those relations, defines them as three-place predicates. Here too we must note the replacement of the one-place predicate 'motion', as used in classical physics for characterizing 'absolute motion, by the two-place predicate 'relative motion'.

By the 'absolute' motion of a body (system) we understand a state by which the moving body differs from those at rest. Such a state must be ascertainable as a special property of the body; let us say in that certain phenomena follow a different course in moving bodies and in stationary ones. Taking bodies to be reference frames, then we understand by the 'absolute' motion of a system S its property that phenomena in S occur according to different laws from those governing a system at rest. Properties are represented by one-place predicates. Thus the statement that a body (system) x moves absolutely can be symbolically expressed as $M(x)$: this predicate's being one-place expresses the absolute character of the motion referred to. Thus, whether a system moves absolutely would have to be ascertained by observing the phenomena occuring in it and especially according to what laws they run.

With motion the concept 'non-motion', that is 'rest', is equally defined. A body that is not moving absolutely is at 'absolute' rest, which again means a special state of the body, that is, a special property of the body at rest, and this we shall denote by the predicate symbol $R(x)$. Here too, the absolute state of rest of a system S is ascertained by observing the laws according to which phenomena in S take their course.

However, we may take the motion of a body (system) also as mere change of place. In that case the motion of a body x can only be ascertained relatively to another body y. Thus motion is represented as a two-place relation and we must define it as a two-place predicate. Motion thus viewed is called 'relative' and symbolized '$M(x, y)$'. We can at once indicate a property of it; for relative motion is to mean only the change of place of the moving objects but none of their changes of state. Therefore $M(x, y)$ and

$M(y, x)$ must be synonymous, and $M(x, y) \equiv M(y, x)$: the two-place predicate 'relative motion' is symmetrical. Further, we note that it makes no sense to say of a body that it moves relatively to itself, so that $M(x, y)$ is irreflexive. We can thus describe relative motion as a symmetrical and irreflexive two-place predicate.

With relative motion the concept of 'relative' rest is likewise defined. If two bodies x and y are not in relative motion, then they are at relative rest. 'Relative rest' we denote by '$R(x, y)$', which also is a symmetrical two-place predicate, with $R(x, y) \equiv R(y, x)$. Since 'relative rest' is defined by means of the irreflexive predicate $M(x, y)$, namely by $R(x, y) \equiv \sim M(x, y)$, $R(x, y)$ too is irreflexive, a fact that must not be overlooked. One is readily inclined to accept the proposition 'body x is at rest with regard to itself' as true and therefore significant. However, the logical analysis of the concept 'relative motion' ($M(x, y)$) indicates as shown that $R(x, y)$ (that is $\sim M(x, y)$), is irreflexive so that the expression 'x is at rest with regard to itself' is senseless.

We shall see presently that physics does not always keep the concepts $M(x)$, $M(x, y)$, $R(x)$, $R(x, y)$ strictly apart, or fails, in its theses, to attend to the logical properties of $M(x, y)$ and $R(x, y)$.

7. LOGICAL ANALYSIS OF THE RELATIVITY PRINCIPLE

According to special relativity theory all translational motions are relative. Motion or rest of a body x can thus be asserted only relatively to some other body y. Symbolising relative motion and rest respectively by the two-place predicates $M(x, y)$, $R(x, y)$, by what formula must we represent the relativity principle?[5]

This principle states that mechanical and electromagnetic phenomena in systems with relative translational motion to each other obey laws of the same form. It is therefore not possible to ascertain the relative translational motion or rest of two systems S_1 and S_2 by observing the course of phenomena in those two systems. Therefore such systems are often called 'equivalent' and the relativity principle may accordingly be put thus: if S_1 and S_2 are in relative translational motion to each other then the two statements 'S_1 is at rest and S_2 moves with regard to S_1 (at speed v)' and 'S_2 is at rest and S_1 moves with regard to S_2 (at speed $-v$)' means the same. This interpretation of the relativity principle is of course imprecise and does not enable us to indicate an unambiguous meaning for it.

Recalling that 'relative motion' and 'relative rest' are symmetrical two-place predicates, with '$M(x, y) \equiv M(y, x)$' and '$R(x, y) \equiv R(y, x)$', what is the meaning of 'S_1 is at rest and S_2 moves with regard to S_1 at speed v'? 'S_2 moves with regard to S_1' is expressed by $M(S_2, S_1)$, with $M(S_2, S_1) \equiv M(S_1, S_2)$. Between S_1 and S_2 there is thus only the one relation $M(x, y)$. What could be the meaning of 'S_1 is at rest'? It is not enough to try to answer the question by saying that the relativity principle states merely that all phenomena in systems with mutual translational motion happen according to laws of invariant form. For if of two systems S_1 and S_2 one is described as at rest and the other as moving with regard to the first, then this account must make sense independently of the assertion that the reverse characterization of the two systems means the same as the one just mentioned. Indeed to say that 'S_1 is at rest and S_2 moves with regard to S_1' and 'S_2 is at rest and S_1 moves with regard to S_2' mean the same is itself unambiguously meaningful only if the meaning of the two parts can be unambiguously specified.

However, logical analysis of the concepts 'relative motion' and 'relative rest', that is $M(x, y)$ and $R(x, y)$, shows that these mutually exclusive contradictory relations can be asserted only as holding between two objects in each case. Yet if only two systems S_1 and S_2 are given and $M(S_2, S_1)$ holds, then we cannot assert relative rest for S_1. For this we should need to indicate a system relatively to which S_1 is to be at rest. S_2 cannot be this system, for if $M(S_2, S_1)$ holds, then so does $M(S_1, S_2)$, that is S_1 moves with regard to S_2. Thus at best we may assert that S_1 is at 'rest' with regard to a third system S_3. One might object that conceivably only S_1 and S_2 exist: would it then be impossible to attribute precise significance to the relativity principle? To be sure, it could in that case be expressed by $M(x, y) \equiv M(y, x)$. However, it can be shown that this representation cannot be sufficient. For this equivalence allows us only to speak of the motion of x and y, while the holding of the relativity principle demonstrably relates to systems in which one has to state of two systems that the one is in motion and the other at rest.

In order to find out the logical conditions for unambiguous interpretation of the relativity principle we shall apply the method of fictitious predicates. Let there be two systems S_1 and S_2 in motion at relative speed v. Assume a fictitious system S_3 and compare the two cases: $R(S_1, S_3)$ & $M(S_2, S_1)$ and $R(S_2, S_3)$ & $M(S_1, S_2)$, that is 'S_1 is at rest with regard to S_3

and S_2 moves with regard to S_1' and 'S_2 is at rest with regard to S_3 and S_1 moves with regard to S_2'. (The third possible case $M(S_1, S_3)$ & $M(S_1, S_2)$ & $M(S_2, S_3)$ may be disregarded for the moment since it does not exhibit the two relations 'motion' and 'rest' applied to the systems.)

If now we represent the relativity principle by a formula like $R(S_1, S_3)$ & $M(S_2, S_1) \equiv R(S_2, S_3)$ & $M(S_1, S_2)$, the relations R and M are indeed applied to S_1 and S_2, but only by relating them to a fictitious system. This 'relation' remains of course largely unspecified in this formula. Empirically, we only have the relative speed v of S_1 with regard to S_2. If we introduce a fictitious predicate S_3 and relate it to S_1 and S_2 by the representation of the relativity principle proposed above, then the formula takes on an empirical content only when we can indicate how it could be tested empirically. We test it by deriving decisive predictions. On further reflection we see that the relation of S_3 to S_1 or S_2 must be fixed more precisely if we are to obtain predictions. In other words, we must choose a metric for the relation of S_3 to S_1 and S_2, so that we can derive predictions with whose help we can test that the relativity principle in the form given above is empirically valid.

The equivalence $R(S_1, S_3)$ & $M(S_2, S_1) \equiv R(S_2, S_3)$ & $M(S_1, S_2)$ certainly meets the usual representation of the relativity principle, for on the basis of the formula we can say: if two systems S_1 and S_2 are in relative translational motion to each other, then it is indifferent whether we take S_1 as at rest and S_2 as in motion, or S_2 as at rest and S_1 in motion. However, we must not overlook that in spite of what remains indefinite in the equivalence, the logical possibilities as to relations between S_1 and S_2 are marked rather more precisely than in the usual representation of the relativity principle which is logically inadequate.

In choosing a metric for unambiguously fixing the relations between the systems S_1, S_2, S_3 we shall opt for that of special relativity, leaving aside for the time being whether its specifications suffice to characterize the relations between the three systems. (In the sequel we shall indeed discuss phenomena that fall within the ambit of general relativity, but we shall find that even in such cases we can assume limiting conditions under which the metric of special relativity can be asserted.) According to this metric the speed of a system causes a certain contraction of the length of objects and a certain slowing down of clocks. These two effects, derivable from relativistic mechanics, differ in an essential characteristic. Changes in length occuring through changes in relative speed leave no 'trace', so to speak, that is no empirically observable data remain from which one might infer the changes in length and speed that have taken place. In contrast, the changes in the running of clocks occasioned by changes in speed leave behind the difference in their settings. Thus the slowing down of clocks appears after the event in the difference in the positions of their respective hands. Let us consider whether from this, with introduction of the fictitious system S_3, we might not be able to derive predictions about the relative speeds or other relations between S_1 and S_2.

Let there be a clock each at the origins of S_1 and S_2. At time 0, when the origins of S_1 and S_2 pass each other, the two clocks are synchronized. The relativistic metric presupposes that clocks in systems in identical motion run alike. Since S_1 and S_2 move at translational speed v with regard to each other, this speed must influence the running of the clocks, according to the metric of special relativity. It is evident that the relativity principle admits indefinitely many assumptions as regards the change in running of the two clocks.

By hypothesis all we are given empirically is the relative speed v of S_1 and S_2. If we introduce a fictitious system S_3 then clearly there are infinitely many possibilities for choosing the relative velocities of S_3 with regard to S_1 or S_2. Let the relative velocity of S_1 and S_2 with regard to S_3 be v_1 and v_2 respectively, then the only restrictions on v_1 and v_2 is that their relativistic sum should equal precisely v. (In this case v corresponds to the relative velocity of S_1 and S_2 measured from S_1 or from S_2).

According to the relativistic metric the two clocks in S_1 and S_2 synchronized at time $t = 0$ of their encounter must after a certain time show a difference in their respective settings. Which of the two will lag behind the other? By introducing a fictitious predicate S_3 we can see that given the relativistic metric this question admits infinitely many answers. If S_1 and S_2 move relatively to S_3 at speeds v_1 and v_2 respectively as specified above, the clocks in S_1 and S_2 will after time t have different settings according to what values we choose for v_1 and v_2. Thus we must further specify the relativistic metric if we are to be able to derive unambiguous predictions as to the differences in the settings of the two clocks.

We shall see later that the uncertainties in the relations of the two clocks in relative motion to each other are logically of the same kind as certain so-called 'paradoxes' of probability theory (e.g. Bertrand's paradox). These seeming contradictions arise because under certain given circumstances various metrical specifications may be assumed, giving different characteristic values for the cases in question. For two moving clocks we thus can derive various results for the settings, according to what metric stipulations are made for the relations between S_1, S_2, S_3. These apparently contradictory results are usually summarized under the heading of the 'clock paradox'. In fact, however, the matter concerns a range of uncertainty in the relations between the systems. As explained several times before, the uncertainty can be removed only by fixing a metric, which is done in the form of introducing fictitious predicates. Which amongst the many metrics turns out to be the most appropriate is decided by the predictions derivable by means of the chosen metric. Let us make this plain in terms of the example of the two moving clocks.

8. THE CLOCK PARADOX

In two systems S_1 and S_2 in uniform rectilinear motion at relative speed v to each other two clocks at the same respective origins pass each other at time $t = 0$ and are synchronized at that instant. The relativity principle is then usually applied as follows. If we regard S_1 as at rest and assume that S_2 moves for a sufficient length of time away fron S_1 at the translational speed v and then returns to S_1 with the same speed in the opposite direction, the relativistic law of clock retardation states that the clock in S_2 must lag behind that in S_1 by a certain amount which shows itself in the corresponding difference in setting. The fact that at the reversal in S_2 the clock undergoes accelerated motion that disturbs its running can be neglected if the time of translational motion forward and back is long enough. The longer the translational motion of the clock in S_2 reliatively to S_1, the larger its lag with regard to the clock in S_1, that is the greater the difference of the settings on return. Thus we can choose a time of translational motion so large that the disturbance of the clock by the accelerated motion on reversal can be made arbitrarily small and then neglected.

If in this way the clock in S_2 lags if we regarded S_1 as 'at rest' and S_2 as 'in motion', it similarly follows that the clock in S_1 lags by the same amount if we take S_2 as 'at rest' and S_1 'in motion'. According to the relativity principle the two 'descriptions' of the same phenomenon are synomymous ('equivalent'). Since however when comparing clocks in S_1 and S_2 on return it is impossible for each to lag behind the other at the same time, we here have a contradiction seemingly derived from the relativity principle or the theses of relativity theory.

It is worth noting that M. v. Laue[6] regards the clock paradox as a genuine and unresolvable contradiction, if the only thing that existed in the world were these two clocks; but in line with the generally prevailing view in physics today, he holds that the contradiction disappears if alongside the existence of the systems S_1 and S_2 we further assume that of inertial systems to which we can always refer states of rest or motion of S_1 and S_2. With this assumption the 'clock paradox' as usually represented cannot in fact be derived. In inertial systems Galilei's law of inertia holds, such systems are free of gravitation. If we now assume that there are inertial systems to which the state of motion of all other systems (bodies) is referred, then 'S_1 at rest' means rest with regard to an inertial system; in S_2, which on reversal is accelerated with regard to the inertial system, a gravitational field arises according to the laws of general relativity with intensity proportional to the distance between S_1 and S_2 which correspondingly acts on the running of the clocks in such measure as necessarily to result in a lag of the clock in S_2 against that in S_1 when the two are compared. If therefore alongside S_1 and S_2 there exist inertial systems to which the states of motion of S_1 and S_2 are referred, then of course the two statements 'S_1 is at rest relatively to the inertial system and S_2 moves forward and back with regard to S_1' and 'S_2 is at rest relatively to the inertial system and S_1 moves forward and back with regard to S_2' no longer mean the same even according to the relativity principle, since they describe different phenomena.

What, however, does this solution of the clock paradox mean? What is the meaning of the assumption that there are inertial systems to which the states of S_1 and S_2 are to be referred? v. Laue is certainly right to denote the clock paradox as 'unresolvable', if one assumes that all that exists are the two systems S_1 and S_2. Moreover we can now indicate in what the unresolvability of this 'paradox' consists. The 'unresolvability' is a logical uncertainty in the way the question is framed: if the two systems are all that exists and their relative speed is all that is known, it is impossible adequately to describe their states of motion without further data. Therefore the phenomena (that is, the running of the clocks) whose course is to be derived fall into a range of uncertainty. The reflection that shows this is quite simple.

If by motion we are going to mean relative motion, then, as we know already, 'motion' is a two-place predicate $M(S_1, S_2)$ which is symmetrical and therefore $M(S_1, S_2) \equiv M(S_2, S_1)$. By 'rest' we can understand only relative rest and if two systems S_1 and S_2 are at relative rest, then here too $R(S_1, S_2) \equiv R(S_2, S_1)$. From this it follows directly that if S_1 and S_2 alone exist and $M(S_1, S_2)$ holds, nothing can be said about the being at 'rest' of S_1 and S_2. Without further assumptions the states of S_1 and S_2 remain uncertain and any attempt to derive predictions concerning the running of the clocks from the relative speed of S_1 and S_2 appear logically impossible because in the absence of further data there are insufficient premisses. However, we already know how phenomena lying within a range of uncertainty with regard to a reference frame can nevertheless be given an unambiguous order testable as to its appropriateness: namely, by means of stipulating a metric through introducing fictitious predicates. The stipulating of such a metric thus amounts to the above assumption of v. Laue, that there are inertial systems relatively to which the states of S_1 and S_2 are to be determined. In inertial systems Galilei's law of inertia holds and clocks at rest in such systems run in a certain way (let us say atomic clocks of a certain kind with a certain frequency). On this metric assumption we can now give additional and unambiguous indications about the states of rest or motion of S_1 and S_2 over and above their relative translational speed, precisely with regard to the inertial systems assumed to exist. Then the question about the running of the clocks in S_1 and S_2 acquires unambiguous sense and can be decided, and, as stated already, the paradox can no longer arise.

Here it turns out that the stipulations of the relativistic metric, that is the principle that light moves at constant speed and the (special) relativity principle, are insufficient adequately to characterize the states of two systems S_1 and S_2 in relative translational motion to each other. Amongst the states of a system there is the running of clocks in it. As to that, if all that exists is the two systems S_1 and S_2, the two stipulations of relativistic metric allow only the quite general and rather imprecise inference that in general there will occur a difference in the settings of the clocks in S_1 and S_2. However, since because with $M(S_1, S_2)$ and the sole existence of S_1 and S_2 nothing can be said about the state of 'rest' either of S_1 or of S_2, we can derive nothing further as to the difference of settings nor, in particular, as to which system has the clocks with residual lag behind those of the other.

Thus the two stipulations of the relativistic metric must be complemented by two additional metric stipulations in order to lead to unambiguous derivations on the difference in the settings to be expected in the case of the systems S_1 and S_2. The above-mentioned statement, generally assumed in physics today, that there are inertial systems to which the states of rest and motion of the systems S_1 and S_2 are to be referred, must be regarded as such an additional metric stipulation. Now metric stipulations are indeed chosen with regard to appropriateness, but from the logical point of view the choice is arbitrary. That the stipulations are appropriate, especially if they concern the introduction of fictitious predicates, is shown in that the correlated fictitious quantities together with observation statements enable us to derive new predictions or perhaps a special simplification of the descriptive system of propositions. If the predictions so derived largely agree with the results of observation, then the metric stipulations, that is the ascription of the fictitious predicates, count as appropriate.

In the case of the two systems S_1 and S_2 in relative translational motion to each other, the metrical stipulation that there are inertial systems to which states of S_1 and S_2 are to be referred, very largely follows given empirical circumstances and therefore is certainly quite appropriate. However, we must not overlook that if we assume only the existence of S_1 and S_2 – and the Lorentz formulae presuppose no more than that –, then to assume that there are further systems, namely inertial ones, means nothing other than stipulating a metric by fictitious predicates. Under the conditions mentioned, to assume that alongside S_1 and S_2 there is an inertial system is a stipulation by which the fictitiously introduced values of the inertial metric are related to the values characterizing S_1 and S_2.

If we are to characterize the running of clocks in S_1 and S_2, this metric stipulation is necessary, as mentioned already, because an uncertainty arises when we characterize the systems. That we must here introduce a third (and of course a fictitious) system if we are to attain to an unambiguously defined way of characterizing the states of systems and thus to decidable predictions as to the settings, this is obvious already in virtue of the 'solution' by means of the complementary assumption of inertial systems (which are precisely the required third system). The logical constraints on the question emerge even more clearly if in choosing and introducing the third (fictitious) system we do not rely on chance empirical data – let us say certain observations suggesting the assumption that there are indefinitely many inertial systems – but deliberately assume a system characterized by an arbitrarily chosen fictitious predicate, from which together with observation statements one derives the predictions.

9. CONDITIONS, UNDER WHICH THE RELATIVITY PRINCIPLE DOES NOT HOLD

The 'contradiction' of the 'clock paradox' is called genuine and unresolvable by v. Laue if we assume two clocks (or, what amounts to the same, the two systems S_1 and S_2) as all that exists: but, as we have explained, it arises from applying the relativity principle to that case in an inadmissibly uncertain form. If $M(S_1, S_2)$ holds and there are only the two systems S_1 and S_2, then it is logically impossible to speak of S_1 or S_2 being 'at rest'. For 'rest' like 'motion' can be asserted only with regard to a reference frame. The relativity principle in its usual form 'S_1 is at rest and S_2 in motion relatively to S_1' $\equiv$ 'S_2 is at rest and S_1 in motion relatively to S_2' can thus not be applied at all to the case of the two systems S_1 and S_2 alone. If one does so nevertheless, one obtains an 'unresolvable contradiction', which here arises from the imprecise form of the relativity principle or from a logically impossible and therefore senseless use of the concept 'rest'.

That the paradox is unresolvable therefore rests on the fact that the relativity principle in its above form cannot be applied to two systems only. It is however conceivable (logically possible) that there might be only two systems S_1 and S_2. This case is all the more deserving of examination since the Lorentz transformations state connections between the co-ordinates of only two systems in translational motion to each other. They thus do not presuppose that there must be more than two systems. What can we derive about the running and settings of clocks in the case of relative translational motion of the two systems S_1 and S_2? We have seen that, in order to obtain a consistent solution for this case, physics assumes that there is a third reference frame of inertial systems. Let us now adopt the view that there are only S_1 and S_2 and let us ask, by what methods we can under these conditions reach unambiguous derivations about the running and settings of the clocks in S_1 and S_2?

Here, as we shall show, the only method leading to our goal is the method of fictitious predicates. In physics one is inclined to interpret the 'clock paradox' as a problem that can be dealt with only on the hypothesis of general relativity theory. This interpretation is certainly possible, since, as already explained, the relativity principle in its customary and frequently used form can simply not be applied to the systems S_1 and S_2 alone. This of course offers the possibility of making complementary assumptions such as to render the problem concerning the running of the clocks in S_1 and S_2 discussable only if the laws of general relativity are adduced. Such is the assumption that there are inertial systems to which the states of rest and motion of S_1 and S_2 are to be referred. However, this amounts to evading the surely possible case of there being only two systems S_1 and S_2. As to that case, the only result obtained to date is that there a contradiction arises if one applies the relativistic metric. Our enquiry has further made plain that in this case the relativity principle can simply not be applied in unambiguous fashion.

However, we know that in cases where the characterization of states is uncertain we can introduce fictitious predicates and thereby stipulate relations between the elements given by observation and the chosen fictitious ones, and so reach unambiguously decidable problems. Application of this method to the systems S_1 and S_2, assumed as all that exists, can proceed, as we shall show, in such a way that the question can be decided within the rules of special relativity alone.

If 'rest' and 'motion' are defined as two-place predicates and the relativity principle is to say something about the state of rest or motion of systems, it can be applied in principle only to domains of at least three systems. Therefore alongside the systems S_1 and S_2 assumed as really existing we assume a fictitious system S_3. S_1 and S_2 are in relative motion to each other at speed v. Then we might represent the relativity principle thus: the two propositions 'S_1 is at rest with regard to S_3 and S_2 is in motion with regard to S_1' and 'S_2 is at rest with regard to S_3 and S_1 is in motion with regard to S_2' mean the same. Using our previous notation we can express this by $R(S_1, S_3) \& M(S_2, S_1) \equiv R(S_2, S_3) \& M(S_1, S_2)$. In what follows we denote the left hand side by (I) and the right hand side by (II), so that the relativity principle can be represented by (I) $\equiv$ (II).

This representation differs from the usual formulation of the relativity principle in that not only the motion but also the state of 'rest' of S_1 or S_2 is now asserted relatively to a system. As to the properties of S_3 the principle presupposes merely that no disturbing forces act in it, nor in S_1 and S_2. Besides, by introducing the system S_3 we can further assume a third case $M(S_1, S_2) \& M(S_1, S_3) \& M(S_2, S_3)$, henceforth denoted by (III). In view of this possibility the above form of the relativity principle (I) $\equiv$ (II) can be generalized as follows: if S_1 and S_2 are in uniform rectilinear motion with regard to each other, then for an observer in S_3 the course of phenomena obeys the same laws whether S_3 is at rest with regard to S_1 or S_2 or is in uniform rectilinear motion with regard to them. This form of the relativity principle is represented by the formula (I) $\equiv$ (II) $\equiv$ (III). We shall now examine in which form and under what conditions we can assert that the relativity principle holds.

Consider the case in which (I) holds, that is $R(S_1, S_3)$ & $M(S_2, S_1)$. The clocks of S_1 and S_3, because of the relative rest of the two systems, run alike and their hands have the same positions. At the instant when S_2 passes S_1, synchronize the clock of S_2 with that of S_1. S_2 moves in translation with regard to S_1 and S_3. According to the relativity principle, systems in translational motion to each other are equivalent, that is, all phenomena occur according to laws of invariant form relatively to such systems. Hence two systems that have the same speed of translation with regard to a third can replace each other. This possibility is exploited by A. Grünbaum[7] when dealing with the clock paradox. He designs a model that actually allows us to examine all the questions here relevant within the framework of special relativity theory alone.

The device that Grünbaum uses consists in introducing new fictitious systems alongside the actually existing systems S_1 and S_2. This enables him to examine the problem of the settings of the various clocks, exclusively in terms of systems in translational motion without having to assume that one of the systems is accelerated on reversal. Still, it seems to me that he does not completely implement his metric assumptions and therefore fails to draw the decisive epistemological conclusions. If the procedure presently to be discussed is interpreted as a pure application of the method of fictitious predicates, this renders visible the logical conditions under which the question as to the running and settings of the clocks in S_1 and S_2 can be unambiguously answered.

S_1 and S_2 pass each other at time $t = 0$ at speed v, at which instant their clocks are synchronized. After a certain time t_1 the two systems are at a distance d from each other. We now introduce a fictitious system S_3 spatially close to S_1 which is to be at relative rest to it. Then S_2 will have moved away from S_3 by a distance $d = vt_1$ in time t_1. This corresponds to $R(S_1, S_3)$ & $M(S_2, S_1)$. According to the Lorentz formulae the clock in S_3 (and S_1) shows time d/v at the instant when the distance between S_2 and S_3 is d, while at the same instant the clock in S_2 indicates the time (d/v) $(1 - v^2/c^2)^{\frac{1}{2}}$. Grünbaum now assumes that at this instant a system S_4 passes

S_2 towards S_3 (S_1) at translational speed — v with regard to S_3 The clock in S_4 is synchronized with that in S_2 at the instant of meeting. As soon as S_4 reaches S_3 (S_1), the clock in S_4 will indicate the time $(2d/v)$ $(1 - v^2/c^2)^{\frac{1}{2}}$, while the clock in S_3 (S_1) indicates $2d/v$. The difference of the settings in S_3 (S_1) and S_4 will thus be $(2d/v)$ $\{ 1 - (1 - v^2/c^2)^{\frac{1}{2}} \}$. The clock in S_4 lags by that amount behind those in S_3 and S_1.

Grünbaum now says, that S_4 is equivalent to the 'returning' system S_2, without accelerated motion arising in the course of the phenomena described. We can therefore interpret the result obtained as showing that on return the clock in S_2 will lag by $(2d/v)$ $\{ 1 - (1 - v^2/c^2)^{\frac{1}{2}} \}$ behind that in S_1. This, as mentioned, corresponds to the case $R(S_1, S_3)$ & $M(S_2, S_1)$. Grünbaum stops at this one example (without of course introducing a fictitious system S_3) and can thus show that the difference between the settings remains the same whether we measure from S_1 or from S_2. Therefore, he thinks he has shown that the 'clock paradox' simply cannot be derived from the hypotheses of the relativistic metric.

However, we have seen that the 'relativity principle' (a basic stipulation of the relativistic metric, which is the metric used in Grünbaum's model), as used in the question of the settings of the clocks in S_1 and S_2, can be unambiguously asserted only if we bring in a fictitious system S_3. The principle than says that $R(S_1, S_3)$ & $M(S_2, S_1)$, or (I), is 'equivalent' to $R(S_2, S_3)$ & $M(S_1, S_2)$ or (II), the two mean the same, or (I) $\equiv$ (II). However, we can show, precisely by pursuing Grünbaum's model, that from (II) we can derive consequences as to the settings of the clocks in S_1 and S_2 different from those derived from (I). Hence (I) and (II) do not make the same statement (that is, they describe two different phenomena), so that the relativity principle in the exact and precise form (I) $\equiv$ (II) does not hold. This is a remarkable consequence arising from the application of the method of fictitious predicates to the case of clocks in translational motion to each other. The divergent results arising on the basis of (I) and (II) can be obtained as follows by pursuing Grünbaum's model.

The derivation when (I) holds we have met already. Now assume (II). According to $R(S_2, S_3)$ & $M(S_1, S_2)$ we must assume that S_2 and S_3 are spatially close and at rest to each other. Let S_1 pass S_3 (S_2) at time $t = 0$ at translational speed $- v$, at which instant the clocks are synchronized in all three systems. These are the conditions that correspond to (II). At time t_1 the distance between S_1 and S_3 (S_2) is $- vt_1 = - d$. The clock in $S_3(S_2)$ then shows the time $-d/ - v = d/v$, and that in S_1 the time $(d/v)(1 - v^2/c^2)^{1/2}$. At this instant let a system S_5 pass S_1 toward S_3 (S_2) at the translational speed v with regard to S_3, and synchronize the clock of S_5 with that in S_1. As soon as S_5 reaches $S_3(S_2)$, the clock in $S_3(S_2)$ indicates the time $2d/v$, and that in S_5 the time $(2d/v)(1 - v^2/c^2)^{1/2}$. Thus the clock in S_5 lags by $(2d/v)\{1 - (1 - v^2/c^2)^{1/2}\}$ behind that in S_3 (S_2). Since according to Grünbaum's procedure S_5 is equivalent to S_1, in the present case the clock in S_1 lags behind that in $S_2(S_3)$ by $(2d/v)\{1 - (1 - v^2/c^2)^{1/2}\}$.

Thus in case (I) the clock in S_2 lags behind that in S_1, and the reverse in case (II). Therefore (I) and (II) are not equivalent and the relativity principle (I) $\equiv$ (II) does not hold. What is logically decisive here is that the relativity principle for S_1 and S_2 can be unambiguously represented only with regard to a fictitious system S_3. One might think that the relations between empirically real systems were independent of the introduction of fictitious relations – in the present case the relations of S_1 and S_2 to S_3. However, as explained, this is so only if the 'relations' between the real systems can be represented with unambiguous certainty without the help of fictitious predicates, so that no domains of uncertainty arise for the characterization of states. Under the assumed conditions this is impossible for the systems S_1 and S_2, which is why we have to introduce a fictitious system S_3 and fix its relations to S_1 and S_2. This amounts to a metric stipulation complementing the relativistic metric. The metric of special relativity leaves the relation between the two systems S_1 and S_2 partly undefined, so that we require complementary metric stipulations if we are to be able to characterize phenomena (the running of clocks) in S_1 and S_2 unambiguously and to represent the relativity principle with regard to S_1 and S_2 in unambiguous fashion. The way to complement the relativistic metric in the above case is by introducing the fictitious system S_3 and fixing its relations with S_1 and S_2.

In order to bring out even more clearly the logical function of the metric introduced by means of a fictitious predicate, let us consider the various metric stipulations possible in the choice of S_3. What is given is always the two systems S_1 and S_2 in translational motion to each other at speed v. In the previously discussed prevailing view of physics today, one assumes that there are inertial systems to which the states of rest or motion of S_1 and S_2 are to be referred. Here then one introduces inertial systems as S_3, a possible metric stipulation for complementing the metric of special relativity for the case of the two systems S_1 and S_2. Since this approach retains the motion away and turning back of one of the clocks, an accelerated motion arises and we cannot assert that S_1 and S_2 are at relative rest to S_3 in the same sense. On these assumptions, that is by introducing this sort of metric for the relations of S_1 and S_2 to S_3, the paradox of mutual retardation between the clocks can simply not occur.

In A. Grünbaum's model there are only systems in translational motion that can stand for each other under certain conditions. In this case S_1, S_2, S_3 can all be inertial systems (since there are no accelerated motions), but it turns out that we cannot now maintain the relativity principle $R(S_1, S_3)$ & $M(S_2, S_1) \equiv R(S_2, S_3)$ & $M(S_1, S_2)$, because we can derive the 'simultaneous' mutual retardation between clocks.

The logical function of introducing the fictitious system S_3 stands out especially clearly if we assume that both S_1 and S_2 might be in relative motion to S_3, which corresponds to case (III) above. This is perhaps the most general form of the possible complementary statement under the assumed conditions.

To obtain a complete account of the logically possible relations, let us now examine what difference in settings of the clocks compared arises in case (III), that is $M(S_1, S_2)$ & $M(S_1, S_3)$ & $M(S_2, S_3)$. Let S_1 and S_2 again be in uniform rectilinear motion to each other as well as with regard to S_3. Take the special case with S_1 moving relatively to S_3 at speed $-v$ and S_2 relatively to S_3 at speed v. (Viewed from S_3, S_1 and S_2 are in relative motion to each other at speed $2v$, viewed from S_1 or S_2 at speed $2v/(1+v^2/c^2)$. (In what follows all quantities are to be taken as relative to S_3, except when otherwise stated). As soon as S_1 and S_2 have moved away from S_3 by $-d$ and $+d$ respectively, systems S_5 and S_4 are to pass S_1 and S_2 respectively, both in the direction of S_3 at speeds $+v$ and $-v$ with regard to it. When S_5 passes S_1, their clocks are to be synchronized, likewise for S_4 and S_2. S_5 and S_4 reach S_3 at the same time. If an observer in S_3 now compares the clocks in S_5 and S_4, what difference in their settings will he detect? A simple reflection shows that in the case assumed the answer is: none.

For if the observer in S_3 at time $t = 0$, S_1, S_2 and S_3 are close together at $x = 0$, has synchronized his clock with those in S_1 and S_2, then the clocks in S_5 and S_4, in the case here described, will both lag behind the observer's by $2(d/v)\{1-(1-v^2/c^2)^{1/2}\}$. The clocks in S_5 and S_4 thus show no difference in setting.

For case (III) we can assume conditions so general that the cases so far discussed, (I) and (II) included, can be derived as special cases. We assume again that S_1 and S_2 are in uniform rectilinear motion to each other. Further, S_1 and S_2 are to move at speeds $-v_1$ and $+v_2$ respectively with regard to S_3. As soon as S_1 and S_2 have moved from S_3 by $-d_1 = -v_1 t$, $d_2 = v_2 t$ respectively, systems S_5 and S_4 pass S_1 and S_2 respectively towards S_3 at speeds $+v_1$ and $-v_2$ respectively. The clock in S_5 is synchronized with that in S_1 at the instant the two systems meet, and likewise for S_4 and S_2. Because $d_1/v_1 = d_2/v_2 = t$, S_5 and S_4 arrive at S_3 together. The clock in S_5 indicates the time $2(d_1/v_1)$ $(1 - v_1^2/c^2)^{1/2}$, that in S_4 the time $2(d_2/v_2)$ $(1 - v_2^2/c^2)^{1/2}$. The difference in setting between the clocks in S_5 and S_4 will thus be given, if $v_1 \geqslant v_2$, by the expression (A) $= 2(d_2/v_2)$ $(1 - v_2^2/c^2)^{1/2}$ $- 2(d_1/v_1)$ $(1 - v_1^2/c^2)^{1/2}$. For $v_2 > v_1$ we must here replace d_2 by $\bar{d}_1$ and v_2 by v_1. If $v_2 = 0$ or $v_1 = 0$, then $d_2 = 0$ or $d_1 = 0$; but then too $d_1/v_1 = d_2/v_2$ so that (A) now yields $2(d_1/v_1) \{1 - (1 - v_1^2/c^2)^{1/2}\}$ or $2(d_2/v_2)$ $\{1 - (1 - v_2^2/c^2)^{1/2}\}$. These are the expressions for the difference in setting for cases (I) and (II). If $v_2 = -v_1$, then $d_2 = -d_1$ and (A) vanishes, which is the special case (III) discussed above, where S_1 and S_2 more relatively to S_3 at speeds $-v$ and $+v$ respectively.

Let us now consider an even more general case. We presuppose the metric of special relativity; as shown already, in the case of the sole existence of two systems S_1 and S_2, this is inadequate for unambiguously describing states (amongst them the running of clocks) by spatio-temporal quantities (measurable quantities). Therefore we introduce a fictitious system S_3 and shall denote the 'state' of S_1 and S_2 relatively to S_3 by v_1 and v_2 respectively. (If v_1 or v_2 are zero, S_1 or S_2 respectively are at rest with regard to S_3). At the instant when S_1 and S_2 pass each other, we synchronize their clocks. We assume that at the same instant S_3 is likewise near to S_1 and S_2 and its clock is synchronized with theirs. After a sufficient lapse of time S_1 and S_2 meet again at S_3, where we assume that their motion has been almost all the time translational at speed v, so that the 'accelerations of reversal' can be neglected. The total path covered by S_1 and S_2 relatively to S_3 between the two meetings is denoted by s. (Because S_3 is fictitious, s is too and cannot be ascertained by measurement.) Let s_1 and s_2 be the corresponding values of that path when measured in S_1 and S_2 respectively. Then $s_1 = s(1 - v_1^2/c^2)^{1/2}$, $s_2 = s(1 - v_2^2/c^2)^{1/2}$ so that

$$s_1(1-v_2^2/c^2)^{1/2} = s_2(1 -v_1^2/c^2)^{1/2} \tag{1}$$

Let S_1 and S_2 traverse the paths d_1' and d_2 respectively relatively to S_3 in the time t between the two meetings. (t also is a fictitious quantity not ascertainable by measurement). Since S_1 and S_2 move relatively to S_3 at translational speeds v_1 and v_2 respectively as assumed, then (neglecting accelerations of reversal) $d_1 = v_1 t$, $d_2 = v_2 t$, from which

$$d_1/v_1 = d_2/v_2 \tag{2}$$

When S_1 and S_2 meet again their clocks are compared and the difference D in their settings is read off. This must satisfy

$$D = (d_1/v_1) (1 - v_1^2/c^2)^{1/2} - (d^2/v^2) (1 - v_2^2/c^2)^{1/2} \tag{3}$$

This can be simplified in view of (2), it being indifferent whether D turns out positive or negative. The relative speed of S_1 and S_2 must satisfy

$$v = (v_1 + v_2)/(1 + v_1 v_2/c^2). \tag{4}$$

The values of D, v, s_1 and s_2 are obtained by measurement, and d_1, d_2, v_1, v_2 are calculated from Equation (1)–(4). These are the paths covered and velocities both relatively to S_3 of S_1 and S_2 respectively. Now S_3 has been assumed only as a fictitious system, which shows especially clearly that introducing it means precisely the stipulation of a metric for deriving the values of the speeds characterizing S_1 and S_2, from the observed values of D, v, s_1, s_2. Note that S_1 and S_2, which move relatively to each other at speed v, can 'in themselves' (that is, relatively to S_3) have quite different speeds v_1 and v_2. On the metric chosen, v_1 and v_2 are determined by Equations (1)–(4), that is they depend on D, v, s_1, s_2, so that beside v, we need the data D, s_1, s_2. The epistemological function of the introduction of fictitious predicates becomes clearer still if we discuss these formulae from various angles.

Choosing as general a metric as in the above example clearly shows the uncertainty in the meaning of the relativity principle for the two systems S_1 and S_2. If we assert the principle in the much too general form $M(S_1, S_2) \equiv M(S_2, S_1)$, we obtain what v. Laue rightly describes as the 'genuine and unresolvable contradiction' of the clock paradox. However, in this form the principle leaves uncertain 'the state of rest' of either S_1 or S_2, and so does not express the usual meaning of the relativity principle, namely that 'S_1 is at rest and S_2 in translational motion with regard to S_1' and 'S_2 is at rest and S_1 in translational motion with regard to S_2' mean the same. With this meaning, the principle can be applied to S_1 and S_2 only with the help of a fictitious system S_3 whose relations to S_1 and S_2 are stipulated. The properties chosen for S_3 or its stipulated relations to S_1 and S_2 have the character of 'fictitious predicates'. If we have chosen an S_3, then, as already explained, we can represent the relativity principle in agreement with the last mentioned interpretation by the equivalence (I) $\equiv$ (II), or $R(S_1, S_3)$ & $M(S_2, S_1) \equiv R(S_2, S_3)$ & $M(S_1, S_2)$. However, the principle is not generally valid in this form, as our further discussion has already shown. Likewise for the previously mentioned more general form of the relativity principle (I) $\equiv$ (II) $\equiv$ (III), which need not be further explained.

If the relativity thus represented, whether in the narrow or wider form, is applied to clocks relatively at rest or in motion to each other, then, as shown above, it turns out that the (special) relativity principle does not hold in the form given by applying the Lorentz formulae. For an observer in S_3 the difference in settings of the clock is given by the general expression (A) above, applicable to all three cases (I), (II), (III), but yielding in general different results for the lag of clocks in S_1 and S_2 in the three cases.

It has been shown that the relativity principle can simply not be applied to just two systems, in fact at least three systems are required for unambiguous application. It then emerges that the three propositions (I), (II), (III) do not mean the same (are not 'equivalent'), for in the various differences in settings of the clocks in the three cases we have observational data that enable us to see empirically that any two of the three propositions are incompatible thus excluding their equivalence (asserted in the relativity principle). If therefore by the relativity principle one understands the equivalence of (I) and (II) or even that of (I), (II), (III), then one can say that our reflections exhibit the fact the relativity principle does not hold for the running of clocks. We shall establish this by analysing our examples further.

In the above example we have assumed that the difference of the setting in S_1 and S_2 is read off after re-encounter, in which case the further data v, s_1, s_2 help us to calculate the speeds of S_1 and S_2 with regard to S_3. In some cases it might however be expedient to fix the velocities of S_1 and S_2 relative to S_3 and then to derive the difference in setting to be expected. We might for example assume S_1 and S_2 have the velocities $v_1 = 0$ and $v_2 = (\pm) v$ respectively relative to S_3, or conversely $v_2 = 0$, $v_1 = (\mp)v$. The relativity principle (I) $\equiv$ (II) asserts that these two cases are equivalent. However, we have already recognized that this assertion is false, since in the two cases the settings for S_1 and S_2 are not the same. Even more interesting would be the possibility of making both v_1 and v_2 different from zero. This corresponds to the assumption that S_1 and

S_2 are in relative motion to each other at speed v and with regard to S_3 at speeds v_1 and v_2 respectively, subject to equation (4) above. What values we choose for v_1 and v_2 in any case is a matter of expediency. It might be that for example D, s_1, s_2 are not or insufficiently known so that in order to obtain predictions we must fix values for the velocities of S_1 and S_2 relative to S_3. According to how far the derived predictions come true, the stipulations as to the values of v_1 and v_2 (that is the stipulated metric) counts as more or less appropriate.

Such possible stipulations, by whose means alone the unambiguous application of the relativity principle to S_1 and S_2 becomes possible, clearly show the metric character of the chosen relation of S_3 to S_1 and S_2. If the relative velocity of S_1 to S_2 has the value v, then $v_1 = 0$ corresponds to case (I) and $v_2 = 0$ to case (II), but these are by no means the only possible metric choices. Let us collectively denote as case (III) all metric stipulations in which the values of v_1 and v_2 are choosen as different from zero. A special instance of (III) would be values of v_1 and v_2 such that $D = 0$; that is, we should have to ask what are the velocities of S_1 and S_2 with regard to S_3 if on re-encounter of the systems their clocks show no difference between their settings. (This corresponds to the case already treated above, where $v_1 = -v$ and $v_2 = +v$.) If one assumes that all relative motion of the systems is translational, excluding disturbing forces (under these conditions the case to be considered does not fall under the laws of general relativity theory), this possibility seems to be excluded given the customary (though inadequate) representation of the relativity principle in relativistic physics. If no gravitational fields or other forces arise, the relativistic principles as usually formulated must produce differences between the settings of clocks in the systems that separate and reunite. Identical settings could arise according to the relativistic law only by the action of 'disturbing forces' (gravitational or other force fields). As against this the exact representation of the relativity principle, which in the case of two systems S_1 and S_2 requires introducing a fictitious system S_3 with corresponding metric stipulations, shows that the difference D of the settings of the clocks in S_1 and S_2 can, if all disturbing forces are excluded (that is, we have purely translational, force-free systems), assume not only positive and negative values but also zero.

Here one might object that the explanatory example speaks of the running or change in running of clocks, but it is surely problematic whether for mechanically made clocks or for periodically recurring phenomena that can be used as 'clocks' one can assume the same or variable running in different systems. It is not unobjectionable simply to suppose that 'clocks' that are not further characterized quite generally run identically in the systems in which they are. As to this we must however point to a hypothesis of the relativistic metric. It makes sense to speak of certain changes in the running of clocks in differently moving systems only if there are periodically recurring phenomena of which we can be sure that they occur in force-free systems and in identical manner for observers in those systems. This presupposes the relativity theory of electromagnetic oscillations produced by field sources identical in kind. If for example in a system S we produce light rays of a certain frequency v, by burning a certain chemical, we assume that in all systems at rest relatively to S the light waves thus produced will have that same frequency.

On this hypothesis we can then view changes in the running of clocks in moving systems as changes in the frequency (and therefore of the period and wave length) of electromagnetic waves produced in corresponding manner. About the running of such electromagnetic clocks we can likewise show that the differences in frequencies (that is of the 'clock settings') in the systems S_1 and S_2 can be unambiguously derived only if we introduce a fictitious system S_3 whose relations to S_1 and S_2 are chosen and stipulated.

S_1 and S_2 are in relative motion to each other at speed v, and with regard to the fictitious system S_3 at speeds v_1 and v_2 respectively. Once again

$$v = (v_1 + v_2)/(1 + v_1 v_2/c^2) \tag{1^1}$$

For an electromagnetic radiation which has frequencies v_1 and v_2 in S_1 and S_2 respectively, we have

$$v_1 = v_2 (1 - v/c)^{\frac{1}{2}}(1 + v/c)^{-\frac{1}{2}} \tag{2^1}$$

A radiation of frequency v in S_3 has frequencies v_1 and v_2 in S_1 and S_2 respectively, where

$$v_1 = v(1 - v_1/c)^{\frac{1}{2}}(1 + v_1/c)^{-\frac{1}{2}}$$
$$v_2 = v(1 - v_2/c)^{\frac{1}{2}}(1 + v_2/c)^{-\frac{1}{2}}$$

From (3^1) and (4^1) we obtain

$$v_1 = v_2 \{(1+v_2/c)(1 - v_1/c)\}^{\frac{1}{2}} \{(1 - v_2/c)(1+v_1/c)\}^{-\frac{1}{2}} \tag{5^1}$$

Equations (1^1)–(5^1) are partly independent from each other. v and possibly v_1 and v_2 can be determined by measurement, and v_1 and v_2 can then be calculated from (1^1), (2^2), (5^1). If for some reason v_1 and v_2 cannot be ascertained, then we can assume for this source of radiation (for example a certain atomic decay) a fictitious frequency v relative to S_3 and therefrom calculate v_1, v_2, v_1, v_2 by means of equations (1^1) (4^1). The predictions further derived with the help of these values then decide whether the value chosen for the fictitious v is appropriate to our purpose. The differences $v - v_1$, $v - v_2$, $v_1 - v_2$ must be viewed as the differences in the running of clocks in S_1, S_2 and S_3.

10. DOMAINS OF UNCERTAINTY AS A PRESUPPOSITION FOR APPLYING THE NEW METHOD

The epistemological presupposition for applying the method of fictitious predicates in physical descriptions is always that there is a certain domain of uncertainty arrising from the attempt of characterizing phenomena by values of measurement. By 'uncertainty' we are to understand by no means merely 'relations of uncertainty' and 'imprecision' in quantum physics. These last, to be discussed in detail later, are indeed special cases of 'uncertainties' that require the stipulation of metric systems by means of introducing fictitious predicates, but here we take 'uncertainty' in the characterization of phenomena much more generally. It can arise even where there are no 'limits of precision' to our procedures of measurement, that is where phenomena can be sufficiently characterized by exact values using the available procedures of measurement; or, what comes to the same, if arbitrarily small changes in the values of the characterizing quantities can be ascertained by measurement. In such cases the 'uncertainty' consists in the measuring procedures being simply inapplicable within certain domains, that is neither accurately nor inaccurately, and yet we are quite able to speak of the occurence ('existence') of phenomena in such domains. As we have tried to show in the examples of the previous chapters, phenomena between which no relation of action is possible fall into such 'uncertainty domains'.

During procedures of measurement action relations are established between the measuring instruments and the objects to be measured (phenomena). Only where transmission of action can be established between the relevant objects (phenomena) can one carry out measurements. Now all empirical transmissions of action spread at finite speed. If for our measurements we use transmissions of action (signals) of the highest possible speed of propagation, then there are no possible action relations between the phenomena within the time span bounded by the sending and return of the 'action' on the one hand and the phonomenon of 'reflexion' of the action on the other. Therefore there are no empirically performable measuring procedures for this domain of phenomena, by which one might characterize relations between phenomena by means of unambiguous values of measurement. If nevertheless one wants to indicate a spatio-temporal order even for phenomena lying within such a domain of uncertainty, one can do it only by stipulation, namely by introducing fictitious predicates. Fictitious time relations between the phenomena concerned might for example be fixed by insisting on invariance under transformation for certain expressions (that represent natural laws). By means of the metric thus chosen exact individual values (not just 'probable values') are assigned to the phenomena that lie in the uncertainty domains, and these values have the character of 'fictitious' values of measurement. Their fictitious character rests on the fact that according to the conditions under which alone the measurements can be carried out, there is in principle no procedure of measurement and observation, by means of which the ascribed fictitious values could be tested. Their correlation is always effected only by way of a calculation to be performed on the basis of the chosen metric stipulations. If we change the stipulations – which is always permissible –, the calculation yields different – and again 'fictitious' – values for the same phenomena. Which of these fictitious ascriptions, that is which of the predicates that fix the metric to be chosen, we decide to adopt is, as previously explained, a question of expediency. The aim is the most complete and simple description possible. What is decisive here is the amplest possible derivation of predictions with the minimum of auxiliary hypotheses, made possible by the chosen metric and further, the confirmation of these predictions by observation.

The step of choosing and assigning fictitious predicates to phenome a that cannot be characterized by using measuring procedures and for this reason lie in domains of uncertainty, this step we can regard as the new method for reaching scientific knowledge; by this method, modern analytic enquiries have enriched traditional methods for gaining scientific knowledge. As regards the question how far we can apply the relativity principle to two systems in relative motion to each other, the new method leads to the result that in that case the principle can be represented unambiguously only if we introduce a third fictitious system with determinations that complement the relativistic metric. Possible forms of such representations are $(I) \equiv (II)$ and $(I) \equiv (II) \equiv (III)$. Of course the relativity principle under the conditions concerned can no longer be asserted as generally valid.

If this remarkable result already makes clear the peculiar nature of the
new method, the characteristic features of the necessary epistemological
analysis of physical concepts, relations and measuring procedures stand
out more vividly still when the 'method of fictitious predicates' is applied to
the description of micro-phenomena.

Introducing fictitious predicates and relating them to values of measurement obtained by procedures of observation (measurement) means the formation of new kinds of concepts, that were not applied so explicitly in traditional scientific enquiry. Within quantum physics this new procedure
developed in such exact form, that there the sometimes repeated application of the step of introducing fictitious predicates occurs by the automatic
application of mathematical operators to expressions for physical quantities. However, epistemologically this does not make these steps different in
kind from the relating of certain fictitious predicates in relativity theory to
phenomena that lie in uncertainty domains as regards measuring procedures. It is not unimportant to recognize this if one is to understand what is
epistemologically novel in this method of modern scientific enquiry.

PART III: PREDICATES AS PROBABILITY QUANTITIES

11. ANALYTIC-DEDUCTIVE INFERENCES INFERENCES FROM CONJUNCTIVE CLASSES OF PROPOSITIONS

In everyday life, mathematics and empirical science, we make statements
and inferences that are explicity accorded only probable validity, where
'probability' is usually opposed to 'certainty'. Enquiry as to what might be
the marks of 'probability' soon showed that this concept was being used
with different meanings. For exact science, whether mathematics or the descriptions of mathematical physics, only those forms of the concept of
probability are relevant in which the probability values ascribed to phenomena, statements (predictions) or inferences occur as values of functions.

For epistemology there arises the task of finding at least the necessary
conditions that must be given if we are to speak of the 'probability' of any
elements. In view of the ambiguity of the word 'probable', it must of course
remain an open question whether there are general necessary conditions
that must be present in every instance where one might speak of 'probability' in its various senses.

That a statement probably holds is based on reference to the holding of other propositions. There is thus a certain relation of derivation between the probable statement and the propositions from which one infers its probably holding. Let us call general inferences of this kind 'probable inferences' and the relation that exists between their premisses and conclusion a 'probability derivation relation' or 'probability deduction relation'. In what then do probable inferences[8] differ from logical analytic deductions?

Merely pointing out that logical mathematical derivations and above all syllogisms are exact inferences while 'probable inferences' are less precise is surely not enough and in a certain sense even incorrect. For, to take an example, derivations of probabilities in the calculus of probabilities are unambiguously exact. Even if in other cases 'probable inferences' are made on grounds that cannot be strictly valid, nevertheless there are strict logical relations of dependence between probability statements of the kind occuring in the exact sciences and the propositions from which they are obtained.

The difference between so-called strict deductive inference and 'probable inference' must therefore be sought elsewhere and certainly in the contrast between absolutely rigid and less valid forms of inference. Here we may use the procedure of mathematical logic and analyse the relations between premisses and conclusions, in order to discover the characteristics of logical structure of analytic deductive inference as well as of 'probable inference', insofar as this last concerns formalisable inferences used in the exact sciences.

Deductive inference is often presented as though it were impossible by means of it to obtain new knowledge, since the conclusion is always asserted already in the premisses. This account is surely correct. The conclusion p_i must then figure amongst the premisses and be connected with other propositions. Consider the individual premisses $p_1, p_2, ...p_n$, then p_i must figure amongst them. Taking the premisses as a class $[p_1, p_2, ...p_n]$, the task of analysis is then to discover the relations amongst its elements. For just as the conclusion is connected with the premisses by a relation (of dependence), so too there must be some relations or other between the individual premisses. These relations are decisive for the strict validity of the inference.

In a formalized calculus schemata and rules for applying them are given for the forms of inference admissible in it. The best known schema used in almost any calculus is that of '*modus ponens*', according to which we can write down a statement 'B' as holding if 'A' and 'B follows from A' are already given as holding. The application of this inference schema presupposes that 'A' and 'B follows from A' both hold, and so therefore does their conjunction. Representing the relation of consequence by ' $\rightarrow$' and conjunction by '&', the connection of the two premisses of modus ponens is given by A & ($A \rightarrow B$).

If we ragard A and B as the individual propositions p_1, p_2 of which the premisses of analytic deductive inference consist, then the individual propositions in the schema are connected by the relations '&' and ' $\rightarrow$', it would not be appropriate to call this formula, in which the individual propositions are linked by various relations, a 'class of propositions' with elements p_1, p_2 (that is, A, B). For the elements of a class are marked precisely by each having the same property in the same way. If as predicate characterizing the elements of a class of propositions we choose the property of each element being related to the others in a certain way, for example by conjunction '&', disjunction '$\vee$' or equivalence '$\equiv$' and the like, then between any two elements of the class only that same particular relation must always hold, for example only conjunction, only disjunction or only equivalence. Thus the conjunction of $p_1, \ldots p_n$, which holds when they are connected only by &, that is p_1 & p_2 & ... & p_n, can properly be regarded as a class of propositions. The predicate common to all its elements is the property of each being linked with any other by '&'. With Carnap [9] we shall call this a conjunctive class of propositions. Similarly we can speak of 'disjunctive classes of propositions', 'equivalence classes' and so on.

A characteristic of a class of propositions thus understood is that it is itself a statement that can be true or false. A conjunctive class is true if and only if each of its elements is, and false as soon as even a single element is false. Let us denote the conjunctive class of the elements $p_1, p_2, \ldots p_n$ by $[p_1, p_2, \ldots p_n]_c$, so that $[p_1, p_2, \ldots p_n]_d \equiv_{df} [p_1$ & p_2 & ... & $p_n]$. We shall also need the disjunctive class of propositions', whose elements are connected only by '$\vee$', so that $p_1, \vee p_2 \vee \ldots \vee p_n$ holds, which we define by $[p_1, p_2, \ldots p_n]_d \equiv_{df} [p_1 \vee p_2 \vee \ldots \vee p_n]$. The predicate common to all the elements of a disjunctive class is the property of each being linked with any other by '$\vee$'. These classes too are statements that can be true or false. A disjunctive class is true as soon as even one of its elements is, and false if and only if all its elements are false.

Looking back to the formula A & ($A \rightarrow B$) which links the premisses of modus ponens, we can show by the transformation rules of the propositional calculus that this is equivalent to the conjunction of A and B, that is A & $(A \rightarrow B) \equiv A$ & B. If for A and B we substitute p_1 and p_2 respectively, p_1 & $(p_1 \rightarrow p_2) \equiv p_1$ & p_2. However, $[p_1$ & $p_2]$ is the conjunctive class $[p_1, p_2]_c$ of elements p_1, p_2. It is characteristic of analytic deductive inference, that its premisses are equivalent to the conjunctive class of individual propositions amongst which the conclusion must itself figure. In saying that the schema of modus ponens allows inference from A & $(A \rightarrow B)$ to B, the relation between the conclusion B and the premisses is certainly one of tautological consequence. Denoting that relation by the implication sign '$\rightarrow$', we have the inference schema A & $(A \rightarrow B) \rightarrow B$. Substituting p_1, p_2 and recalling that the premisses are equivalent to the conjunctive class, we can replace $[p_1$ & $p_1 \rightarrow p_2)] \rightarrow p_2$ by $[p_1$ & $p_2] \rightarrow p_2$. Here we recognize a common characteristic of all analytic deductive inference forms (we need not here consider so-called 'transfinite' inferences). In such inferences the premisses are always equivalent to a conjunctive class amongst whose elements figures the conclusion, so that we can denote these inferences as 'inferences from conjunctive classes of propositions'. As a further characteristic of such inferences we might regard the circumstance that the relation between premisses and conclusion is always that of tautological implication. This follows from the property of any such inference forms, that its premisses can be replaced by an equivalent conjunctive class amongst whose elements figures the conclusion. Thus one obtains the inference schema $[p_1$ & $p_2] \rightarrow p_2$, in which the relation between premisses and conclusion necessarily is tautological implication.

12. INFERENCES FROM DISJUNCTIVE CLASSES OF PROPOSITIONS.
PROBABILITY INFERENCES

If we can characterize analytic deductive inference as occuring from conjunctive classes of propositions, we can raise the purely formal and more general question whether there might not be 'inferences from disjunctive classes of propositions'? It might be thought that nothing could be inferred from a proposition $p_1 \vee p_2 \vee \ldots \vee p_n$ so that inference from disjunctive classes is simply impossible. However it would be premature to assume this. For in everyday science we often infer from premisses amongst which there are disjunctive classes of propositions. These are 'probability infer-ences'.[10]

Given propositions p_1, p_2, ... p_n disjunctively linked, that is $p_1, \vee p_2 \vee ...$ $\vee p_n$ holds, our everyday comment as to the holding of the individual prop-ositions often is 'perhaps p_1 holds' or 'perhaps p_2 holds' and so on. Here we infer from the given disjunctive class to the 'uncertain' holding of an indi-vidual proposition. This uncertainty alone is sometimes denoted as 'proba-bility', although in the present case numerical discrimination of degrees of probability is out of the question without further determinations.

As with inference from conjunctive classes we can ask for the relation be-tween premisses and conclusion in derivations from disjunctive classes. In analytic deductive derivations, that is those from conjunctive classes, there is, as we saw, only one kind of relation of consequence between premisses and derivable conclusions, namely tautological implication. What are the corresponding relations for derivations from disjunctive classes? Or, to ask more cautiously, what relations between premisses and derivable proposi-tion are possible in such inferences?

Consider how, starting from a disjunctive connection of individual prop-ositions, we can infer to the 'probable' holding of one of these propositions. It turns out that under the same conditions this is possible in various ways. If the question concerns the probability with which we are to expect a cer-tain outcome when throwing a single die, the point is to infer from a disjunction of six individual propositions to the 'probable' holding of one of them. Let p_1, p_2 ... p_6 stand for 'the upmost face is 1', the 'upmost face is 2' and so on, then the premisses are given by the disjunctive class $[p_1, p_2 ... p_6]_d \equiv [p_1 \vee p_2 \vee ... \vee p_6]$. How, from this, can we now infer, say, the 'probable' holding of p_6? At first there is only one answer: without further stipulations nothing definite about the 'holding' or probability of p_6 can be derived. It is clear at once that the premisses (the disjunctive class $[p_1 \vee p_2 ... \vee p_6]$) must be complemented by a stipulation that is not further fixed in general, if we are to infer the 'holding' of one of the class elements.

We might for example stipulate that every individual proposition should have equal 'probability', from which together with the disjunctive class we infer that p_6 holds with probability $\frac{1}{6}$. The conclusion now actually has an unambiguous relation to the premisses. From the disjunctive class $[p_1 \vee p_2 \vee ... \vee p_6]$ and the stipulation that each element holds in equal measure, we can deduce with certainty that the probability of p_6 is $\frac{1}{6}$.

However, the stipulation of equal 'probability' for each of the possible cases is arbitrary and it is quite conceivable that we might fix different probability values for the propositions characterizing the individual cases. Indeed, we might fix no prior (*a priori*) probabilities for the possible cases or the propositions characterizing them, but to stipulate that the average frequency with which each case has occurred in a large number of actual cases is to be assigned as the degree of probability to the respective individual propositions.

We see that there are indefinitely many possibilities of choosing a system of probability values, which then enable us to infer from the disjunctive class unambiguous probabilities for the occurence of the class elements (that is the individual propositions). Without some such arbitrarily chosen system the relation of consequence between the premisses (that is the disjunctive class) and the conclusions that may exist remains uncertain. Now 'uncertainties' in the relations between elements are something that we have already met and we know how such uncertainties can be removed: this can be done only by stipulating a system of relations for the elements in question. Such chosen systems are called 'metric' systems and a metric is always chosen with regard to appropriateness, as explained several times before.

Therefore, in order to remove the uncertainty in the relations of consequence between disjunctive classes and their elements, we must stipulate a metric system, which we shall call 'probability metric'. Such a metric must be chosen whenever conclusions are to be derived from disjunctive classes of propositions. Hence we can generally denote inferences from disjunctive classes as probability inferences. Now we can answer the question what relation of consequence is possible between the disjunctive class given as premiss and such conclusions as might be derived: these are 'relations of probable consequence', which have special properties. The comparison of such a relation with the corresponding one between a conjunctive class and its elements, clearly illustrates the logical syntactic connection between absolutely certain deductive inference and probable inference.

In analytic deductive inferences, characterized as from conjunctive classes, there is only one relation of consequence between premisses and conclusion: namely tautological implication, holding between a conjunctive class and each of its elements. In contrast, between a disjunctive class and its elements, indefinitely many kinds of relation of consequence are possible, corresponding to the indefinitely many metric systems that we might choose in order to fix a 'relation of consequence' between premisses and conclusion in such cases. Between a disjunctive class $[p_1, p_2 \ldots p_n]_d$ and its elements p_i indefinitely many 'relations of probable consequence' are possible.

The question now arises, whether the various relations of probable consequence that can exist between a given disjunctive class and its elements are independent or mutually dependent. Further we might ask whether the 'relation of probable consequence' might not have some logical relation or other to tautological implication, that is the relation of consequence between a conjunctive class and its elements. Examining these questions will certainly help to clarify the logical character of the 'relation of probable consequence' and therefore of probable inference.

13. SYSTEMS OF PROBABILITY METRIC AND THEIR ORDER

The unambiguous and complete characterization of probability inferences or relations of probable consequence is made more difficult (if not impossible) by the circumstance that such inferences are often applied without prior indication of precise probability values for the derived propositions. With such unformalisable inferences we need not here concern ourselves. For exact empirical description the only suitable probability statements and inferences are those that make it possible to ascribe exact numerical probability values to phenomena. Recent theories of probability try in this precise sense to give definitions of the concept or relation of probability, and here it turns out that for formalisable systems of probability metric one can indicate an order expressing the functional dependence of systems from each other.

For empirical descriptions two different kinds of probability concepts have always been used and the difference in their significance is noticeable: namely 'a priori' and 'statistical' probability. The first is defined by the classical probability calculus as the 'ratio of favourable to possible cases'. To this it has been objected that the definition is circular, since it presupposes 'equal probability' for the individual cases in the domain of all possible cases. It is certainly correct that 'a priori probability', however defined, presupposes the 'elementary probability' of the individual cases as an undefined concept. Here it is indifferent whether the individual cases are given equal or different probability values. In every instance the probability values are assigned by stipulation to the individual cases in the domain of all possible cases. The elementary probabilities of these individual cases are presupposed as an undefined and directly intelligible concept. In this sense *a priori* probability can indeed be defined without circularity. Thus these elementary probabilities fixed by stipulation count as '*a priori* probability values', and so do all further probability values derived from them by the rules of the probability calculus.

As against this, the concept of 'statistical probability' presupposes no undefined elementary probabilities for any case. What counts as statistical probability of a case is the 'statistical frequency' with which it usually occurs. The number of relevant cases occuring (in a long sequence) and the total number of all cases that have happened is obtained by empirical counts. The ratio of these two numbers counts as the 'statistical probability' of the case in question. It remains quite undetermined, what initial probabilities belongs to the possible individual cases. The concept of an elementary probability, independent of any experience, assigned to the individual cases and therefore '*a priori*', does not belong to the constitutive hypotheses of 'statistical probability'. Statistical probability is particularly suitable for empirical descriptions, since statistical values are obtained from experience (counting, measurement) and easily tested by observation. For *a priori* probability values the conditions for empirical testability must be selected in each individual case. Only by being thus interpreted does the ascription of *a priori* probability values to phenomena become an empirically testable description. The fact that on the basis of this interpretation of *a priori* probability values their testing proceeds by means of statistical counts or measurements, does not abolish the logical difference between the '*a priori*' and the 'statistical' concepts of probability. The logically decisive difference is that *a priori* probability presupposes the undefined elementary probabilities of the individual cases independently of any experience, while statistical probability exclusively relates numbers obtained from experience (counting, measurement), without presupposing an undefined probability concept whose meaning is already known.

Now we can show that '*a priori*' and 'statistical' probability are two extreme cases in a series of those probability concepts according to which 'probability' means a relation between numbers. From quite simple problems we can see that under certain empirical conditions it is expedient to stipulate within what range the probability values are to lie for the cases in question, while at the same time enlisting statistical empirical counts and measurements for ascertaining the exact values. If, for example, we know that there are twenty spheres in a container, some black and some white, but at least five of the latter, then it is appropriate to stipulate for the probability of drawing a white or black sphere corresponding *a priori* 'limiting values'. In the present example it would not be inappropriate to stipulate that these probabilities should be $\geqslant \frac{1}{4}$ or $< \frac{3}{4}$ respectively. A more precise 'adjustment' of these values could then easily be undertaken by a statistical count in a large number of draws.[11]

From the possibility of such cases we can see that alongside purely *a priori* and purely statistical probability, in empirical descriptions one may use probability values determined partly by elements of either kind. To these values correspond probability concepts that must be distinguished from the two pure concepts, as well as from each other according to the measure in which they are determined by either.

Thus in the domain of exact probability ascriptions there are infinitely many probability concepts constructible from the same syntactic elements and arrangeable in a logical order. The English mathematician W. E. Johnson [12] was perhaps the first to indicate the mathematical functional connection between probability concepts insofar as they are defined in terms of *a priori* or statistical determinations. Johnson's function has been generalised and perfected by R. Carnap[12]. Let w and k be the number of favourable and all possible cases respectively, s and s_i the number of all cases that have happened and those of them whose probability is in question (both empirically counted), then the logical syntactic connection of all probability concepts defined by means of these four elements is given by $G(w, k, s, s_i) = (s_i + \lambda w/k)/(s + \lambda)$, where λ is arbitrary and can have any real value from 0 to ∞. For $\lambda = 0$, G represents probability as 'relative frequency', that is statistical probability, while for $\lambda = \infty$, G tends to w/k which is the '*a priori*' probability as ratio of favourable to all possible cases.

The G function indicates a continuous order for all possible systems with an exact probability metric. We saw above that inference from disjunctive classes becomes possible only after choosing a probability metric. The propositions in a disjunctive class are the cases whose probabilities are to be inferred. These elements may all be known, or an unknown number of additional propositions may also belong to the class; they may be more or less alike and either finite or infinite in number. In all these different instances closer examination will show which probability metrics are applicable at all and which is the most appropriate; that is, what values of λ for given disjunctively connected propositions will enable us to obtain probable inferences that are the most appropriate for the requirements of description.

The arbitrary choice of a probability metric for making probable inferences occurs in terms of choosing a value for λ in G. It can happen that for certain values of λ we obtain probable inferences (or probability values) that not only deviate from the probable inferences in statistics or *a priori* probability calculus, but even are diametrically opposed to them. However, this too is quite in tune with the logical possibility of fixing arbitrary relations of probable consequence between given propositions (that is for a given disjunctive class).

Choosing a probability metric means introducing a fictitious predicate system, that is a system that fixes certain relations between the given propositions of a disjunctive class. Before further examining how systems of probability metric in the form of fictitious predicates are introduced into physics in order to establish relations between physical quantities, let us mention certain guidelines often used in empirical practice (by no means only in exact scientific enquiry) when a probability metric is chosen for the sake of obtaining readily testable empirical descriptions or predictions.

The principle generally presupposed when one puts forward laws of nature is the so-called 'principle of induction'. In its most general form this proposition states that if an event B under like circumstances has always followed an event A, then if these circumstances recur, so will B given A. As has long been recognized in epistemology, the principle of induction is not justifiable on logical grounds but is a practical rule, which it is indispensable to follow or presuppose both in everyday life and in empirical enquiry. According to that principle, probability as the expectation that some event will happen must be taken as the average frequency of its occurrence, the value being regarded the more certain the greater the number of observed sequences that have yielded the same frequency for the event concerned.

The 'principle of induction' if used for describing probabilities is however not just a formal syntactic rule demanding that 'probability' must exclusively be taken as 'statistical'. Indeed, we have seen already that syntactically there can be infinitely many systems of probability metric and therefore probability concepts. However, in any given actual case one of these must be chosen, and for this there are no formal syntactic rules, any selection rule having to be imported into the formal systems from without. As repeatedly mentioned, the choice is governed by expediency: in the case of empirical description it is to yield the most far-reaching though always testable predictions. From this quite non-syntactic point of view the principle of induction likewise demands that for the individual cases the systems of probability metric be chosen in such a way as to make the derived probability values always testable as statistical values. From the fact that in the end we can test probability values only by statistical methods (counts), it follows by no means that probability in empirical description could only be understood as 'statistical'. The best example for cases in which empirical probability statements are made by means of the *a priori* concept (which is non-statistical) are the predictions in games of chance (for example roulette). Here one presupposes 'equal prior probability' for the individual possible cases and the probabilities for the various cases are likewise derived *a priori*. Nevertheless, probability predictions in games of chance are always tested statistically.

If then we have grasped the syntactic form a probable inference or the logical connection between systems of probability metric, we must now ask under what conditions and in what way probability concepts are used in empirical description.

14. THE CHARACTERIZATION OF STATES BY CONJUNCTIVE AND DISJUNCTIVE CLASSES OF PROPOSITIONS

In order to describe phenomena and states, natural science ascribes to them metrical quantities, whose mutual relations (or that of their changes) indicate the order governing the sequence of phenomena. Now we have already met instances where such ascriptions are in principle not possible by way of measurement. Since we want to have a descriptive order for such domains too, we there characterize phenomena by assigning to them fictitious quantities or fixing fictitious relations between empirically given data and fictitiously introduced predicates. This method of introducing fictitious predicates into empirical description thus relates quantities obtainable by measurements with other 'quantities' in principle inaccessible to measurement.

There are various reasons why measuring procedures cannot be applied to any phenomena. One reason of this kind we have already recognized in the finite speed of propagation of the transmission of action. This leads to uncertainty for the time relations of certain phenomena, which can be eliminated by assigning fictitious time values or fixing fictitious time relations. We have already tried to explain how these fictitious predicates are to be related to data obtained from measurement and how we can thereby obtain predictions.

However, it may happen that this impossibility of characterizing states (phenomena) by measurable quantities arises from what in empirical enquiry are usually called irremovable imprecisions in measurement. For measurement always means establishing a transmission of action between measuring instrument and state to be measured. Since because of the finite speed at which action is transmitted it may be the case that no such transmission can occur between empirical states (phenomena) and measurement is therefore impossible, it is plain that the possibility of performing a measurement depends on the laws governing the transmission of the measuring action. Amongst these laws is the relation between the spatio-temporal order of magnitude of the measuring process on one hand and of the phenomena to be measured on the other. It is well-known that we cannot perform precise measurements with instruments that are too 'coarse'. By the 'fineness' of a measuring instrument or the 'precision' of a measuring procedure we must understand the relation between the orders of magnitude of procedure and phenomena to be measured. Evidently this relation can no longer change (that is the precision of measurement cannot be refined) if the two orders of magnitude here being related both have the smallest possible empirical values: amongst these last are the quantities characterizing charge, mass and size of elementary particles, wavelength of cosmic rays and so on. If there are thus limits to the precision of measurement it is clear that under certain conditions we obtain 'uncertainty ranges' for the characterization of states by metrical quantities.

Of course one thing must here not be overlooked as regards imprecision of measurement: if it were merely a matter of obtaining by measurement one single quantity characterizing a state, for example a single measurement of position, then one can always choose a measuring procedure whose order of magnitude is small enough as against that of the phenomenon to be measured. That is, we can always determine a single quantity by measurement with quite sufficient accuracy, although this is as yet of little use for

physical description. For to characterize a state only by one value of measurement is in principle insufficient to lead to predictions, whereas precisely
that is the proper goal of any physical description.

We derive predictions by means of laws of nature, as soon as we know a
'sufficient' characterization of a state. A law of nature functionally relates
the characterizing quantities. In simplified form we might express the content of a natural law thus: if of two (or several) quantities p and q characterizing a state the one, say p, undergoes a certain change, then q changes in
such and such a way. If a law of nature is expressed as a differential law,
then it expresses a functional relation between 'differential' (arbitrarily
small) changes in the characterizing quantities.

Thus we can establish a law of nature only where states and phenomena
are characterized by at least two metrical quantities, or more precisely,
where at least two such are required. A law of nature is indeed a relating of
characterizing quantities or their changes and this simply requires at least
two quantities. Quite generally let us call 'sufficient' any such characterization of states and phenomena required to obtain predictions by means of
laws of nature. As a necessary feature of a sufficient characterization we
have recognized the assignment of at least two descriptive quantities to the
states or phenomena in question.

Further analysis of statements in which empirical science usually em-
ploys metrical quantities to characterize phenomena to be measured leads
to two forms of description that differ specifically both in form and meaning. If for example a material point is characterized by simultaneous values
of position and momentum, the indication of these two metrical quantities
already counts as a sufficient characterization of the particle's state. Denoting the propositions assigning the values of position and momentum by q
and p respectively, then the material point can be characterized by the conjunction $p \& q$. (Strictly, we should have to indicate that they hold simultaneously at time t, perhaps by writing $[p \& q]_t$, about which more presently.)

Suppose now that to characterize the state of a mass at time t we need not
just the position q but a number of position values $q_1, q_2, \ldots q_n$, and not
just p but a number of momentum values $p_1, p_2, \ldots p_m$. Thus, to specify the
'position' of the mass we might have to indicate where its parts are, and likewise for its momentum, the momenta of its parts. The state of the mass at
time this then represented by $[p_1 \& p_2 \& \ldots \& p_m] \& [q_1 \& q_2 \& \ldots \& q_n]$.

Here the p_i and q_i can be taken as elements of conjunctive classes of propositions. The expressions in square brackets have the form of conjunctive classes which are in turn conjoined in the whole and may thus be themselves regarded as elements of a conjunctive class. Now it is worth mentioning that all sufficient characterizations in classical physics, relativity theory included, are given only in the form of conjunctive classes of propositions. Purely conjunctive combination of the assigned values of quantities can thus be regarded as the basic epistemological presupposition of the descriptive form of classical physics, or of the attendant forms of concepts.

For classical physics and relativity theory it is a self-evident presupposition that an empirical state at a given time can be completely and sufficiently characterized by a conjunction of certain ascribed individual values. Even if relativity theory has recognized that these values cannot always be obtained by measurement, nevertheless the method proposed for such cases, namely introducing fictitious predicates or stipulating a corresponding metric, does not go beyond this presupposition. For phenomena between which action cannot be transmitted and which therefore fall into uncertainty ranges as regards the characterizing quantities in the observational system concerned, the chosen relativistic metric stipulates individual values which are however linked conjunctively with the other values in the total description of states in precisely the same manner as this occurs in descriptions by means of values obtained by measurement throughout.

Since in classical physics and relativity theory states (phenomena) are always characterized by conjunctive classes, it follows that in these physical domains all derivations, in particular of predictions, occur as inferences from conjunctive classes of propositions. (Let A denote the change in certain quantities of a state, and B the resulting changes in certain others, then $A \rightarrow B$ expresses the law that certain changes A are usually attended by changes B. If A and B both indicate a conjunction of characterizing quantities (change of quantities), then A & B is a conjunction of the propositions ascribing the individual values. If A is given in the form p_1 & p_2 & ..... & p_n, then with $A \rightarrow B$ we can derive the prediction B, let us say in the form q_1 & q_2 & ..... & q_m which thus follows from the conjunction of the premisses A and B, since, as previously shown, A & $(A \rightarrow B)$ is equivalent to A & B. Every individual description p_i and q_i can here be derived with analytic certainty. All such deductions are thus inferences from conjunctive classes whose elements ascribe individual values to phenomena).

Purely formally we can go on to ask what should be understood by a proposition in which individual values ascribed to phenomena are linked disjunctively. Take the state of a mass sufficiently characterized by indicating the simultaneous values of position and momentum: it is conceivable that for the 'position' of the mass we indicate a spatial domain, somewhere within which the mass is located. A spatial range can be conceived as consisting of a number of individual values. Let Q be the ascription of a spatial range q_1, q_2 ..... q_n the ascribed individual positions within the range Q; then the ascription of the range of position to the material point expresses that it must be at one or the other, that is at $q_1 \vee q_2 \vee ..... \vee q_n$. Whether we think of Q as dissected into finitely or infinitely many partial positions is irrelevant to the epistemological step taken in going from a description by conjunctive to one by disjunctive classes. What is decisive is to recognize that ascribing disjunctively linked values of a quantity to phenomena means characterizing them by a range for that quantity.

In this way we can consider a partial or even exclusive characterization of phenomena by means of ranges for quantities. It is useful to adopt the symbol '$\triangle$' to denote a value range. The ascription of $\triangle Q$ to a mass thus means that it is at one of the partial positions q_1, q_2, ... q_n of the spatial range $\triangle Q$. To express that this occurs at a time t, we apply the index t so that $\triangle Q_t \equiv [q_1 \vee q_2 \vee ..... \vee q_n]_t$. Similarly one may conceive that in characterizing phenomena by other quantities one uses ranges of quantities, for whatever reason. Thus we may think of ranges for momentum, energy and time, if necessary with specification of whether the domains concern the same time or place and so on. Denoting momentum, energy, time by p, E, t respectively, the corresponding ascriptions of intervals thus are given by $\triangle p_t \equiv (p_1 \vee p_2 \vee ..... \vee p_n)_t$, $\triangle E_q \equiv (E_1 \vee E_2 \vee ..... \vee E_n)_q$, $\triangle t_q \equiv (t_1 \vee t_2 \vee ..... \vee t_n)_q$. The disjunction of individual propositions has been called its 'disjunctive class'. Thus characterizing by means of ranges may be called characterizing by disjunctive classes. The possibility of using these for empirical description becomes interesting when we have conditions under which this way of providing it appears to be appropriate.

15. PRESUPPOSITIONS OF MEASUREMENT IN CLASSICAL AND RELATIVISTIC PHYSICS

We have said that classical physics and relativity theory characterize phenomena by ascribing to them exact individual values conjunctively linked in description; now we must ask on what epistemological presuppositions this mode of description is based.

Classical physics already distinguished between precise and less precise values of measurement, and the endeavour has always been to improve the precision of measuring instruments and values measured. Where this met with all but irremovable difficulties, one devised procedures for obtaining corrections by repeated measurement or by observing external circumstances. By means of such formulae we can calculate, from actually measured values, new ones that count as the exact 'values of measurement'. By 'exact' values we are to understand those for which the (always present) imprecision of the measuring procedure lies between limits so narrow as no longer to fall within the domain of observable differences.

The question how far we can obtain exact values of measurement becomes more important in view of the form of differential laws used for empirical description. As already mentioned, such laws state that for differential changes in one quantity certain such changes in other quantities will generally occur. By 'differential' changes we must understand 'arbitrarily small' changes. The entire system of statements of classical physics then rests on the assumption that arbitrarily small changes in the quantities characterizing states (phenomena) can in principle be ascertained by measurement; or perhaps more precisely: the smallness of a change is no obstacle in principle to its being measurable.

Beside this first presupposition of the system of description of classical physics, taken over by relativity theory, there are further conditions presupposed in classical physics for assigning characterizing quantities to phenomena. As already explained, to characterize states sufficiently we must always assign to them two quantities at least. By way of example, we have mentioned the state of a mass, sufficiently characterized by indicating the two quantities 'position' and 'momentum', that is we can derive predictions if the state is thus characterized (at a given time). The characterizing values or their changes must be ascertained by measurement. Here classical physics presupposes that in every case we dispose of procedures enabling us to ascertain the required values of measurement independently of each other. Since to 'measure' is to transmit action between instrument and phenomenon to be measured, measurement of the individual quantities establishes for each of them a chain of actions between instrument and 'object' (phenomenon, state) to be measured. Since sufficient characterization requires the determination of at least two quantities, at least two actions must be transmitted between measuring instruments and the state to be measured. One of these action chains can disturb the state in such a way that the conditions for setting up the second no longer exist. If sufficient characterization requires finding more than two quantities, we can imagine even more complicated circumstances concerning the disturbance of a phenomenon by a measuring process. The disturbance can occur in such a way that further action chains towards ascertaining the required values can no longer be established.

Classical physics and relativity theory presuppose as self-evident that all measurements can be carried out without disturbance, however small the changes in quantities. It is precisely these two presuppositions, first that arbitrarily small changes in characterizing quantities are always measurable and second that measurement of a quantity never disturbs the state to be measured in respect of the others quantities required for sufficiently characterizing it, that allow us to represent the relations of dependence between changes in quantities in the form of differential laws. The differential changes connected with each other in the laws signify here changes in the quantities themselves; that is to say: in the differential laws of classical physics and relativity theory, characterizing quantities (like length, time, mass, momentum, energy and so on) always appear as differentiated with respect to another characterizing quantity. Here every quantity, both those being differentiated and those with respect to which one differentiates, have a physical dimension, expressible in the 'length-time-mass' system. That in physical description one might use differentiable quantities that cannot be interpreted as quantities in that system and therefore have no physical dimension in the usual sense, is a possibility not yet envisaged in the conceptual system of classical-relativistic physics.

Before examining further possible forms of differential laws, let us point out another presupposition that determines the form of many differential laws in classical physics.

Over and above the two cited assumptions that relativity theory has adopted from classical physics, the latter makes yet a third assumption that transgresses the conditions of possible experience. Today we know that to assume the measurability of arbitratily small changes in value of several quantities required to characterize states sufficiently is a requirement going beyond experience. Likewise for the assumption that such quantities can be measured independently; that is, without mutual disturbance: critical epistemological analysis shows that under certain conditions this requirement too goes beyond experience. Both assumptions demand conditions which, in crucial empirical cases that do occur, cannot be realized. Of this kind too is the presupposition of classical mechanics that there are actions that are transmitted at infinite speeds. For example mechanical actions like gravitation, impact by rigid masses and the like are supposed to spread infinitely fast. This enables one, in the corresponding laws, to represent relations between changes in quantities independently of time. If these laws are nevertheless represented as differential laws in which the characterizing quantities are differentiated with respect to time, as though the actions were propagated in time, that is at finite speed, without simply jumping arbitrarily large spaces, then these differential expressions are empty. Such laws are known as 'laws of pseudo-proximate action'; they are 'laws of action at a distance' differentiated with respect to temporal domains. Action-at-a-distance laws make statements about transmissions of actions for which infinite speeds of propagation are presupposed.

Relativity theory abandons this presupposition of classical mechanics, and therefore does not contain laws of pseudo-proximate action. Of course, from the principle that all action is transmitted at finite speeds of propagation there arises a possibility of domains in which events have no mutual relations of action. We have already discussed that for events in such domains a fixed spatio-temporal order becomes possible only if we introduce fictitious predicates (relations) that relate the events concerned to others, the values of whose quantities can be determined by measurement, in such a way that the partially fictitious system of relations allows us to derive testable predictions.

A very similar logical step (namely, introducing fictitious predicates) becomes necessary for empirical description if the other two presuppositions of classical physics mentioned above are given up. As we saw, relativity theory takes over from classical physics the assumption that arbitrarily small changes in characterizing quantities are measurable under the conditions required for sufficient characterization of phenomena, and the assumption that thus measuring any quantity leaves phenomena undisturbed as regards other quantities. Accordingly, relativistic differential laws have the same epistemological character as genuine laws of proximate action in classical physics (for example classical electro-magnetism). The new logical step that becomes necessary if one wants to go on obtaining descriptive laws and predictions while giving up these two presuppositions accordingly consists in introducing into empirical description differential expressions that can no longer be interpreted as arbitrarily small changes in the quantities characterizing phenomena, or as relations between such changes. The form of the law of proximate action that is typical for part of classical physics and for the whole of relativity theory is thus given up.

16. DESCRIPTION BY RANGES OF QUANTITIES

The interpretation of physical differential expressions as denoting arbitrarily small changes in the quantities characterizing phenomena or relations between such changes may be taken as the basic presupposition for the content of laws of proximate action. If these laws are to state something empirically testable, the arbitrarily small changes in quantity connected in laws of proximate action must be ascertainable by measurement under the conditions concerned. However, we must not overlook a restriction here. In relativistically characterizing phenomena one partly assigns fictitious quantities to them, whose values and changes in value are not accessible to measurement. However, this does not alter the character of laws of proximate action, so long as by linking the fictitious system of relations with measurable ones we can derive sufficiently many predictions testable by observation. Moreover, the predictions deduced must be of the same form as propositions derived from a system of laws of proximate action set up without use of fictitious predicates.

Eliminating the assumption of actions transmitted at infinite speed results in excluding laws of action at a distance and of pseudo-proximate action, as well as in generalizing the form of laws of pseudo-proximate action to cases where fictitious predicates must be introduced; but dispensing in addition with the assumption concerning arbitrary precision of measurement results in much more far-reaching consequences still.

As soon as it is impossible to interpret differential changes in quantities as arbitrarily small but measurable changes in the quantities characterizing phenomena, it becomes problematic what could be meant by connecting differential quantities in laws of nature. It therefore becomes problematic, likewise, how far differential laws can state anything at all about empirical events. Thus the form of laws of proximate action is called in question.

However, the purely mathematical form of differential expressions offers enormous advantages for the representation of a system of propositions. As mentioned earlier, classical mechanics for example has not renounced the representation even of laws of action at a distance as differential laws with respect to time, although in such laws differential expressions with respect to time have no physical meaning. In relativity theory the presuppositions for setting up such pseudo-differential laws are no longer given, but then in what sense do differential statements remain possible if, once the other classical principles are given up as well, differential expressions are no longer to mean arbitrarily small changes in the characterizing metrical quantities?

In laws of proximate action, measurable quantities (or relations between them) always figure as differentiated with respect to another such quantity. What such a law states is always this: 'if such and such quantities (quantitative expressions) undergo arbitrarily small changes, then such and such other quantities will suffer differential changes according to the functional relations given by the law'. The presupposition that arbitrarily small changes of any quantity can always be ascertained by measurement without disturbing the other characterizing quantities, may be taken as the epistemological condition making it possible to formulate and test laws of proximate action and therefore for their having a content that makes empirical sense. As already mentioned, the requirement that 'arbitrarily small' changes in quantities be measurable counts as fulfilled if in a given case changes are measurable in all those quantities that cannot be neglected with regard to the order of magnitude of the calculated values of measurement to be tested. It can happen, for example, that if we repeatedly measure the same quantity we obtain, under the same conditions, values so little different from the derived value to be tested, that we assign a negligible order of magnitude to these deviations as compared to that value. This kind of relation between value to be tested and deviations of the results of measurement often arises from the 'inaccuracy' of the instruments or procedures used. Indeed, the procedure establishes an effect-relation between instrument and phenomenon to be measured. The ratio of the orders of magnitude of procedure and phenomenon respectively may be viewed as a measure of accuracy: the smaller the first and the larger the second in the ratio of the two orders of magnitude, the greater the accuracy of measurement.

Now it is conceivable that under certain conditions we dispose only of measuring procedures whose order of magnitude is approximately the same as that of the phenomenon (states) to be measured. In that case these measuring procedures are subject to a limit of accuracy below which one simply cannot go. In the assigning of values this is expressed by not ascribing precise values to the 'objects' concerned but saying 'they have about such and such a value' or 'a value lying in such and such a range'. The latter mode of assigning 'values of measurement' to measured phenomena (states, objects) is of decisive importance for empirical description in conditions where the accuracy of measurement is in principle restricted.

The change of an exact value may be represented as the difference between two values: the quantity changes by an amount $x_1 - x_2$. Whether the quantity is differentiable depends on whether at all relevant values this difference can be assumed to converge to zero, which corresponds to the interpretation that arbitrarily small changes of the quantities required for sufficient characterization are measurable without mutual disturbance. If we give up this assumption, this means in mathematical terms that $x_1 - x_2$ does not converge to zero, but remains above or at some finite value d, and the interpretation just mentioned can no longer hold.

If $x_1 - x_2$ cannot fall below d, then we can no longer set up differential laws in the traditional way, that is as laws of proximate action. However, ascertaining the empirical mutual dependence of changes in quantity is at the same time a presupposition for obtaining predictions, since laws of proximate action express relations of dependence between arbitrarily small changes in quantities, so that such laws permit derivation of predictions expressing the sequence of states within the domain of arbitrarily small changes in spatio-temporal quantities. Such predictions about continuous changes of state or of the quantities sufficient to characterize them, I have elsewhere called 'second-order predictions'.[13] A finite irreducible value for the difference $x_1 - x_2$, that is for ascertainable changes in the characterizing quantities under the conditions required for obtaining predictions, would thus make it impossible to set up laws of proximate action and therefore second order predictions.

What kind of laws remain that can be set up under these conditions? If in the decisive cases we must take the observed values only as finite differences in values of quantities, is epistemology any longer justified in characterizing phenomena (states) by assigning to them exact values or relations between quantities? For, indeed, the ascribed exact values or their differential changes are now in principle unmeasurable in the conditions of the decisive cases. What further sense is there in wanting to characterize states by ascribing exact values to them?

To answer this, let us recall in what form states can be characterized by 'values of measurement', if for any reason individual values cannot be obtained by measurement. Here it is indifferent whether exact values might be obtained under special conditions, while being excluded only under certain other conditions. Rather, we shall now examine the general question whether and in what form we might characterize states by measurement-based numerical values even when exact values for the descriptive quantities are not obtainable, for whatever reason.

In general we may say that when measuring imprecisely we can assign numerically 'imprecise values of measurement' only if the imprecision possesses a numerical 'measure' whose definition for individual measuring procedures is vital. Where we have to ascribe individual values of a quantity to phenomena, we often indicate a so-called 'plus or minus range' as the measure of the imprecision, it being presupposed that the (unknown) exact value lies somewhere within that range. What is worth noting about this way of describing by means of imprecise values is that one relates an unknown fictitiously assumed individual value to a range that can be intuitively exhibited either on the object measured or on the scale of the instrument.

In this way we come nearer to answering the question what numerical elements are required to characterize phenomena by assigning numerical values in imprecise measurement. First of all we must have a 'range' bounded by numerical values, where 'range' is to be understood quite generally as a system of values, for example individual numbers arranged according to some system, or a continuous sequence between two values and the like. However, if a range of values is available, this is not yet enough to characterize a phenomenon (state), since a range is a system of values and therefore signifies a multiplicity of values: is one of these to denote a phenomenon and the rest be empty? Or is the whole set of values in the range to characterize a phenomenon, or is the range to denote a multiplicity of phenomena corresponding to the many values in it? We shall see that all these possibilities are discussed in recent physics in connection with description by means of ranges.

In whatever way we assign a range (used because exact measurement is unperformable) to phenomena, this will make unambiguous description possible only when the range is given the meaning of an ordered system of values. If we do not indicate a metric ordering the values of the range, then, clearly, we can say nothing about phenomena taking place within the domain marked off by the boundaries of the range: without choosing a metric we cannot assign characterizing numerical values to the phenomena concerned. Thus we must additionally indicate an ordering system of relations by which the individual values are related to each other and to the whole range. Because individual values are not measurable within the range, these relations, that is many-place predicates, can be chosen only arbitrarily and thus have the character of fictitious predicates.

In what follows we shall examine what governs the introduction of fictitious predicates, or metric systems defined by them, in the empirical description of ranges.

17. DEFINITION OF PROBABILITY QUANTITIES BY MEANS OF FUNCTIONS

In classical physics and relativity theory, states (phenomena) are characterized by assignment of individual values of metrical quantities, the assigning propositions being conjunctively linked. As against this there are cases in quantum physics, where the propositions assigning individual values are disjunctively linked in the description. In the former case we can say of every individual value ascribed in the conjunction of propositions that something measurable and thus really existing corresponds to it, which is certainly not so for all values assigned in a disjunction of such assignments. Here, the values obtained or to be obtained by observation are related to fictitious values to which by definition nothing observable or really existing corresponds. Such descriptions by disjunctively linked value assignments are used wherever we can characterize phenomena only by ranges of values of the descriptive quantities.

In this the range is interpreted as a system of individual values and each of these as characterizing one 'individual phenomenon', but since only one amongst the individual cases (states) can actually occur under the conditions in question, the 'characterization' of the remaining cases involves using fictitious predicates, related to each other or to those values of the range that are to be obtained by measurement (counting) and chosen with a view to their appropriateness. If, for example, we stipulate a finite number of elementary partial ranges (partial values) for a given 'total range', where elementary parts are used in the characterization of 'fictitious' individual phenomena, then it may be expedient to assign to each 'fictitious' phenomenon characterized by individual values the ratio of elementary to total range as a 'characterizing value'. Under other conditions it might be expedient to assign to individual phenomena ratios of partial to total range that are not uniformly distributed but depend on the characterizing values in some functional way. Thus it might further the obtaining of predictions if to phenomena that do or can occur in different partial ranges of a spatio-temporal range we assign values of the ratio that depend functionally on the way the partial ranges are arranged in the total range, by means of different corresponding factors of proportionality.

The stipulated division of a range into a certain arrangement of partial values indicates a system of individual values of which, so to speak, some are measurable while others are purely fictitious or denote 'fictitious phenomena'. The functional values of the relations stipulated between the values of an ordered range – for relations can be functional connections of various kinds –, which values are then assigned as characterizing values to the phenomena concerned (partly empirical and real, partly fictitious), we shall call quite generally 'probability values'. The functions whose values are probability values we call 'probability functions' and say that they define 'probability quantities'. The way in which probability values are introduced into physical description when the data of measurement are characterized as values of ranges, shows that probability values are defined as values of functions. The arguments of such functions (probability functions) are individual values (partial values of ranges) some of which are measurable and others purely fictitious. Thus we can say in general that a physical probability quantity is always a function of observed values (based on counting or on measurement) and fictitious values.[14] Later we shall consider a more detailed dissection of the arguments of a function defining a probability quantity, that is a function whose values are interpreted as physical probability values. At present let us further examine the relations and differences between the two forms of description – namely by exact values of measurement and by ranges of values.

18. DESCRIPTIONS BY MEANS OF PROBABILITY AND
METRICAL QUANTITIES

So far we have discussed only the case where limited accuracy of the measuring procedures might force us to characterize phenomena (states) by whole ranges. The conditions and circumstances under which we infer to this limitation in principle and are forced to establish a new form of description (namely by range or probability) are certainly fundamental. However, we must not overlook that interpreting certain data of measurement as ranges or partial values of ranges becomes possible only with the help of arbitrary stipulations, by means of which a chosen system of values is introduced in such a way that observed values are interpreted as partial values of it. Leaving aside appropriateness, the procedure of interpreting individual measured values as partial values of an arbitrarily chosen system can be understood as the general possibility of selecting in principle, for any measured values, functions whose arguments take these values amongst others. If the functions are chosen on the pattern on which we define physical probability quantities, then the arguments would always include not only measurable values but also additional ones of merely fictitious significance (denoting 'fictitious phenomena').

Generalizing currently usual terminology, the procedure of viewing individual values of measurement as suitably given argument values of chosen functions (probability functions) might alternatively be regarded as the possibility of defining for any measurable quantity a corresponding probability quantity. We shall leave it open whether it is appropriate to apply this kind of probability description generally – that is, wherever we characterize phenomena by measured values. Here our aim is to become familiar with the probability mode of description in all its epistemological consequences.

From this point of view it is worth noting that not only can individual measured values be ranged into a chosen system of values, thus becoming argument value of a 'probability function' (whose values are probability values), but conversely such probability values can suitably be interpreted as exact 'individual measurable values' assigned to characterize 'phenomena' that are of course in the first place 'fictitious'. This procedure too may have advantages for physical enquiry, because, assuming 'fictitious' phenomena to be characterized by probability values interpreted as individually measured, forces us to look for further quantities characterizing these phenomena, which will always lead to a broadening of the descriptive theory. Modern physics actually uses both procedures: as enquiry might dictate, individual measurable values might be ranged into a chosen value system defining certain of their relations as 'probability values' which play a part in characterizing the phenomena to be described; sometimes, however, enquiry does not stop at assigning probability values but on occasion interprets some of these as 'measurable values' that are to characterize 'phenomena'. Before we examine the corresponding logical mathematical operations, let an example from physics explain the use of such concept formations.

If corpuscular rays pass through diaphragms arranged in a certain way, it is well known that interference patterns result on an intercepting screen. On the one hand we can record the impact of corpuscles on the diaphragm, on the other the interference bands or rings arising on the screen from the distribution of corpuscles suggest that it is waves that traverse the diaphragms. Thus physical description ranges measurable values like positions of impact of individual corpuscles, width and intensity of interference bands and the like, as arguments of suitably chosen functions into a system of values amongst which there are fictitious argument values as well. The functions thus set up then describe a fictitious phenomenon, namely a fictitious wave phenomenon. From the argument values, those based on measurement included, and the functions, we can in each case derive the amplitude of the fictitious wave. The amplitude values are however not assigned as characterizing values of measurement to some actual phenomenon, but the square of the amplitude, which signifies a certain relation between measured and fictitious arguments, is defined as a probability quantity. The values of this quantity indicate the probability of the distribution or states of the corpuscles and are assigned to the actual individual phenomena, in this case the individual corpuscles traversing the diaphragms and hitting the screen. In this way we obtain a probability description for paths, distribution, momentum and so on of impinging corpuscles.

Now the amplitudes of the fictitious waves need not be interpreted merely as probability quantities whose values are assigned to the actual individual emphirical phenomena, but can also be viewed as 'measurable quantities' characterizing 'phenomena' (say, energy waves). In this case too the domain of arguments of the wave functions consists of the system of fictitious values within which the individual measurable values are ranged.

Only now the amplitudes thus obtained characterize actual waves (energy waves) whose properties are described by exact individual values (even if some of these are fictitious).

Once this second step is taken, we must of course ask what are the further properties of the new 'phenomena' which are now characterized by values themselves assigned by the first step of the method as probability values to certain actual individual empirical phenomena. Examining the properties of these 'phenomena' may lead us to derive values that can be assigned as measurable values to certain observational data. Processes initially characterized as fictitious can thus under certain circumstances be regarded as actual empirical processes. Of course this step is justified only if, from the laws assumed for these phenomena, we succeed in deriving the regularity of the other phenomena described by probability quantities, in such a way that these last can then be sufficiently characterized by means of exact individual values of measurement alone. In the cases concerned this would mean that probability quantities and laws had been eliminated from the description.

In the two steps mentioned we first bring in fictitious predicates to characterize phenomena and then try to replace fictitious values and their relations by individual measurable values and their relations. This can be well illustrated in the experiments on interference of corpuscular rays. First we have individual values of measurement such as thickness and distance of the diaphragms, position of impact or momentum of individual corpuscles, width and intensity of the bands and so on. Since they do not suffice to derive (or predict) the states of individual particles for arbitrarily small spatio-temporal domains, these measured individual values are supplemented by fictitious ones; that is, they are ranged into a fictitiously assumed value system. Between the individual values of this domain, functional relations are fixed; that is, we define the fictitiously supplemented system as the domain of arguments of a chosen function, in our example a wave function. It is so chosen that the square of the amplitudes (these are interpreted as ranges within which lie the individual values used for characterization) of the fictitiously introduced waves can be interpreted as certain relations between the measured values characterizing corpuscular phenomena and the additional fictitious individual values. These numbers, which belong to the values of the chosen wave function, are the probability values assigned to the individual corpuscular phenomena.

Following this first step which characterizes phenomena by probability quantities (we may call it quite generally 'first quantisation'[15]), we have in our example a second step aiming at characterizing phenomena by means of exact individual values alone. To this end we now no longer interpret the amplitude of the wave marked by fictitious predicates as a probability quantity(that is, as a range for certain values) whose individual values are to characterize the states of the individual particles, but the values of the amplitude are interpreted as exact individual values of measurement for which we assume the empirical existence of corresponding phenomena, in our case specific waves. This step, of viewing as measurable quantities expressions previously interpreted as probability quantities, forces us to examine the phenomena characterized by such quantities and now assumed as empirically existent, especially with regard to such further properties as might be testable by observation. It is in the sense that de Broglie in particular has assumed the existence of waves (de Broglie waves, guide waves, pilot waves), interpreting their amplitudes as measurable values directly characterizing actual phenomena, while from another angle they are construed as mere probability expressions (ranges of values). This interpretation raises questions as to the speed of propagation, energy and other properties of the waves assumed as actually existent, queries that did not arise when the amplitudes were interpreted in terms of probability. For on that last interpretation these waves, their amplitudes and other properties are merely relations between the values that are assigned to corpuscular phenomena by measurement and the values that characterize them fictitiously. On this view it is obviously senseless to ask whether something physically real corresponds to the quantities characterizing 'probability waves', nor is there any sense in demanding that the properties of these waves and thus their actual existence be tested by observation (measurement).

From this possibility of double interpretation of expressions consisting in relations between fictitious and measurable values, first as ranges and then as individual measurable values, it has been inferred that we have here two appearances of the same phenomenon: depending on the conditions under which we observe them, actual empirical phenomena appear now as corpuscular, now as wavelike. On this view we must correspondingly apply the concepts either of corpuscular or of wave description. This view is most clearly expressed in the complementarity.thesis of Niels Bohr. This may be advantageous for carrying physical enquiry further: but then the thesis is merely a rule as to what forms of concepts we must apply under certain conditions in order to describe the observed data, if we wish to succeed in obtaining predictions. However, it would be erroneous and incompatible with the principle of contradiction if the theses were to be construed as stating that the phenomena concerned are themselves both corpuscular and wave-like. That such an ascription is logically untenable is shown by contrasting the characterization of phenomena by measured individual measurable values on the one hand with that by value ranges on the other. To the first corresponds description in terms of wave concepts, with the assertion that the waves actually exist empirically. The second or 'corpuscular description' assigns value ranges to phenomena and therefore logically excludes the possibility that exact individual measurable values in decisive cases of characterization could have an empirical meaning. The further pursuit of this conceptual contrast stands out particularly clearly when we apply the method of fictitious predicates in physical enquiry.

19. UNCERTAINTY DOMAINS AS PREREQUISITE FOR THE METHOD OF FICTITIOUS PREDICATES

Where we characterize phenomena by the use only of exact individual measurable values, we can sufficiently characterize by means of a conjunctive class. From the view that exact individual values are always sufficient it follows further that all partial states, down to differential spatio-temporal domains, as well as complexes of states must be sufficiently describable in terms of conjunctively linked ascriptions of such values. Perhaps the most perfect example for this mode of description is that using wave concepts, where all quantities characterizing the waves are held to be measurable, down to arbitrarily small changes. Wave fields are complexes of wave states. Under the conditions mentioned the differential changes of state in a wave field also count as in principle measurable. It is further characteristic for this mode of description that in the expressions describing arbitrarily small changes of state the individual measurable quantities (or their relations) directly co-ordinated with states always appear differentiated with respect to other individual quantities likewise deemed to be measurable. Such a differential expression has empirical physical significance only if the arbitrarily small changes in the quantities to be differentiated as well as in the quantities with respect to which we differentiate, are in principle equally amenable to measurement.

Descriptions of state by means of ranges of quantities are based on assumptions not compatible with those just mentioned. This is especially obvious where in describing phenomena one uses ranges of quantities and wave concepts. Here the quantities characterizing wave states are interpreted not as expressions denoting empirically real waves, but as ranges of individual values that in no way characterize 'waves' themselves. If for example we adopt a wave system to describe the distributions of position and momentum of corpuscles and the wave amplitudes are interpreted as ranges for the individual values of position, then the quantities that characterize such waves are exclusively elements of fictitious systems of relations between the individual values characterizing possible states of the corpuscles, including in particular the position values marking the possible points of impact. Here only the corpuscles are assumed to be actually existent but not the 'waves' which are now used merely as an ordering system of concepts for describing the distribution of corpuscles. In such cases too the 'waves' are characterized by quantities such as frequency, wave length, amplitude, speed of propagation and so on, in analogy with empirically real waves; but this must not blur the fact that in real waves the characterizing quantities are individual measurable values assigned to the real wave phenomena themselves, while for the distribution 'waves' the analogus quantities are functions of precisely such measurable values as have nothing to do with characterizing the waves themselves, being assigned to phenomena of quite a different kind (in our case, to corpuscles and their states.)

Accordingly, where amongst differentiated quantities there are some that are not measurable values characterizing real phenomena but are interpreted as ranges of possible individual measurable values, there differential laws mean something quite different from what they mean in cases in which quantities that directly characterize phenomena are differentiated. Physical description always aims at representing laws of nature in the form of differential laws. However, if under whatever conditions there is a limit to what can be ascertained in the way of changes in quantities, so that these changes can be determined only in the form of finite ranges, then differential expressions relating differential changes in the characterizing measurable quantities lose all empirical meaning. In order that we may nevertheless give to descriptive laws the form of differential laws in these cases too, we assign systems of new 'individual values' ('value ranges') to the observed data, with only some of these values counting as measurable and the rest as fictitious, individual values complementing the former. In such domains of individual values we define functions whose values then signify relations between individual values of the domain (range) as fixed by the function. Let us recall that the values thus set up are called 'probability values' and the corresponding functions 'probability functions'. We further say that such functions define 'probability quantities'. Physical description requires that probability functions be differentiable if possible: such are the wave functions in our above example, used for describing the distribution of particles on the intercepting screen (insofar as the squares of their amplitudes are interpreted as probability quantities). From this we can understand what new meaning attaches to differential laws if to characterize states sufficiently we must assign probability quantities to them. The laws then contain expressions for probability quantities, differentiated with respect either to measurable quantities or to further probability quantities. Such a law states for example that a differential change in such and such a measurable quantity will occasion a differential change, according to that law, in the probability that other measurable quantities characterizing the state concerned will assume such and such values. In the case where a probability quantity is differentiated with respect to some other one, the law states for example that a differential change in the probability that a state characterized by such and such values of measurement will occur is attended by a differential change, according to that law, in the probability that a state characterized by such and such other values of measurement will occur.

The difference in meaning between the two kinds of differential laws (namely classical and probability laws) presupposes an epistemological difference between the two forms of concept used for description. In order to grasp the mutation in concepts supervening with the application of the method of fictitious predicates it is not unimportant to realize that classical differential laws (laws of proximate action represented in differential form) and probability laws are logically incompatible both as to presuppositions and content. This is surely clear from our account concerning the presuppositions on which we can set up laws of proximate action and the fact that we are forced to resort to probability differential laws only where these presuppositions are not satisfied. However, in order to make the difference between the two kinds of concept forms even clearer, let us further pursue what are the consequences for modes of description if we introduce 'probability quantities'.

Amongst the presuppositions for setting up differential laws in classical form there is the assumption that all metrical quantities used to characterize phenomena are differentiable. Physically, as we already know, this means that arbitrary small changes of any quantities, under the conditions required for sufficient characterization of states, are measurable. It is decisive for understanding physical probability descriptions, that we only ever assign probability quantities to phenomena (states) if the conditions for measuring arbitrarily small changes in the metrical quantities concerned are not fulfilled; what corresponds to this in formal terms is that under the conditions that do hold the metrical quantities concerned are not differentiable. More accurately, we ought perhaps to say that no empirical physical sense attaches to the expressions denoting differential changes in the metrical quantities.

This clearly shows that the presuppositions for applying classical laws of proximate action on the one hand and probability laws on the other hand are logically incompatible. The former presuppose that to all differential expressions there corresponds a sense testable by empirical physical means (by measurement of the characterizing quantities). Where these conditions are not fulfilled (as for example wherever under conditions of sufficient characterization finite limits exist as to what is measurable in the way of changes in quantities), it becomes logically impossible to set up laws of proximate action. More precisely: if under such conditions one sets up 'laws of proximate action' after all, such 'laws' or the differential expressions occuring in them cannot in principle have an empirically testable meaning. It is under just these conditions that differential probability laws are set up, and on further analysis they turn out to be establishable only if we presuppose that changes in characterizing quantities within certain ranges are not measurable, so that nothing can be said about such 'changes'. This very clearly illustrates that the logical conditions to be satisfied by laws of proximate action on the one hand and probability laws on the other are incompatible. From laws of proximate action we must be able to derive statements (predictions) about arbitrarily small spatio-temporal domains, that is about arbitrarily small changes in the metrical quantities used for sufficient characterization, and these statements must be in principle testable and therefore have empirical content. For differential probability laws, on the contrary, there must be finite spatio-temporal domains, that is finite ranges for these metrical quantities, within which no changes testable by measurement can be derived for the characterizing quantities under the conditions concerned. In other words, for laws of proximate action, changes in the quantities concerned retain their exact mutual relations even in differential domains, while description by differential probability laws presupposes that changes in these quantities remain undefined as to their relations in differential domains. If nevertheless here too an ordering relation is to be indicated for the characterizing quantities or their changes, we know already what method must be applied: we can only stipulate relations between observed and fictitiously introduced values, in such a way that when the relations are linked with observation statements we can derive testable statements. This is the 'method of fictitious predicates'.

We remember that the presuppositions of the relativistic metric make transmission of action between phenomena impossible under certain conditions: phenomena in such a spatio-temporal domain are subject to a so-called 'uncertainty of temporal sequence', removable only by stipulating which phenomena occuring in such domains are to count as 'simultaneous'. The fictitious 'simultaneity' thus stipulated, which fixes the time relations 'earlier than' and 'later than' for the other phenomena undefined as to their relative temporal sequence, is a many-place (namely three-place) fictitious predicate. The fictitious character of the relation so chosen consists in that we can in no way ascertain or test the stipulated simultaneity (or the other time relations thereby fixed) by means of observation (measurement). Nevertheless it becomes possible, by combining these fictitious time relations with observation statements, to obtain predictions of observable phenomena. If for example we stipulate a fictitious 'simultaneity' for phenomena in undetermined temporal sequence according to the relativistic metric, then by connecting statements about the stipulated temporal relations of masses in motion with statements about conservation of momentum and energy which are already tested by observation, we obtain predictions about the changes of masses in motion, which under certain conditions are indeed accessible to testing by observation.

We must not overlook that to assign time values on the basis of stipulated fictitious time relations presupposes that the phenomena concerned are undetermined as to temporal sequence with regard to all possible measuring procedures. Such uncertainty of values to be assigned must be presupposed in general if the method of fictitious predicates is to be applied. This is no less evident when we introduce probability quantities, which is also done by stipulating certain fictitious relations. The difference in the ways in which the relativistic metric and probability metrics introduce fictitious values lies merely in this: in the former, individual values of measurement are complemented by values that are fictitious but otherwise of the same form as the others (for example, both fictitious and measurable quantities must here be continuous and differentiable); in the latter, observed values are indeed ranged into a system of individual values some of which are fictitious, but the fictitious predicates defining the probability quantities are here one-many or many-many relations, so that both in form and meaning the quantities they define are in principle different from any kind of metrical quantity.

20. THE DEFINITION OF PROBABILITY FUNCTIONS BY MEANS OF OPERATORS

In order to underline the epistemological character of the concept of 'probability quantity' as used to characterize phenomena in physics, we must get to know not only the comparison between it and the concept of metrical quantity but also the new logical step by means of which the former concept is set up and assigned to phenomena. Having recognized these steps, we shall find it easy to grasp the formal mathematical methods and concepts of modern physics to which it is often impossible to assign connections between phenomena in empirical reality.

As mentioned earlier, under certain conditions we introduce fictitious quantities that have the form of metrical ones into empirical descriptions. Such expressions of fictitious quantities (fictitious time quantities, in the example of the previous section) have the same physical dimensions as the metrical quantities which they were introduced to complement. The fictitious simultaneity stipulated for phenomena of undefined temporal sequence is a time value of the same physical dimension as time values measured by clocks. Accordingly, we stipulate for these fictitious quantities the same kind of properties as those attaching to the metrical quantities concerned. In the example mentioned these might be the properties of being continuous and differentiable. From this it follows that in such cases the same formal operations can be applied both to fictitious and to the corresponding metrical quantities.

Now it is conceivable that under some conditions or other we complement values obtained by measurement by fictitious quantities such as are not of the same kind as those values, differing either in form or in some special properties. Thus it is conceivable that under certain conditions we can measure only values that we cannot assume to be continuous. We can, however, range such measured values into a value system in which there are fictitious continuous values (that is, not measurable in any way). Such fictitious complements become even more obvious if they introduce values that are physically dimensionless or to which attaches a physical dimension quite different from that of the other values used for defining the fictitious quantity.

For example, the values measured under certain conditions for the momentum of corpuscles may be complemented into a value domain by means of assuming 'possible' (empirically quite unascertainable) values of momentum. Further, it is possible to assign arbitrarily chosen and physically dimensionless 'probability values' to the individual momentum values. Such an assignment would for example be achieved by stipulating that the individual momentum values (if we have chosen a domain with a finite number of values) have each the same probability. Here, the momentum values of the domain are stipulated to be so related that the numerical values (probability values) assigned to the individual phenomena are dimensionless.

However, it is also possible to assign probability values to individual phenomena, namely the probability that a phenomenon is to be characterized by such and such an individual value of a certain metrical quantity (say, momentum), in functional dependence from another metrical quantity characterizing the phenomena concerned. Thus we might choose a function that assigns the probability for a particle having a certain momentum as depending on the place of impact. If the domain of impact is divided accordingly into surface parts, then the spatial quantities characterizing them, such as extent, position, shape and so on may on occasion themselves be denoted as 'probability quantities' (in our example, the probability that an individual particle has a certain value of momentum). Here the defined probability quantity has a certain physical dimension, namely that of a length or a power, logarithm and so on of a length, but this 'physical pseudo-dimension' of the probability quantity is not of the same kind as that of the metrical quantity for whose values that probability quantity was defined. (In our example this measurable quantity is momentum, of dimension mlt^{-1}, that of the probability values assigned to the momentum values is of dimension l or a power and so on of l.)

The property that underlines the fictitious character of the function defining a probability quantity is its being continuous and differentiable. As already mentioned, probability quantities are defined in empirical physical description where we cannot ascertain that the metrical quantities characterizing phenomena are continuous under the conditions required for sufficient characterization, nor therefore can we in such cases significantly assume that they are; nor are they differentiable. In order that here too we are to be able to describe in terms of testable differential laws, we define new quantities, namely probability quantities, as continuous and differentiable functions of the metrical quantities that are partly not so.

In the description, probability values are assigned to the characterizing 'exact' measured values and therefore to the phenomena to be described. Assigning an exact measured value to a phenomenon without giving the probability value corresponding to that value is in these cases a description that is in principle empirically untestable and therefore empty. On the other hand we must require that in the probability description we can test by observation that the assigned probability values are continuous and differentiable, otherwise the assertion that probability quantities undergo differential changes would be just as empirically empty as the assertion that the measured values themselves do.

The empirical testing of probability values assigned to phenomena is always carried out by statistical evaluation (for example counting in a large number of instances and forming average values). To the probability quantities being continuous and differentiable corresponds the arbitrarily close approximation of statistically obtained (observed) values (average values) to the asserted probability values in a sufficiently large number of cases. We now note that because certain metrical quantities used as arguments of the function defining a probability quantity are not differentiable, this form of differential description requires the introduction of fictitious differentiable arguments, which can have a physical dimension or may be dimensionless parameters. Analysing the structure of functions defining probability quantities will exhibit the steps in which physical probability functions are defined by the introduction of fictitious arguments. This method can then be developed into a mathematical operation consisting in the application of operators of a certain form. The application of such operators, occasionally called "quantisation"[16], is equivalent to the definition of corresponding probability functions for the metrical quantities concerned. In this we must not overlook that applying these operators may result in the appearance of further expressions in turn interpretable as continuous and differentiable 'metrical quantities'. This makes it possible to apply to expressions partly obtained by quantisation the same kind of operators a second time, in which case we speak of "second quantisation"[17].

There exists of course a purely formal logical possibility to continue in this way. By applying the corresponding operators to the relevant expressions we can define probability functions for metrical quantities. In this way one arrives at differentiable expressions which can in turn be interpreted as continuous 'metrical quantities'. This has recently produced the possibility for applying corresponding operators and so on. Naturally, physical description uses these formal possibilities only as far as this furthers its purposes and the expressions obtained can be interpreted in an empirically testable way.

From this point of view it is also worth considering the extreme mode of description that would have to be applied if we could not ascertain by measurement differential (that is, arbitrarily small) changes in any of the metrical quantities characterizing states. In that case we should have to define probability quantities for every metrical quantity, and any probability quantity could be differentiated only with respect to another such. To characterize this mode of description ('the probability field') further, we also need to analyse the structure of probability functions.

21. EMPIRICAL AND FICTITIOUS EXISTENCE
OF PREDICATES

Keeping in mind that wherever we can sufficiently characterize phenomena by differentiable metrical quantities it is superfluous and pointless to introduce probability quantities into description, we can understand attempts at characterizing by means of differentiable metrical quantities (or differentiable functions of such quantities) alone, where previously an account in physically testable form could be given only with the help of probability quantities, the point being to exclude the 'less perfect' mode of description that uses these last. Now probability quantities too are defined as differentiable functions of metrical quantities, but amongst the latter there must be some which, under the conditions required for sufficient characterization, cannot be assumed to be continuous and differentiable. However, this raises the question of how to define probability quantities as differentiable functions if amongst their arguments must figure non-differentiable quantities, that is, ones that change only by discrete steps; for after all, a function is differentiated with respect to its arguments. (In saying that probability laws are to have differential form we did not mean that they must. For example, they could be set up on the basis of statistical counts alone, without regard to whether the function that mutually relates the counted values is differentiable. However, differential probability laws enable us to make much more far-reaching derivations, which is indeed why differential laws are used at all in physical description; moreover, the setting up of such laws reveals especially clearly the step that effects the ranging of only discontinuously variable values of measurement into a fictitiously complemented value system: it is the complementing that enables us to define differentiable functions in the argument domain concerned.) Probability functions are defined as continuous and differentiable. Amongst their arguments there are measurable quantities of which under certain circumstances we cannot assume that they are continuous. What scope is there for defining a differentiable function by means of quantities that can assume only discrete values in such a way that those quantities are arguments of the function and the differential changes of the function's values are empirically testable? It is here not enough merely to examine the formal mathematical possibilities for defining differentiable functions under the conditions concerned. There are obviously many such possibilities, but in defining a function for the purpose of physical description we must take into account what empirical content attaches to the expressions defined or used for definition. From this point of view we must examine more precisely the form and significance of the quantities by which physics characterizes empirical states (phenomena).

As elementary and no further analysable metrical quantities of physics we adopt 'length', 'time' and 'mass'; all other metrical quantities are defined in the form of relations between elementary ones. Such a relation is then called the physical dimension of the quantity defined. If now we assume that under certain conditions a quantity a (say, momentum) is not continuous, this does not exclude our assuming continuity for some partial quantity defining a (for example, the time 't'). If then we define a probability quantity by means of a continuous function $F(a, t)$ with the discontinuous argument a and the continuous argument t, it may be possible to differentiate with respect to t. This way of defining physical probability quantities is often used. In its simplest form, the many-place function F functionally connects individual argument values with the class of so-called 'possible' argument values. This class is 'fictitious', that is, it contains mainly values assigned to phenomena that do not occur. The values of F (namely the probability values) in this case relate occurring phenomena (that is, empirically real ones) in functional dependence to fictitious phenomena, so that, even in this simple form, F appears as a fictitious predicate.

A second mode of definition uses the possibility of admitting new 'fictitious' quantities (with a fictitious value domain) amongst those arguments for which a probability function indicates a continuous relation. Such quantities (for example amplitudes of fictitious waves) have a physical dimension and we can use them, alongside discontinuous ones, as continuous arguments in the setting up of probability functions that thus may be differentiable. In such cases certain expressions formed by means of fictitious quantities have distributions of individual phenomena (more precisely, distributions of measured values characterizing individual phenomena) assigned to them as observational data, in order to make testing possible. Thus the interference patterns on the screen of the diaphragm experiments are interpreted as distributions of particles and, being observable data, are assigned to certain expressions formed by means of the amplitudes of the fictitiously introduced probability waves. What is noteworthy here is that such observable (measurable) particle distributions count as continuous with respect to space but not time, in the usual experimental arrangements. Thus no measurable differential changes in distribution can be assigned to differential time changes which clearly illustrates the fictitious character of the continuous quantities characterizing the probability waves.

The third way of defining differentiable probability functions of metrical quantities that are assumed discontinuous, may be viewed as an extreme logical generalization possible in this setting. In this case we admit a physically dimensionless and continuous parameter (or system of such parameters) amongst the arguments of the probability function to be defined. The other arguments, whether they are empirical metrical quantities, or fictitious, have a physical dimension. If amongst the arguments of a probability function there is a dimensionless parameter, this expresses the measure in which the probability of individual phenomena depends on the individual quantities (which we call 'metrical quantities' quite generally) characterizing them. If a probability function is differentiated with respect to one of these parameters, then the differential expression states that the probability for such and such a metrical quantity assuming certain values changes in such and such functional dependence from the change in that parameter. In this case each metrical quantity is assigned a probability function, amongst whose arguments there is a continuous parameter (or a system of these). If such a function is differentiated with regard to such a parameter, the differential coefficient states the probability for this or that metrical quantity assuming certain values or undergoing certain functional changes with the parameter concerned. What values, value domain and type of change are chosen for the parameter concerned depends on what is appropriate: the aims that condition the choice of value system for the parameter are simplicity of description and the possibility of reaching predictions in a simple manner.

This mode of defining probability quantities illustrates that statements about the existence of fictitious predicates (for example, 'there is such and such a probability that...') are never about empirical and actually existing objects or their observable (measurable) predicates. This further brings out the epistemological difference between hypothetical empirical and fictitious statements of existence. If the existence of some empirical phenomena is asserted in the form of a hypothesis (for example, that some particles or waves and the like exist), then we always presuppose that the phenomena concerned can be sufficiently characterized by measured values, that is by exactly measurable individual values: that the characterizing values can in principle be determined by measurement counts as necessary condition for the empirical existence of the phenomena. That measured values satisfying this condition might be predicated of phenomena only in the form of a hypothesis does not alter the epistemological character of the (empirical) existence ascribed to the phenomena. Therefore we rightly say that hypotheses are verifiable; that is, a testing procedure will decide whether what they say is true or false. Here too it is epistemologically indifferent whether hypotheses are completely or only partly verifiable. What matters is only that hypotheses assert the existence of phenomena that are characterized only by exactly measurable individual values in logically the same way as with phenomena whose (empirical) existence is asserted on the basis of actually performed measurements (observations).

In contrast, fictitious predicates are always ascribed by stipulation. Whether an object (phenomenon) has a fictitious property or whether an asserted fictitious relation holds is in principle not ascertainable by measurement. Indeed, fictitious relations (many-place predicates) are introduced precisely where it is in principle impossible to determine anything by measurement about the law-like relations between phenomena. That is why for phenomena lying within such uncertainty ranges (for example of temporal sequence or measurement) there are many and in principle even infinitely many possibilities for arbitrarily choosing and stipulating fictitious ordering relations that are all epistemologically 'equivalent', since it is in principle impossible to find out anything by measurement about the empirical validity of the predicate statements concerned. In the sequel the ascription of a fictitious property or the choice of a fictitious relation or system of relations (for example a system of probability metric) may of course turn out to be inappropriate, since fictitious statements (for example about 'simultaneity' of phenomena in uncertain temporal sequence) can be linked with other observation statements and predictions be derived from them. If these last mainly fail to come true, the stipulated fictitious relations count as inappropriate and might accordingly be modified. It would however be misleading if we were to conclude from this that it was after all decidable by experience which system of ordering relations held between phenomena in uncertainty ranges, no less than with connections where the descriptive laws relate only measurable quantities and whose validity is thus decided by measurement. The logical difference between the two modes of description, that is, by empirical and fictitious predicates, let us repeat, consists in the first always stating relations between quantities that are in principle measurable and the second between quantities some at least of which are not accessible to measurement and can therefore be introduced or modified only by stipulation. This is true even if the fictitious predicate statements when linked with observation statements lead to predictions that are not confirmed and the fictitious system of relations is rejected as inappropriate. The uncertainty range, in which the phenomena to be ordered lie, continues to exist unchanged if the stipulated fictitious order turns out to be inappropriate for the phenomena. This is decisive for the fact that in every case an order for the phenomena in the range can be given only by stipulation. For the uncertainty in the relations between phenomena states that nothing can be determined by observation or measurement as regards certain properties of, or relations between, the phenomena concerned. The fictitious character of predicates thus introduced then appears to be based on the uncertainty in principle of the phenomena as regards the relevant properties and relations. Since, as we said, this uncertainty remains, however satisfactory stipulated fictitious systems of ordering relations may subsequently prove, phenomena of this kind can be described only if we introduce (stipulate) fictitious predicates.

From this point of view we gain a better grasp of the endeavours in physics to exclude probability descriptions.

22. THEORIES OF MEASUREMENT AS A PRESUPPOSITION OF EXACT-CONTINUOUS
AND OF PROBABILITY DESCRIPTION

Physical description uses probability quantities wherever it is in principle impossible to ascertain arbitrarily small ('differential') changes in metrical quantities under the conditions required to characterize phenomena sufficiently. Probability quantities are defined by functions amongst whose arguments figure the relevant metrical quantities. It follows that every physical theory using probability quantities for description must indicate a corresponding theory of measurement that has to show precisely under what conditions a continuous change in the relevant metrical quantities is not measurable, which then enables us to use these last as discontinuously variable arguments of a probability function to be set up. From this point of view we can say that a physical probability description (and therefore also the probability functions defining the quantities concerned) receives its empirical physical sense only relatively to a corresponding theory of measurement. Accordingly the definition of a physical probability quantity is always framed with a certain theory of measurement presupposed. From this connection between probability quantities, metrical quantities and theory of measurement certain conditions result that physical theories must generally satisfy.

The theories of classical physics, amongst which we here count the theories of relativity, do not contain theories of measurement that indicate limits to accuracy of measurement. Thus they give no limits for the ability to ascertain differential changes of metrical quantities under the conditions required for sufficient characterization of phenomena. (Relativity theory does indeed base the introduction of its metric on a theory of measurement, namely that measurements of lengths and times can generally be carried out by means of signals. From the theses of this theory it follows that the temporal sequence of certain phenomena is uncertain, which is a prerequisite for introducing fictitious time relations. However, the relativistic theory of measurement does not concern accuracy limits of measurement). In physics, indicating limits for the accuracy of measurement is a prerequisite for introducing and defining probability quantities. Because of this limitation to accuracy and the attendant consequences, one school of thought in theoretical physical enquiry sees the use of probability quantities as a defect in physical description and therefore tries not to use in descriptive laws expressions that are interpreted as probability quantities: rather, the only expressions to be used are those that directly characterize individual phenomena as continuous metrical quantities.[18] In the light of our new insights we can indicate what condition every experiment must satisfy, if it is to yield a description of the course of phenomena testable without the use of statistical data. It is this: each physical theory must contain a corresponding theory of measurement that can actually be carried out. This condition is necessary because when we define probability quantities as elements of physical description we presuppose in principle that there are certain actually existing conditions of observation (measurement). Differential laws express relations between differential changes in the 'describing' quantities. If these last are all of them metrical, then we must demand that arbitrarily small changes in all of them should be measurable, under the conditions required for sufficiently characterizing phenomena. In that event we say that the metrical quantities are continuous, but if the demand is not satisfied, then we denote them as discontinuous. In order to obtain differentiable descriptive quantities even in these cases, we here define probability quantities as continuous relations (functions) between metrical quantities some at least of which are discontinuous.

If therefore, in physical laws, expressions construed as probability quantities (in the sense explained above) can no longer be so construed (because of existing conditions of measurement), but must be understood as expressions for metrical quantities interpretable as continuous under the crucial conditions, then this 'reinterpretation' can be given only if one indicates a new theory of measurement. One must indicate measuring procedures, with whose help one can measure the arbitrarily small changes in the relevant metrical quantities with sufficient accuracy in practice, under the conditions concerned. Similarly for the case where new physical theories are set up in order to describe by means of continuous metrical quantities alone phenomena whose course could hitherto be described only with the help of probability quantities. For the new metrical quantities which may occur in such theories one must always indicate a theory of measurement that shows how one is to measure the arbitrarily small changes of metrical quantities required for a sufficient characterization of phenomena under precisely the crucial conditions. Failing such a theory of measurement for the new laws, it becomes doubtful how to derive testable predictions from them and therefore how far such laws describe empirically real phenomena at all. It is instructive to clarify these circumstances in terms of an existing theory. However, let us first give a summary of the elementary preconditions for applying the method of fictitious predicates.

23. THE EPISTEMOLOGICAL CONDITIONS FOR APPLYING THE METHOD OF FICTITIOUS PREDICATES

We have said that stipulating an order for phenomena by means of fictitious predicates becomes necessary wherever it is impossible to attain sufficient characterization of phenomena, by measurement, whatever the reason. Thus we must know the precise conditions under which in the case concerned we can or cannot measure, if we are to be able to apply the method of fictitious predicates. This naturally holds not only for cases in which under the decisive conditions it is impossible to measure changes in quantities within specified uncertainty ranges, but also for cases where these ranges arise from the finite speed of propagation of the transmissions of action used for measuring. It follows that description by means of fictitious predicates presupposes that we indicate a theory of measurement that is actually practicable and yields the necessary appearance of uncertainty ranges. Only on that precondition can we obtain unambiguous empirically significant propositions by applying the method of fictitious predicates.

Thus in relativity theory the propositions derived for phenomena in uncertain temporal sequence by stipulation of fictitious time relations receive empirical content only if we indicate that measurement of time and length are to be carried out by signals, that is, transmissions of action propagated at finite speed. From this theory of measurement first we learn within exactly what domains (ranges) phenomena are in principle not characterizable by quantities of measurable value, these being the uncertainty domains. Secondly, the theory of measurement by signals shows in what way the stipulated fictitious values and observed values can be combined into an ordered value domain.

In just the same way physical indications gain empirically testable content only through a corresponding theory of measurement; they too link observed values with fictitious ones. The thesis that measurement of position, momentum, energy, time occurs by means of action relations between corpuscles and electromagnetic waves again shows the limits below which phenomena under the decisive conditions can no longer be characterized by arbitrarily small but measurable value changes. Here too the theory of measurement tells us which values are obtained by measurement and in what way they can be complemented by fictitious values to make up an ordered value domain. The probability relations (functions) defined in these domains receive their testable meaning only in view of what the theories of measurement indicate as to the kind of observed values that can be obtained under the individual conditions or as to when uncertainty domains necessarily arise.

Since we know how applying the method of fictitious predicates is connected with indications of theories of measurement, we can now adopt a firm position as to theoretical attempts aiming at the exclusion of differential probability laws from physical description.

L. de Broglie[19] has lately set up what he calls a 'theory of double solution' whose goal is to view 'guide waves', that are normally interpreted as statistical distributions of particles, as actual continuous waves, according to a certain formal mathematical modification. Epistemologically these waves would be real phenomena of the same kind as electromagnetic waves (energy waves) in classical electromagnetism. The new wave equation given by de Broglie in his theory is to allow us to predict individual phenomena exactly (for example individual paths of particles) even where on the usual interpretation of wave mechanics only statistical predictions can be made.

The decisive step towards this solution de Broglie sees in the new 'non-linear' form of his wave equations, which he introduces as general wave form with components that can be shown to be the traditional linear wave equations.

For describing all those phenomena whose description in quantum physics requires the dual conceptual form of 'wave-corpuscle', according to the complementarity principle, de Broglie assumes a general non-linear wave function u which can formally be split into two parts u_0 and v.[20] u_0 is a nonlinear wave function that assumes large values in very small domains (so-called 'singular domains') but falls off rapidly outside these. Thus u_0 marks the very small domains (and their motion) corresponding to particles. v is a linear wave function of the same form as the 'guide wave' ψ of wave mechanics, except for a factor of proportionality, so that $v = C\psi$. Now de Broglie puts $u = u_0 + v$, the two partial waves not being independent, precisely because of non-linearity. Thus there is a continuous mathematical dependence between u_0 and v, to which a continuous causal dependence between particle and 'guide wave' is to correspond. In contrast, traditional wave mechanics lacks causal dependence between corpuscle and the linear guide wave ψ; rather, the latter merely indicates possible statistical distributions of the particles.

This discontinuous causal dependence between corpuscle and guide wave in traditional wave mechanics shows up not least in that, because of the statistical interpretation, differential changes in the guide wave function do in principle not represent changes of states in arbitrarily small times. Rather, what corresponds to differential changes in the quantities (fictitious or probability ones) characterizing the guide wave functions are differential changes in the distribution of corpuscles or of the individual values characterizing them (individual states). Such differential changes in distribution however occur only if we have observed a sufficiently large number of individual phenomena; but the multiplicity of individual phenomena is only ever given in a finite (sufficiently long) time. It follows that differential changes in probability quantities characterizing states can be construed only as statements about a multiplicity of states, whose occurrence is spread over a finite length of time.

It is precisely this feature that distinguishes differential changes in probability quantities from differential changes in metrical quantities as they appear in classical differential laws (laws of proximate action). The latter changes concern changes that occur within arbitrarily small times and are in principle measurable. For differential changes in probability quantities this is logically impossible. The empirical content of such a change is merely that the individual values to be observed in a finite time characterizing states occuring in that time now have a different average value (distribution). While therefore differential changes in metrical quantities indicate changes of state in arbitrary small (differential) times in the classical sense, differential changes in probability quantities denote changes in distribution that in principle will not occur until some finite time has elapsed.

If therefore a theory, like de Broglie's above, aims at excluding description by probability quantities and putting in its place a so-called 'causal' account (that is, by means of continuous, and differentiable metrical quantities), then along with the characterizing metrical quantities concerned we must be given a corresponding new theory of measurement. Failing this, that is, if the traditional theory of measurement is retained, then amongst the continuous and differentiable 'metrical quantities' exclusively involved in the new account there must be some whose differential changes are in principle inaccessible to measurement. These 'hidden' metrical quantities ('hidden parameters') can at best be interpreted as probability expressions in their turn, otherwise the new theory floats in thin air, that is, its statements are in principle untestable by observation.

It is undeniable that in the 'theory of double solution' the wave u_0 describing particle motion and the 'guide wave' v depend on each other in a formal mathematically continuous way, but how can this dependence be tested by observation (measurement)? The answer decides whether the expressions newly introduced by de Broglie are empirically interpretable as differentiable metrical quantities or once more only as probability expressions. In the latter case the differential changes in de Broglie's 'hidden' wave quantities would merely mean changes in the distribution of average values occurring in finite time, which is precisely what he wants to exclude by means of his theory. It is in this light that we must judge de Broglie's attempt to demonstrate that his theory is empirically testable. In this connection he points to the possibility that by means of the wave u one can describe refraction and interference phenomena of photons as continuous processes, as with other particle waves[21]. However, as we already know, this would mean that the laws describing these experiments (Young's on refraction and those on the scattering of corpuscular rays), namely the wave equations for u or u_0 and v, had the form and meaning of classical laws of proximate action. Thus we should have to view the 'quantities' whose differential changes are related by the functions u, u_0 and v as continuous metrical quantities arbitrarily small changes in which must be measurable under all conditions required for predictions. In particular it would have to be possible accurately to measure arbitrarily small changes in the metrical quantities figuring in u, u_0 and v (such as speed of propagation, wave lengths, frequencies, amplitudes, energies) precisely in those cases where we are to describe or predict the motion of individual particles (that is, their individual paths) in arbitrarily closely neighbouring spatio-temporal domains. However, the measurability of the relevant characterizing values could at best be based on a new theory of measurement, deviating from the traditional one. Such a theory would have to indicate de Broglie's 'theory of double solution', if statements on continuous changes in particle motion derivable from it are to be testable and therefore have empirical content. The required theory of measurement would have to indicate procedures precisely for the case of changes in individual particle motions in arbitrarily small spatio-temporal domains, in such a way that the predicted differential changes (derived from the new theory) in the characterizing measurable quantities can be captured with sufficient practical accuracy. Failing such a theory of measurement, it becomes doubtful how far the 'theory of double solution' has new empirical content as against traditional wave mechanics statistically interpreted. For if we retain the current theory of measurement in wave mechanics for the 'theory of double solution', then here too there are restrictions on accuracy for measurements that can actually be carried out in the decisive cases. This necessitates the appearance of uncertainty domains for which we must in any case stipulate a probability metric if we are to obtain testable predictions. The propositions derived from the new theory can again at best be verified only in a statistical sense. The new content of the theory could then at most consist in specifying a new (modified) statistical metric as against the existing wave mechanical statistics. However, the laws of the theory would again be only differential probability laws, against de Broglie's view. Thus the 'theory of double solution' is subject to the general condition valid for any theory in which one tries to exclude probability quantities from physical description: the theory must indicate a theory of measurement corresponding to the (newly introduced) continuous metrical quantities whose functional relations are to replace the traditional probability quantities; and the theory of measurement must allow the testing by practical measurement of predicted differential changes in state for arbitrarily small spatio-temporal domains.

New theories that aim to replace probability description by a continuous deterministic one (using laws of proximate action with the help of continuous metrical quantities only) can thus attain their goal only by indicating a new theory of measurement that would have to differ in principle from those currently adopted in quantum physics. This difference concerns the current rule that, under the conditions required for sufficiently characterizing states, the metrical quantities concerned are not measurable down to arbitrarily small changes. If this rule is retained in the theory of measurement belonging to a theory newly set up, it follows that its expressions and predictions are again only statistically interpretable and testable; or else, since a corresponding practicable measuring procedure is lacking, the metrical quantities figuring in the theory assume the character of 'hidden quantities' ('hidden parameters'), that is, expressions totally beyond testing, which makes the theory empirically empty.[22]

That changes in the characterizing metrical quantities may under certain circumstances not be captured in arbitrarily small ('differential') domains results in the individual values of those quantities remaining uncertain in the ranges concerned, as previously mentioned. This uncertainty can be removed only by stipulating fictitious relations between observed values and introduced fictitious values. In the case of quantum phenomena the relations (fictitious predicates) are so defined within the established value domain that states are characterized by disjunctively linked value ascriptions, that is, by a disjunctive class of propositions. This is one of the main logical differences between the ways classical and relativistic physics on the one hand and quantum physics on the other characterize phenomena. If there are no restrictions on empirically coping with differential changes in characterizing metrical quantities, then states can be sufficiently characterized by conjunctively linked value ascriptions. Only in that case does a complete, continuous, deterministic description become possible. If under any conditions differential (arbitrarily small) changes in the characterizing metrical quantities defy measurement, we can fix only individual values for the uncertainty range, with co-ordinated probability values; as we know already, this always takes the form of disjunctively linked value ascriptions characterizing states.

Now it is worth noting that the two ways of characterizing states (by conjunctively or disjunctively linked value ascriptions) can occasionally be used several times in alternation in the description of phenomena. This partly coincides with alternative applications of wave and corpuscle descriptions in certain domains of micro-phenomena. The epistemological form of this mode of description by two mutually exclusive conceptual systems, which corresponds to sufficiently characterizing states by conjunctive and disjunctive classes of propositions, we shall discuss later. Here we merely observe that the conjunctive case might involve only empirically catchable values, or also these in stipulated connection with introduced fictitious values (for example, in the case of relativistic stipulations, fictitious time values for phenomena in uncertain temporal sequence). As against this, the disjunctive case always requires a stipulated connection between empirically catchable and introduced fictitious values. Characterization by probability quantities is one of this last kind.

The recognition that using probability quantities in physical description rests on theories of measurement and that the elimination of probability quantities therefore requires an indication of appropriate theories of measurement capable of being implemented in practice, allows us to take a further logical step by means of which the concept of 'probability quantity' is introduced into physical description as a universal conceptual tool.

24. THE PROBABILITY FIELD

Physical probability quantities are introduced where, under the conditions required for sufficiently characterizing phenomena, it is impossible to observe the continuity of measurable quantities. In such cases we define a continuous differentiable function (relation), to whose arguments belong the measurable quantities assumed as discontinuous. Functions of this kind define probability quantities by stipulating a probability metric which ranges the measurable quantities concerned into a number system (of 'possible' or purely 'fictitious' values of measurement). The values of the probability function fixing the probability metric then count as the individual values of the probability quantity defined. What characterizes probability functions (for example functions of probability waves) are parameters with whose help we fix the relation between arguments that is to count as a probability quantity. For the parameters there are no rules as to what physical dimension, if any, they should have, they can be dimensionless. Let Ψ be the function stipulating a metric, a_i its arguments associated with measured, possible or fictitious values, λ the parameter figuring as argument (in the case where one parameter suffices; more complicated cases with systems of parameters involve no new logical steps); then the function representing the metric may be written $\psi\,(a_i\,,\lambda\,)$.

We are free to choose what relations between values of measurement we wish to define as probability quantities (with the help of parameters). Leaving aside for the moment the custom in physics of introducing probability quantities only where, under the conditions required for sufficiently characterizing phenomena, it is impossible to observe continuity in the metrical quantities concerned (for example in so-called 'conjugate' quantities), we can consider the possibility of defining a probability quantity for every metrical quantity by choosing a suitable probability function. The function defining the probability metric can be introduced in such a way that it assigns a precise probability value to every individual value of the metrical quantity concerned. This would mean that under given conditions of measurement the probability of obtaining a certain individual value is such and such. In this sense we can conceive of defining a probability function for every physical metrical quantity. This ranges the individual values of a metrical quantity into a metrical system in such a way that they belong to the argument values of the probability function whose values can be understood as numerical probabilities for obtaining the individual values of measurement.

The probability functions chosen for the individual values of measurement can be defined as differentiable (as indeed happens in wave mechanics). Still, if we fix probability functions for all physical metrical quantities in the sense explained, then the new question arises, how far descriptive laws can in this case assume the form of differential laws, since they now contain relations only between probability quantities. This problem is perhaps best understood if the possibility of field description using only probability laws (in the sense explained above) is logically analysed.

If we characterize physical states and their changes by means of a field description, then (in the classical mode of description) all field positions and their changes are unambiguously marked by metrical quantities and their changes. If, as above, we now assign to each metrical quantity a differentiable probability function and therefore probability quantity, all field positions and their changes would be marked by such quantities and their changes. The differential laws describing the field would then be differential probability laws without exception, and such a field will be called a 'probability field'.[23]

The conceptual scheme of the 'probability field' thus conceived is a logical extreme attained by pushing the quantum physical concept of 'probability quantity' to its logical conclusion. What comes nearest to this amongst quantum physical theories is, I think the 'q-number mechanics' of Dirac.[24]

Here the quantities characterizing a state, including time, are defined as 'q-numbers'. From our viewpoint the functions defining q-numbers are probability functions by means of which the values of metrical quantities, to be obtained by measuring, are ranged into systems of probability metric. We can view this as an assigning of a probability quantity (in the form of a q-number) to each 'metrical quantity'; the field positions then appear exclusively marked by probability quantities (q-numbers). A field described by such quantities (or their functional connections) alone would then be probabilistic in the sense explained.

If the quantities marking the field positions of a probability field, which are all of them probability quantities, are defined by differentiable functions – which need not be the case –, then the differential laws describing field changes relate differential changes in probability quantities only: the differential laws then contain only probability quantities differentiated with respect to other such. Since, as shown, probability quantities are always defined by functions, in the differential laws defining a probability field we always differentiate functions with respect to other functions. According to the rules of differential calculus this requires that each function be differentiated with respect to its arguments as well. Amongst the arguments of physical probability functions there are always metrical quantities too. If now we extend the quantum physical conditions for introducing probability quantities – that is only where arbitrarily small changes in relevant metrical quantities (for example 'conjugate' quantities) are unmeasurable under the conditions required for sufficiently characterizing phenomena –, to take the descriptive form of the probability field, it follows that the probability functions describing the field are in general no longer differentiable with respect to the metrical quantities appearing as arguments. For in the probability field all metrical quantities without exception have probability functions assigned to them, amongst whose arguments these quantities figure. In particular this holds for the metrical quantity.'time' presupposed as neither continuous nor differentiable. For every metrical quantity we thus define a probability quantity. However, after what has been said, this seems justifiable only if under the conditions stated we cannot measure arbitrarily small (differential) changes in the metrical quantities: these latter here count as neither continuous nor differentiable so that we cannot differentiate with respect to them.

Accordingly, differential laws as currently used in wave mechanics can no longer occur in probability fields (these laws often contain probability functions or quantities differentiated with respect to metrical quantities, above all 'time', which assumes these last to be continuous). For metrical quantities appear in probability fields only as arguments of probability functions and are taken to be discontinuous ('time' included) under the conditions required for sufficiently characterizing phenomena.

Thus probability field functions (to consider this logically extreme mode of description) cannot be differentiated with respect to metrical quantities. Nevertheless the descriptive field laws can be differential here too, since a probability function presupposes a probability metric for certain values to be obtained by measurement. In the metric the values of measurement are ranged into a number system (of so-called 'possible' or purely 'fictitious' values). At the same time the probability function defines a continuous relation between the values of the metric. (It is precisely such relations that we call 'probability quantities'.) In setting up such relations we may use continuous and differentiable parameters which will figure amongst the arguments of the defining probability function. Let us call such parameters 'scattering parameters'. They can be introduced as single fictitious quantities or as functions thereof, for example as fictitious vectors in a 'space' of infinitely many dimensions. In any case, they are selected for expediency, as we know already. In complicated cases probability quantities may be defined with the help of a system of parameters. If we choose a fictitious vector in a space of infinitely many dimensions as a parameter, then for any metrical quantity we can define corresponding co-ordinate system with infinitely many axes where the vector's components along the axes yield a value system of partly fictitious, partly possible values of measurement. For the latter, with suitable choice of the vector and co-ordinate system in agreement with conditions of observation and observed values, we obtain a discontinuous order so that we cannot differentiate with respect to them. If in addition we adduce the fictitious vector components or their numerical values, we do, however, obtain a value system in which we can define differentiable expressions representing the differential changes of the vector. From the vector's values or their changes we can derive for the possible values of measurement probability values or their changes.

In the usual differential probability laws of quantum physics, probability functions are often differentiated with respect to time. Thus they state the change in probability of obtaining such and such a possible value of measurement for a differential change in time. However, in the extreme case of probability fields we cannot differentiate a probability function directly with respect to time, since time like any other metrical quantity is here to be discontinuous. Thus the probability functions assigned to metrical quantities in the probability field (for example functions of fictitious vectors) can be differentiated only with respect to continuous and differentiable parameters. In every case it is possible to introduce the parameters, used in the functions defining probability quantities (such as certain vector components in space of infinitely many dimensions), as continuous arguments of these functions, that is, arguments with respect to which we can differentiate. Therefore the probability functions in a probability field can be differentiated only with respect to parameters with whose help the functions define continuous relations between the values of measurement. (These relations then are probability quantities, as previously explained). This determines form and content of differential laws in a probability field.

The laws here state in every case that to differential changes in probability quantities (assigned to one kind of values of measurement) here correspond certain differential changes of another probability quantity (assigned to some other). The resulting predictions are tested by observing (measuring, counting) only distributions and their changes. Relations between differential changes in probability quantities (the only relations expressed by differential laws in probability fields) thus depend on differential changes in the parameters used to define the quantities concerned. If (statistical) observation shows that the change of distributions do not on average agree with the derived probability values, then we must choose different scattering parameters and in addition perhaps other functions for defining probability quantities, that is a new probability metric.

The 'probability field' thus conceived is a possible mode of a description and is to be understood only as such. We are by no means asserting that physics must use this mode or even that it would be expedient. We merely wish to show that the concepts of 'probability quantity' and 'differential probability law' used in physics can be logically generalized and that this leads to the logically extreme form of the 'probability field'. However, individual physical enquiry must decide what mode is most suitable for empirical physical description.

25. PROBABILITY DESCRIPTION AND INDETERMINACY OF PHENOMENA

In a sense we may reckon probability quantities amongst the quantities that characterize physical states. They are defined by fictitious predicates that are stipulated relations between measured and fictitious values. Characterizing states only by metrical quantities on the one hand and characterizing them with the additional help of probability quantities on the other are two kinds of procedure that are currently applied depending on prevailing conditions of observation, nor does empirical enquiry proceed in a uniform manner here. Within the same domain of phenomena states are described now by means of continuous metrical quantities alone, now with the additional help of probability quantities. The unsatisfying aspect of this procedure is the logical incompatibility of the two conceptual systems used, a fact especially evident in the various prior assumptions or in the meanings of the two forms of differential law.

If we describe in terms only of metrical quantities, the differential laws (in their most perfect form 'laws of proximate action') express changes of state in arbitrarily small spatio-temporal domains, while we must presuppose these changes to be in principle measurable if the laws are to have empirical content. Here we can always indicate an unambiguous relation between two states which is always a differentiable functional connection exclusively between continuous metrical quantities. This kind of connection between states (especially temporal sequence) we call strictly deterministic changes of state, so that strict determinacy of changes of state or of the course of phenomena must be presupposed if we are to describe a sequence of phenomena with the help of continuous and differentiable metrical quantities alone. By 'strictness' of determinacy (causal dependence) we must understand that the changes of state are derivable and empirically testable within arbitrarily small spatio-temporal domains (more precisely: domains of quantities). What further marks this kind of description is that sufficient characterization of states is always given by value ascriptions that are conjunctively linked.

As against this mode of description by means only of differentiable metrical quantities, the assigning to them of probability functions presupposes that under certain conditions we cannot measure any changes of the characterizing metrical quantities within certain finite domains (quantity ranges). This presupposition can be expressed in two ways: firstly, that the states themselves are no longer precisely specifiable within certain quantity ranges, or that within certain spatio-temporal ranges there can be several possible values for a state; secondly, that within certain quantity ranges nothing at all can be said about changes of state and their relations, so that to observable orders of phenomena above these domains we cannot unambiguously assign states or their changes within the critical quantity range. The indeterminacy of states and of their changes in those domains here too allows us to state several possible values for the same state.

Whether in such cases we express the conceptual presuppositions for characterizing states in the first or second manner, ascriptions of characterizing values are in any case given in disjunctive connection. It follows that within the critical spatio-temporal domains every characterization of states by quantities must be fictitious in kind. Accordingly the process is such that to an observable order of phenomena, or to the corresponding observed values, one assigns a system of fictitious values. Observed and fictitious values are thus ranged into a comprehensive system in which one defines relations by means of which numerical relations, namely probability values, are assigned to the possible values of measurement, and these assigned values are testable by statistical counts or measurements of distributions. What is, however, logically decisive for this kind of probability description is that, within the spatio-temporal domain for which a fictitious value system is stated, it is in principle impossible to make, for individual phenomena, predictions testable by measuring characterizing values. This we must presuppose in order to apply the method explained, and to be able to interpret and test as merely being statistical the statements derived by applying it. This amounts to saying that processes in these spatio-temporal domains necessarily remain uncertain or indeterminate.

In the case of probability descriptions as applied in quantum physics, the indeterminacy of states and their changes in certain domains counts as that 'uncertainty' for value ascriptions which is a logical precondition for apply-ing the method of fictitious predicates. The indeterminacy of states thus presupposed shows especially clearly that the two concept systems cur-rently used for description in microphysicus are incompatible. As men-tioned, description with the help of continuous and differentiable metrical quantities presupposes strict determinacy of all changes of state in arbitrar-ily small spatio-temporal domains. This much holds even where in order to characterize states we must adduce fictitious value ascription, because let us say, certain phenomena are in uncertain sequence; so long as the ascribed quantities have the form of continuous and differentiable 'measu-rable quantities', whose values are ascribed to states along with the other values of measurement and linked with these conjunctively. In such cases too description presupposes strict determinacy of changes in state, not-withstanding the fact that here a part of the states and their changes is characterized by fictitious values. The decisive critical work of strict deter-minacy in the phenomena described is that arbitrarily small changes in state be measurable and therefore the 'metrical quantities' differentiable, these last alone being used to characterize sufficiently the states to which they are directly assigned. Laws that put only such quantities into differenti-able functional relation allow unambiguous derivations within arbitrarily small spatio-temporal domains, in the sense of which definite singular state will follow a given exact and definite individual state.

In contrast, as shown above, description with the help of probability quan-tities necessarily presupposes indeterminacy for states and their changes within certain value ranges. In such domains of quantities it is in principle impossible to characterize states sufficiently by means of exact individual values of measurement, that is by such value ascriptions conjunctively linked. For such cases one therefore gives a characterization by means of a system of disjunctively linked ascriptions of individual values. In systems accordingly set up by introduction of fictitious values one can define proba-bility functions that assign probability values to the individual values of measurement. Only such ascriptions can be tested under the conditions ex-plained, but probability values can in principle not be tested by observing one individual case alone. However, for description by laws of proximate action it is precisely changes of state in differential spatio-temporal do-mains (more generally, quantity domains) that count as those 'individual cases' to be predicted and tested, and the very conditions for the occurrence of such individual cases and therefore the predictions that they will occur can in principle not be unambiguously derived from a probability law. Rather, propositions (predictions) derivable from probability laws alone that concern individual cases always concern a set of individual cases and can be tested only by such a set. Thus every prediction derived from a probability law leaves it open (undetermined) whether an individual case will or will not occur. The occurrence under certain actual conditions of an individual case predicted with a certain probability no more confirms the probability law from which the prediction was derived than the non-occurence of the case would refute it. As regards the individual case, the probability law is compatible both with its occuring and with its not occuring. This means that probability description presupposes that individual events are undetermined or indeterminate under all conditions given for observation (measurement).

This logical opposition in the presuppositions for describing on the one hand by means of continuous metrical quantities alone and on the other hand with the help of probability quantities in addition stands out particularly clearly when phenomena of both kinds of description are used, as it were, in turn. In physics one usually refers to this as the 'double nature' of the relevant phenomena, which under certain conditions are described as corpuscular and under others as undulatory. Without further explanation, however, the contrast of corpuscle and wave is not enough to mark the unbridgeable logical opposition of the two modes of description. For both corpuscular and wave phenomena may on occasion be described by means of differentiable metrical quantities alone, and likewise one can conceive conditions under which both types of phenomena can be sufficiently characterized using only probability quantities.

What is rightly felt as an unbridgeable logical 'break' in quantum physics or wave mechanics when we go from corpuscular to wave description, consists in the special step here taken: characteristic quantities hitherto construed as continuous metrical quantities are now interpreted as arguments of a probability quantity; or, conversely, expressions hitherto used as probability quantities are now interpreted as differentiable metrical quantities characterizing certain phenomena. An example of the first case is the transition from specifying a corpuscle by simultaneous position and momentum – which at first count as differentiable metrical quantities for all cases – to specifying the particle by means of wave functions that are viewed as probability functions whose values are assigned as probability values to individual position and momentum values. The quantities characterizing the probability wave are not metrical quantities assigned to actual phenomena, but rather indicate average distributions of the values of measurement (of position and momentum) characterizing the corpuscles.

Conversely, probability quantities (say, those marking a probability wave) may be assigned as exact metrical quantities to an actual phenomenon. One might for example take quantities originally characterizing a fictitious ('probability') wave, supplement them by further expressions and finally interpret them as metrical quantities describing a real energy field. Such transitions from a description solely by differentiable metrical quantities to one by probability quantities, or conversely, receive unambiguous sense only if a corresponding theory of measurement is indicated, as we know already. If in a given case the conditions of the theory of measurement are indicated, that already decides whether the domain of phenomena concerned can be described only by differentiable metrical quantities. However, it frequently happens that performable measurements can be indicated only for limited partial domains of a connected domain of phenomena, while measuring procedures and results point partly to the assumption of continuous quantities and partly to the view that only average values (distributions) can be ascertained. Currently available results of observation are such that we cannot assume that it would be expedient for enquiry to apply just the one mode of description throughout. There are indeed attempts to unify the form of description and concepts, witness de Broglie's 'theory of double solution' discussed above, but it has so far proved impossible adequately to confirm these theories by observation, not least because there is no corresponding theory of measurement.

In examining the logical foundations of physical knowledge we must take into account that current physics applies description by metrical and probability quantities in turn in somewhat irregular fashion, so that we cannot at present give a formally unified theory in the domain of quantum phenomena. There is indeed some measure of uniformity in the devices of method applied in microphysics whenever arbitrarily small changes in quantities describing states are in principle unmeasurable under the conditions required for sufficient characterization. The discrete values that are measurable are then always ranged into systems of fictitious values – for example matrices assigned to metrical values –, in which we define probability functions whose values indicate for possible individual values of measurement what the probability is of obtaining them under the given conditions of measurement. This is a special form of the method of fictitious predicates as applied in uniform fashion in quantum physics, where it gives a uniform epistemological character to the differential probability laws. Still, we cannot deny that attempts to eliminate probability quantities from physical description and to use only differentiable metrical quantities in characterizing phenomena, are based on notable if insufficient empirical data. That is why at present the logical opposition explained above, namely that between the manner outlined of applying fictitious predicates for forming concepts or describing phenomena and that of applying only continuous and differentiable metrical quantities or relations between them, makes it impossible to attain theories with uniform concept formation in the relevant domains of phenomena. We shall later examine what directions remain open, given the current state of enquiry, for attempts to reach comprehensive theories with uniform concept formation.

26. THE LOGICAL MEANING OF QUANTISATION. FIRST AND SECOND QUANTISATION

One of the most remarkable epistemological steps in modern theory formation is repeated 'quantisation'. As originally introduced into our description, by means of the hypothesis 'that under special conditions action is transmitted only in integral multiples of a least action, 'quantisation' might simply have been viewed as the assumption of a special form of empirical events. Even when applied to the most varied domains of phenomena, such as corpuscular motion and light, 'quantisation' did not at first have to be construed as anything other than the assumption that phenomena take a quantised course.

A somewhat wider interpretation suggested itself when attempts at describing quantum phenomena with the help of continuous metrical quantities led amongst other things to the result that the differential laws (wave equations) thus set up had to be eked out with additional arbitrary (that is, empirically quite unjustifiable) rules if values with empirical meaning were to be derived, and these were found to be testable only statistically. This was the more striking since, beside these attempts by L. de Broglie and E. Schrödinger to derive quantum phenomena from continuous wave processes, the description of the same phenomena was given by W. Heisenberg using different forms of concepts: to the measurable quantities required to characterize states sufficiently, he assigned 'matrices', which logically must be viewed as systems of values that are partly possible and partly purely fictitious, or as functions defined in such systems. The values obtained by matrix operations are probability values which indicate for individual possible values of measurement how probable it is that we should obtain them under certain conditions of measurement. These operations, defining a probability function for each metrical quantity and thus ranging the values of measurement into a partly fictitious value system, show with special clarity that the method of fictitious predicates is being applied.

It might be thought that the above-mentioned attempts at representing events in atomic domains by means of wave processes succeed in avoiding the need to use fictitious predicates (probability functions) in description. Wave-mechanical theories obtain differential equations for wave phenomena. The expressions marking wave states are differentiable, as are the symbols ('metrical quantities') figuring in the wave equations. It, looks therefore as if the differential laws express something about arbitrarily small changes in state thus obviating the need for using probability functions in empirical descriptions and thus for introducing corresponding fictitious predicates. However, on examining more closely how de Broglie and Schrödinger set up their wave equations or under what conditions we can derive testable predictions from them and how these are tested, we see clearly in this analysis the steps by which, even in this wave description, fictitious predicates defining probability functions are introduced and probability values are assigned to the metrical quantities characterizing states.

The de Broglie-Schrödinger theory starts from equations of classical mechanics, that is, equations by which classical mechanics would describe corpuscular processes or states. The theory now modifies these equations, formally replacing the metrical quantities (state quantities) such as momentum, energy and so on by expressions in which certain operators (for example the gradient operator) are applied to an arbitrarily defined function to whose arguments the relevant metrical quantities belong: the result is a wave equation. In the operators used, Planck's constant h occurs as a coefficient, so that applying the operators amounts to a quantisation, which is usually called 'first quantisation'. The Schrödinger function ψ, to which the gradient operator for example is applied with the coefficient h

(the resulting expression then being put in place of classical momentum), has as its arguments 'space co-ordinates' which are, to begin with, assumed to be continuous and differentiable; they are therefore 'fictitious' under certain conditions of measurement, since under the conditions required to characterize states sufficiently we can measure only discrete position values (space co-ordinates).

Agreement between values derived from the differential equations and observed values and therefore empirically testable content for these laws is achieved precisely by applying the operators. The resulting wave equations admit only a discrete sequence of solutions, to which corresponds a discontinuous selection of states (precisely the empirically possible ones). To differential changes in the values of the ψ function there then correspond changes in the probabilities that the individual values of measurement are obtained when we measure under certain conditions.

This reading of the de Broglie-Schrödinger equations, which imposed itself because of the conditions under which characterizing quantities can be measured and statements derived from them empirically tested exhibits the empirical content of wave and matrix mechanics as completely coinciding. In both cases, though using formally different procedures, we define probability functions and assign them to the characterizing metrical quantities. Epistemologically it is indifferent whether we represent quantities of state by matrices and thus from the outset assign probability functions (and therefore probability quantities) to the characterizing metrical quantities, or whether we replace metrical quantities in the classical equations by expressions in which quantising operators are applied to arbitrarily defined fictitious functions of metrical quantities, resulting in expressions signifying probability functions. The common logically decisive step consists in the assignment, to the characterizing metrical quantities, of probability quantities defined in the form of fictitious predicates as functions of metrical quantities and perhaps of other parameters as well.

Moreover we know already that by introducing Planck's constant interpreted as a lower bound of accuracy of measurement under certain conditions, we are forced to define probability quantities and assign them to the metrical ones. If now we denote the use of Planck's constant in concept formation and description as 'quantisation' in general, then we can give a more precise account of the epistemological significance of 'quantisation' as such while still concerned with 'first quantisation'. The original interpretation of it as a description of discontinue step-like events continues to hold only for those special cases in which we can as it were illustrate the discrete actual changes of state ('jumps') by means of models. In by far the majority of cases, quantisation cannot thus be illustrated and discontinuity of states can mean only that for obtaining certain values of measurement we can derive only probability values and indicate only statistical connections for individual phenomena. In that case all we can understand by 'quantisation' in general is merely the defining and assigning of probability functions or quantities in the sense explained above.

This epistemological aspect of 'quantisation' stands out even more clearly in so-called 'second quantisation'.

If quantisation merely meant step-like changes characterized by metrical quantities, it would have to remain unintelligible why quantised sequences of phenomena can be subjected to further quantisation. Repeated quantisation is carried out by repeated application of an operator to expressions of a certain kind, exactly as with repeated application of differential operators. In that case we can no longer justify the general view that 'quantisation' means changes in state occurring in smallest finite changes in quantities. Rather, repeated application of quantising operators can then mean only the formation of probability functions depending on other such. More about this later.

The epistemological problem about de Broglie-Schrödinger waves lies in this: what sort of reality do they have? They do not transmit (transport) energy or momentum as is always the case in mechanical or electromagnetic waves, which thereby become empirically real. We might thus think that de Broglie-Schrödinger waves are a purely mathematical construction used in physical description merely so that we can represent the connection of phenomena more conveniently. This view is reinforced by the fact that the domain of phenomena for which the wave theory was set up can be described in just as close agreement with observation by means of matrices, a description quite unrelated to wave phenomena. On the other hand we must not overlook that certain results of observation, such as refraction and interference of corpuscular rays, have led to the setting up of wave mechanics from which predictions can de derived that are testable (even if only statistically).

When people began to enquire how far quantum physical laws can be reconciled with the theses of relativity theory, they changed their view as to the reality of de Broglie-Schrödinger waves. The Schrödinger equation, which counts as fundamental in wave mechanics, turns out not to be invariant under Lorentz transformation; that is, when we go from one reference frame to another in relative translation to the first, the Schrödinger equation changes its form. From the start attempts were made to modify the Schrödinger equation in such a way as to make it satisfy the relativity principle. It turned out quite in general that every proposition about the distribution (probability) of particles in a spatial domain derivable from a relativistic wave equation always makes statements not only about distribution densities but also about a flow occuring in the same domain. The expression that occurs in a relativistic wave equation and is to be understood as marking flow and distribution density can however assume positive or negative values for the components ('probability components') giving the distribution density (probable arrival) of particles at individual places, because under relativistic requirements the equation must be continuous and differentiable. Yet only positive values of the expressions concerned (which have the form of Schrödinger's ψ functions) can be interpreted as probability densities.

These general conditions or features of wave equations satisfying relativistic principles are taken into account in Dirac's theory. His wave equation is invariant under Lorentz transformation and yields positive values for the expression representing distribution density. The physical interpretation of distribution density expressions for particles (the probability of their arrival in certain spatial domains) as derived from Dirac's wave equation however shows that the probability predictions and therefore the wave equation itself hold only for particles with half values for spin. What then is the equation holding for particles of zero spin or integral angular self momentum? Further research by de Broglie, Klein-Gordon and others led to the setting up of general wave equations from which ψ-functions (densities) could be found even for these last mentioned particles, but it turned out that these 'density functions' are in general not confined to positive values. So it was no longer possible to interpret them as probability functions, that is, to interpret their values as probabilities for the arrival of particles at individual positions in space.

Here physical enquiry took the step, mentioned several times above, of interpreting a probability quantity – namely ψ – as a metrical quantity (quantity of state). If the values of a density function can be positive and negative, the general interpretation of these values as distribution densities is no longer possible, since negative values can then have no meaningful content, as has been stated. However, we do know physical density functions whose positive and negative values have empirical meaning, namely the functions describing electric charge of current densities in a field. Because of the positive and negative values assumed by the Ψ density functions in relativistic wave equations, the functions have been assigned to an electrically charged continuum (field) in the sense that the values indicate electrical charge and current densities at individual positions in space.

The values of the function ψ were first interpreted as probability values, now an attempt was made to construe them according to the relativistically modified wave equations as values of measurement characterizing the field points of an electrically charged continuum. In that case, however, we can no longer say that relativistic wave mechanics describes nothing real and that wave expressions merely indicate average distributions of particles in the various spatial domains. Rather, de Broglie-Schrödinger waves in their relativistic form must be understood as actual energy waves that have empirical reality in the same sense as electromagnetic and mechanical waves.

What is worth noting in the epistemological steps that have led to this interpretation of relativistic Ψ-functions – namely that they mark charge and current densities of spatial domains (field points) –, is that Schrödinger has obtained his wave equation by means of (a so-called 'first') quantisation of certain quantities of state occuring in classical equations. First quantisation indeed consisted in replacing classical quantities of state by expressions obtained by applying certain operators (whose coefficients contain the constant h) to the function ψ. This is a constitutive pre-condition for wave mechanical equations in their relativistic form as well. Since the density function here represents time-dependent continuous vectors counting as differentiable quantities of state of an energy-wave field, the (first) quantisation carried out to yield these functions simply cannot mean the assumption that the course of phenomena is discontinuous. For if relativistic waves are interpreted as energy waves, the Ψ-functions can no longer be understood as probability amplitudes of particles for individual places, since the relativistic equations do not now contain anything that points to the presence of particles.[25] First quantisation, which is required for obtaining the relativistic wave equations too, can therefore mean nothing other than the introduction of fictitious predicates (relations). Of course, the possibility of interpreting the fictitious value systems, chosen in the way explained above, as probability functions, is at first lost in relativistic wave mechanics and it becomes doubtful what is to be understood by first quantisation.

However, further analysis of the quantities characterizing the assumed electric continuum shows that it is in principle impossible to measure experimentally the field energy of the individual field domains (spatial domains) with accuracy. This seems to preclude the possibility of relating relativistic wave equations (originally construed as description of an electric continuum) to the relevant results of observations. What is observed in the decisive cases are corpuscular phenomena, which cannot be satisfactorily characterized by means of expressions purely deductively obtainable from the relativistic wave formulae. Derived statements about arbitrarily small changes of the quantities required sufficiently to characterize field points turn out to be untestable and therefore empty.

In order to obtain predictions about observable phenomena (especially corpuscular ones) in relativistic wave mechanics too, we carry out once more the step of ranging quantities of state into a fictitious value system. The expressions interpreted as quantities of state of field points of the assumed electric continuum are ranged into a wave system represented by matrices. To the continuous function ψ which in its relativistic form is intended (as has been mentioned above) to mark continuous states (charge and current densities) of field points, a matrix is assigned. The corresponding feature in formally carrying this out is again the application of quantising operators to certain expressions. This step is called 'second quantisation'[26]. To the positive and negative values that the relativistic function ψ can assume, which mark the charge of spatial domains, there now correspond positive or negative charge units whose integral multiples are the only obtainable results of measurement. Since all measurable changes of energy and momentum at field points now occur only in integral multiples of positive or negative units of charge, differential laws about field changes are once again possible as statements only about differential changes in the average distribution of units of charge in spatial domains. Thus second quantisation too proves to be a ranging of quantities of state ('metrical quantities') with their individual values into a fictitious value system represented by matrices, in which probability functions are so defined that their values give the probability for the possible individual values of measurement of changes in energy, charge, momentum, position.

The fact that the quantities of state figuring in, or derivable from, relativistic wave equations must also be subjected to quantisation (again by use of quantising operators) if we are to obtain testable statements, shows with especial clarity the epistemological significance of quantisation as such. As regards the description of corpuscular phenomena (corpuscular rays), both quantisations are required in order to obtain testable statements in all the relevant observational domains. The two quantisations always mean the choice of a fictitious value system into which possible observed values of certain kinds are ranged. In this we find the following connection between the two fictitious value systems that are introduced in definite methodical sequence.

In first quantisation measured values (values of state) as obtained under the conditions of classical physics – namely values of position, momentum, energy, time –, are ranged into fictitious value systems in which probability functions (ψ -functions) are defined. By relativistically modifying these expressions the value systems of arguments and values of ψ -functions are extended: values now appear that can be interpreted as marking electric field states (which may be possible values of measurement). However, these connections between quantities could count as a description of phenomena by continuous and differentiable quantities of state only if measuring procedures are indicated for the quantities of charge, energy and momentum marking the assumed electric continuum. These methods of measurement should make it possible to capture arbitrarily small changes in the field quantities under the conditions required to characterize field states sufficiently. Analysis of the theory of measurement here too shows that such measuring procedures are in principle impossible. Measuring under the conditions concerned produces uncertainty ranges inside which nothing can be obtained as regards changes in field quantities. The uncertainty ranges attendant on the theory of measurement arise for quantities of state of the electric continuum, and that with logical necessity, if we may so put it. For these quantities of state were indeed defined on the presupposition of first quantisation, that is, application of quantising operators to certain expressions marking phenomena. Even first quantisation presupposed for all measuring procedures concerned that it is impossible to capture arbitrarily small changes in metrical quantities under the conditions required to characterize states sufficiently.

In first quantisation we have already noticed that the quantising procedure is in general a definition and assignment of probability quantities. In wave mechanics especially, quantisation proceeds by application of quantising operators to functional expressions, for example to Schrödinger's - function in first quantisation. Probability quantities thus defined can therefore be called quantised functions. We must not overlook here that the 'quantities of state' occuring in relativistic wave mechanics are merely modified forms of already quantised expressions of the original de Broglie-Schrödinger wave theory. On this view it is presupposed that relativistic quantities of state are probability functions, in spite of their apparent interpretation as continuous quantities of state in an electric continuum.

If now the relativistic ψ-function and therefore the further characterizing relativistic quantities are quantised a second time, so that from the quantised relativistic equations we can again derive only probability statements, the quantised relativistic quantities of state reveal themselves as 'probability functions of probability functions', which clearly underlines the logical significance of the quantising procedure. Where we use quantising operators to quantise expressions characterizing states, we thereby assign probability functions to the quantities of state. A second quantising of already quantised expressions – for example of the relativistic ψ-function obtained by modifying the already quantised original ψ-function – means assigning probability functions to probability quantities (that in their turn are defined by probability functions). Second quantisation thus yields probability functions of probability functions. Let us call probability functions defined by quantising metrical quantities and quantities of state that directly characterize phenomena 'first-order probability functions'. Correspondingly, let us call those obtained by a second quantisation 'second-order probability functions'. The quantities they define will accordingly be called 'first-order' and 'second-order probability quantities' respectively.

Statements derived from the quantised relativistic wave equations express corresponding changes in second order probability quantities when any changes take place in the other quantities figuring in the laws. The latter quantities can be continuous metrical quantities – today most theories here include time [27] – or they can be first or second order probability quantities. Thus the most general form of statements derived from the quantised relativistic differential equations can be put as follows: differential changes in the probability, that measuring certain quantities under such and such conditions will yield such and such average values, will produce correspond-

ing differential changes in connection with measuring other such quantities, according to the functional relation involved.

If 'quantisation' meant only assuming step-like changes in states or quantities in least finite amounts, repeated quantisation would remain unintelligible. However, let us recall that we must quantise wherever we cannot
capture arbitrarily small changes in quantities required to characterize
states sufficiently, so that uncertainty ranges arise with regard to the determination of quantities: this shows that applying quantising operators is
logically nothing other than stipulating a probability metric, that is, ranging possible values of measurement into a fictitious value system, the values
of the domain so obtained being used as arguments of probability functions
correspondingly defined. Thus construed, 'quantisation' always means defining probability functions and assigning them to characterizing 'quantities', which can be metrical ('direct' quantities of state) or themselves probability in kind. In this way we find for repeated quantisation the unforced
explanation that it defines probability functions of probability functions,
the higher order probability quantities so defined being assigned to those of
order one below. Statements obtained by repeated quantisation then have
the general meaning that a certain probability changes in such and such a
dependence on changes in another.

There is a notion that quantisation always gives a 'continuum' – whether
substantial, or space, or a field, or even time – a 'granular' or corpuscular
structure; or, what comes to the same, that it assumes that states change in
step-like fashion by least finite amounts (for example of matter or energy):
this notion gives an inadequate reading of quantisation, usable perhaps under specially restricted conditions, but insufficient as a significant interpretation for the general application of quantisation, and particularly its repetition.

27. THE QUANTISATION OF ELECTROMAGNETIC FIELDS

The cases so far discussed concern the setting up of relativistic wave equations for corpuscular rays. It turns out that we must here perform two
quantisations in order finally to arrive at equations satisfying relativistic
principles and allowing the derivation of testable predictions in the decisive
cases. First quantisation concerns quantity terms of classical mechanics:
quantising operators of a certain kind are applied to arbitrarily defined

functional expressions (for example the ψ-function) and the classical terms are accordingly replaced by the expressions thus quantised. This operation turns the classical equations into the de Broglie-Schrödinger wave equations. By correspondingly modifying the expressions and equations of this wave theory we obtain equations satisfying relativistic requirements. The equations now seem to represent continuous changes at the field points of an electric continuum, but, in just those cases that are decisive, the statements that we can derive from them turn out to be in principle untestable by observation. To obtain derivations we need a second quantisation, that is the assignment of a probability function (in matrix form) to the relativistic ψ-function or the connected application of corresponding quantising operators on the quantities of state of the 'electric continuum'; the statements thus derived are of course necessarily statistical, because of quantisation.

However, the initial phenomena can be electromagnetic field changes as described by Maxwell's wave equations. A first quantisation of these equations is superfluous because the field is continuous, so that here the relativistic modification of Maxwell's equations can be carried through independently of any quantisation. It is not until electromagnetic waves, for example light waves, interact with matter that we meet phenomena describable only by means of quantised expressions. This quantisation of Maxwell's or of relativistic field equations corresponds to the second kind, here applied in the absence of the first. The fact that atoms struck by light absorb or emit photons (that is, finite least amounts of light energy) can be described only if we replace field quantities by corresponding matrices, which precisely amounts to second quantisation. As we know already, this means the ranging of the possible values of the quantities marking field states into a fictitious value system in which a probability function is defined whose values are assigned as probability values to the possible values of measurement (state values). The quantisations of corpuscular and electromagnetic field phenomena respectively differ in that for the former a second quantisation defines a 'second-order probability functions', while for the latter it defines only a 'first-order probability function' (since first quantisation is absent). In any case quantisation defines and assigns probability functions (probability quantities).

It becomes even plainer that quantisation by introduction of fictitious value systems in which probability functions are defined is a particular mode of applying the method of fictitious predicates when we compare the presuppositions of this procedure with the requirement of relativity theory.

If an atom interacts with an electromagnetic field with for example a photon being absorbed by the atom, we can interpret this as transfer of a certain amount of energy or momentum from field to atom. Such transfers always occur in integral multiplies of a definite, finite least amount or emitted. In the field, however, energy and momentum are distributed continuously. A quantised transfer from field to atom therefore presupposes that the amounts of energy and momentum distributed in the field move at infinite speed to the place where the transfer occurs, that is, where atom and field are in contact. However, infinite speeds of transmission occur only with forces acting at a distance. Quantisation of electromagnetic fields, in the sense that only quantised amounts of energy or momentum are transferred when field and matter interact, thus seems to presuppose that we assume the existence of action at a distance.

However, the relativity principle bars transmission of action at speeds greater than that of light and thus excludes the possibility of action at a distance. Hence the quantisation of field events seems to rest on presuppositions contradicting relativistic principles. We must, however, remember how quantisation is carried out and what it means logically. In every form of quantisation we define least quantities with the help of the constant h. In the case of electromagnetic fields, the continuous field quantities are replaced by matrices. Epistemologically, a matrix representing a physical quantity of state signifies a system of infinitely many values, some being possible values of measurement, others fictitious (not ascertainable by any kind of measurement). By this quantising procedure we obtain amongst others a matrix for the total energy of the field. Thus, the field quantities marking individual field points, as well as the total field energy, are represented by matrices. These physically interpreted matrices are subject to the rules of matrix algebra. In general, matrix multiplication is not commutative, so that their order in a product is not interchangeable. Non-interchangeable matrices representing physical quantities of state are subject to the rule that they cannot both be measured exactly at the same time.

For the relations between the matrices marking individual field points and the matrix expressing the total field energy it is characteristic that the matrices are not interchangeable, from which it follows that uncertainty ranges arise for changes in the relevant quantites. From the changes in quantities marking field points the change of the total field state cannot be calculated, nor can the changes at individual field points (spatial domains) from the change in total energy. This is also expressed by saying that we cannot determine how much the individual volume elements contribute to the total energy, or that we cannot 'localize' the field energy in a quantised field.[28]

From this it follows that quantisation asserts no interaction between total field and individual field points, nor is this presupposed. The presence of quantised amounts of energy and momentum at field points necessary for interaction between field and atom is in principle not derivable from the given total field energy, rather their presence at points where field and matter happen to interact is quite inexplicable and must be accounted fortuitous. That, in a continuous field, action is transmitted at speeds greater than that of light or that forces act at a distance could be asserted only if within quantum physics some law could be indicated, perhaps only a law of action at a distance, that would allow derivation of energy or impulse transfers at individual field points from the total field energy distribution. Conversely, this law would equally allow us to calculate that change of total field energy from changes of state occuring at the individual field points and ascertained perhaps by measurement. Such laws – whether of the form of action at a distance or proximate action – exclude uncertainty ranges for all quantities marking states. Since quantisation necessarily presupposes such ranges – for otherwise no 'quantisation' would be required –, it excludes the possibility that in the domain of quantised phenomena there could be laws that express the exact differential relations between the changes in the 'interacting' states or in the quantities marking them.

Thus according to the laws of quantum mechanics the total state of a field and the 'states' of individual field points are not in the kind of interaction in which it would make sense even to speak of a speed of transmission. On the quantum mechanical presuppositions, total field state and individual states of field points remain largely undetermined as to their relations of 'causal' dependence, in particular in that it is logically senseless to speak of a velocity of propagation of action in the quantised field. A transmission of action, or its velocity, make sense only where from differential changes in some metrical quantities marking a state we can through functional dependence infer the differential changes of other such quantities. Where, because of uncertainty ranges, such inference is barred in principle, we can no longer sensibly speak of propagation of action at whatever speed.

Therefore we cannot rightly object that quantum mechanics assumes relations between electromagnetic field (photons) and matter that involve transmission of action at speeds greater than that of light. To describe interaction in terms of continuous field energy concentrating at the point where field and matter coincide, in amounts specified by the laws of quantum mechanics, which for an extended field means infinite speeds of transmission of action, is to distort the content of those laws. Since, as explained, they make it in principle impossible. to ascertain what are the quantative relations between energy amounts at individual field points and the total field energy, even if the latter is known, there can be no experiments or observations as to how amounts of field energy 'concentrate' themselves at a field point, that is, travel to that point. Thus it becomes senseless to speak of a speed of propagation of energy.

This quantum-mechanical description of energy transfer between electromagnetic field and matter merely observes the transfer of energy quanta; form and content of these laws leave it quite undetermined and inexplicable, how the quantised amounts of energy appearing at fields points are related to the total field energy, being silent in particular as regards the propagation of action in the sense of something starting from a field point and spreading continuously over the whole field – the very thing that would have to be presupposed if we wish to assert that action is transmitted at some speed.

Because of the presupposition of uncertainty in principle as to the way action spreads in a quantised field, it is certainly unjustified to object that quantum mechanics presupposes in these fields the transmission of action at speeds that are greater than that of light or even infinite (action at a distance). On the other hand, we recognize that the uncertainty that quantum description requires in certain circumstances for the causal connection between total field state and the states of the individual field points, compels us to define probability quantities for the uncertainty domains in which possible values of measurement have been ranged into fictitious value systems, the individual values of these functions indicating the probability for individual possible energy states at field points: failing this, it will be impossible to arrive at testable statements about these otherwise unknown causal 'relations'. Such a probability metric is fixed precisely by quantising the electromagnetic field, that is, by assigning matrices of a certain type to quantities of state. Interpreting uncertainty ranges of the quantised field as containing transmissions of action at speeds greater than that of light amounts to importing certain regularities into domains in which according to quantum mechanics no regularities are to be ascertainable.

Even if this contrast between quantum mechanical and special relativity presuppositions, namely that transmission of action at speeds greater than that of light are impossible, cannot be derived, there is of course an incompatibility between the essentially more general presupposition as to causal dependence of the course of phenomena in quantum physics on the one hand and in strictly deterministic systems on the other, relativity theories. amongst them.

PART IV. EMPIRICAL-FICTITIOUS KNOWLEDGE

28. ACTIVE AND FICTITIOUS CAUSALITY

The strictest form of causal determinacy of events must be seen in relations expressed by laws of proximate action. What is usually said to mark sequences of states described by such a law is that from it we can derive arbitrarily small transitions of state, that is, arbitrarily small changes in state for arbitrarily small spatio-temporal domains. This is tied to the presupposition that changes in state are propagated in space at finite speed, for only then is it possible to make such derivations.

We know that laws of action at a distance assert changes in state independently of time, so that derivations from such laws as to changes in arbitrarily small times are empty; therefore such laws are called 'laws of pseudo-proximate action'. Their peculiarity of yielding physically empty expressions for certain differential derivations follows from the underlying presupposition that the actions described by the laws are propagated at infinite speed. In spite of this assertion that actions are to occur instantaneously over arbitrarily large spaces, we cannot here speak of jumps in transmission: the form of concepts used in laws of action at a distance readily permits mentioning arbitrarily small transmissions and changes in state, even if these laws start from the empirically unrealizable presupposition that actions are transmitted at infinite speed. These laws may be viewed as a degenerate extreme form of laws of proximate action. Phenomena described by them can be regarded as strictly determined and it is improper to equate the assumption of actions spreading at infinite speed with the assumption that the course of phenomena in these domains is lawless, merely because the preconditions are empirically unfulfillable.

Certain physicists[29] tend to adopt this last interpretation. According to them, certain quantum-physical presuppositions about the lawlessness of changes in state within certain ranges, leading to the introduction of fictitious value systems and to the definition of probability functions, are equivalent to the hypothesis that we here assume laws for forces acting at a distance or for action at a distance. They base this 'identification' on the remark that in both cases one makes empirically unrealizable or untestable and therefore 'inexplicable' presuppositions for the spatio-temporal domains concerned. However, comparing the form of deductions obtained from probability laws and laws of action at a distance respectively, we see that to assume that under certain conditions phenomena are ruled by chance cannot mean the same as the setting up of some laws of action at a distance for the same domains.

Where we assume that within certain ranges quantities change by chance, we may at most assign probability values to the individual values of measurement, by stipulating a probability metric: only a probability value can then be derived for each possible individual case. That the probability function whose values count as probability values is defined with reference to 'all possible cases' (for example the probability of certain values ôf state at field points with reference to the values of the total field state) naturally does not mean that we here assume special 'forces acting at a distance' between all possible cases and individual ones (say, between the total field and individual field points). The relations connecting the probability values of the various phenomena are arbitrarily stipulated by the probability metric and it is a grave misunderstanding to try to construe these relations as 'action relations', that is, causal relations based on the action of physical forces. We shall presently enquire what kind of 'causality' is to be understood by the ordering of phenomena stated in probability laws.

As regards characterizing phenomena by a probability order, let us at present merely point out that chance in the sequence of cases presupposed for describing phenomena in uncertainty ranges by differential probability laws means just the assumption that between the states concerned or changes in them there can be no forces acting or transmissions of action for which a speed of propagation can be assumed in any sense. This clearly reveals the logical difference between probability laws that presuppose the absence of actions relating the phenomena (changes in state) described on the one hand and laws of action at a distance on the other.

The latter assume action relations between changes in state, in the sense of the propagation of the actions of forces. Even if this is assumed to happen at unattainable infinite speed, still a law of action at a distance expresses a one-one relation between states, enabling us to derive the occurrence of an individual state (phenomenon) that excludes the possible occurrence of other individual phenomena. This deductive form is typical for derivations from laws of proximate action and laws of action at a distance: a certain change in state is followed by one and only one phenomenon (state) and not possibly by some other. Because this kind of statement may be derived from either kind of law, those of action at a distance (for example the classical law of gravitation) likewise belong to strictly deterministic forms of description.

Quite in contrast to this, propositions derivable from probability laws express a one-many or a many-many relation between the phenomena described. Thus, according to a probability law, a given change in state (or perhaps several possible such changes) may be followed by one of several possible and mutually exclusive phenomena, though we cannot unambiguously predict that one of them will occur. The derived proposition has the form 'the change is followed by this event or that event or...', and the probability metric chosen for this form allows us to derive certain probability values for the individual events: nothing else is derivable, which is why we say that probability laws indicate no strictly determined order of phenomena.

The difference in content between strictly deterministic description by laws of action at a distance or proximate action on the one hand and nondeterministic description by probability laws on the other consists in assuming effective forces for the former (the action spreading at a specified speed) and explicitly presupposing that there are no such forces for the latter, so that in that latter we cannot sensibly speak of a speed at which changes in state (that is, the actions of forces) spread.

However, even probability laws indicate a certain order in the course of events. If by 'causality' we understand quite generally an empirically testable or ascertained order of phenomena, then probability laws too express a certain 'causal dependence' of sequences of states. The difference as to strictness of the orders indicated by the two types of laws respectively is readily seen after what has been said. Elsewhere[30] I have distinguished, though from a somewhat different point of view, between the stricter and less strict forms of causality as being of 'second order' and 'first order' respectively. It is there shown that an order of phenomena indicated by statistical laws falls under 'first-order causality'. It might be appropriate here to specify the two orders of causality in terms of the statements about propagation of action or with regard to the introduction of fictitious value systems that make it possible at all to indicate a probable order of events.

Where causal connection between phenomena is asserted by means of indicating a transmission of action, as in all laws of action at a distance and proximate action, let us use the term 'active causality'. Where the order of phenomena is fixed by means of introducing fictitious relations (fictitious predicates) we shall speak of 'fictitious causality'. Thus, probability orders of phenomena are always based on an introduced probability metric. This mode of description presupposes that we exclude the propagation of actions and so therefore of changes in state in uncertainty ranges at speeds given by a law.

The above mentioned division according to orders of causality does not coincide with the distinction between 'active' and 'fictitious causality'. The two divisions overlap, and we need not here consider the various conceptual differences. Incidentally, the term 'fictitious causality' can be applied not only to probability orders of phenomena in the above sense, for we must note that the boundary between active and fictitious causality must be drawn as the presuppositions of physical systems might dictate. Uncertainty about action relations between phenomena arises not only in uncertainty domains of measurement, but also for phenomena between which transmission of action at speeds less than or equal to that of light is impossible on relativistic assumptions. Indeed, we know that such phenomena are in uncertain temporal sequence, and for these we can do no more than stipulate a fictitious time order, so chosen that they are ranged into a uniform system along with events having an observed time order. The uniform causal order thus obtained is however in part fictitious, as we just saw. We are therefore justified in calling the resulting relations between phenomena 'fictitious causal relations', though we must not overlook that in the relativistic case mentioned a fictitious causal relation is a one-one relation between phenomena.

The decisive difference between active and fictitious causality is thus given by whether the order of phenomena (states) is determined by transmissions of action at empirical speeds of propagation or, in their absence, by an at least partly fictitious stipulated order.

29. GREATEST POSSIBLE AND LEAST POSSIBLE PHYSICAL CONSTANTS

Stipulating fictitious causal relations for phenomena becomes necessary wherever uncertainty ranges arise as to the order of phenomena. Modern physical enquiry or theory formation typically attains to domains or ranges of phenomena in which it is in principle impossible to observe or measure anything as to the spatio-temporal order (that is, causal order) of phenomena. On examining more closely under what conditions uncertainty ranges arise in the domain of empirical phenomena, we find that this is necessarily connected with the appearance of a special kind of constant in the formulae. Wherever we indicate finite limits for any quantities marking states or changes in them, in the sense that according to the existing empirical laws of nature those states (phenomena) are impossible that would have to be described by values beyond those limits, there uncertainty ranges arise even within finite domains of values. Phenomena in such ranges cannot be sufficiently characterized by measurement.

Limiting values for metrical quantities (quantities of state) above or below which we cannot go we call 'greatest possible' and 'least possible' empirical constants respectively. An example of the former is the velocity of light c in relativity theory, and of the latter Planck's constant h in quantum physics. Our epistemological analysis has shown clearly that introducing them in both cases leads necessarily to the appearance of uncertainty ranges even in finite value domains. In the case of h as interpreted by Heisenberg this was easy to see, since on his view h is the lowest possible limit of uncertainty for measurements under conditions required for sufficiently characterizing states.

To obtain testable propositions about changes in state within uncertainty ranges, one applies the method of fictitious predicates. In the case of the empirically least possible constant h the (many-place) fictitious predicates chosen for description of such changes define systems of probability metric. Epistemological analysis of quantum physical and wave mechanical probability concepts has clearly shown that there are various ways of choosing these statistical fictitious systems, the criterion being expediency.

However, logical analysis of the relativistic thesis that c has the character of a greatest possible empirical constant, shows up quite similar logical features as regards the appearance of uncertainty domains caused by that empirical limit: just as h fixes unconditionally valid conditions for measuring procedures, so does c. It is important to realize that introducing empirical constants that are to count as limiting values, whether least or greatest possible, necessarily fixes such conditions. For h as construed by Heisenberg this was obvious; the analogous consequence for c here likewise forces us to apply the method of fictitious predicates.

The relativistic theory of measurement starts from the fact that 'measurement' always consists in establishing a transfer of action (action relation) between instrument and object to be measured. Empirical transmissions of action travel at finite speed and there are cases of measurement in which the speed of the transmissions involved must be taken into account when working out the result. Transmissions of action used for purposes of measurement are called 'signals'. If the speed of light c counts as empirically greatest possible limit for the speed of propagation of actions and for expediency's sake we use signals having that speed, it follows, that, in the measurement of the time order of phenomena, ranges of uncertain temporal sequence must arise. That is, there must then be spatio-temporal domains such that between phenomena within them only transmissions at speeds greater than c would be possible. The temporal order of such phenomena is in principle unobtainable by measurement, since c is supposed to be the empirically greatest possible speed for the spread of action. From the uncertainty of temporal sequence of phenomena of the kind mentioned it follows that they are likewise uncertain as regards other metrical quantities, since on relativistic measuring theory other quantities too are measured by signals, which therefore presupposes time measurements.

If therefore introducing c as empirically greatest possible limit results in the appearance of uncertainty ranges as regards certain characterizing metrical quantities, this analogy with the case of h goes further still. For in the relativity case too we can obtain testable propositions about the temporal order, or causal relations of phenomena in ranges of uncertain temporal sequence, only if we apply the method of fictitious predicates. If we wish to obtain testable predictions there is no other way than to stipulate a fictitious order for events of the kind concerned (for example, a fictitious simultaneity for chosen pairs of events in uncertain temporal sequence). The fictitious time values (time relations) thus assigned to phenomena are in principle untestable by measurement. Only when connected with statements about phenomena characterized by values and relations that are measurable can we perhaps attain to testable predictions from the fictitious assignments. The empirical coming true or otherwise of such derivations (or, more generally, their agreeing or otherwise with empirical data) then decides whether the stipulated fictitious relations (predicate systems) are appropriate or not.

Introducing greatest and least possible physical constants thus always involves a determination of measuring theory, namely in the form that within finite value domains of the metrical quantities marking phenomena there must appear uncertainty ranges. Such constants, being limits, are therefore universal and occur in the formulae concerning the totality of physical phenomena. If then in the descriptive laws, which express empirical findings, such constants appear, this means that these findings themselves are determined in form by the conditions that descriptions involving these constants must obey.

30. THE NEW FORM OF KNOWLEDGE

Since Galileo founded the classical method of natural science, cognitive procedures and concept formation in the exact empirical sciences have been developed by empirical enquiry in individual sciences acting in concert with epistemological criticism. Natural sciences specify ever more precise means for obtaining and testing their findings and epistemology sees one of its main tasks as that of indicating unambiguously and exactly what we must understand by 'knowledge', given the state of scientific enquiry at the time. That such essays in critical epistemology can in their turn influence scientific enquiry in its formation of concepts and theories is well attested by famous examples of the past as well as of most recent times.

If, following the methods and theories of modern physics, we too attempt to characterize the knowledge aimed at in current natural science, what comes to the fore in contrast with traditional concepts of empirical knowledge is the new characteristic feature of an indissoluble link between empirical data and fictitious elements. As to what we should understand by knowledge, opinions past and present differ in many ways, not only amongst epistemologists but also amongst natural scientists themselves. Even the creators of relativistic and quantum physics took the epistemologically realist view about matter and its constituents, that what we try to know in empirical science are really existent 'objects' persisting through time, their properties and their relations. Where observational measurement encounters difficulties we ought by no means to infer that these objects do not exist, in principle it remained possible to give a complete, exact and unambiguous account of objects (whether we call them substances, masses, corpuscles or whatever) and their relations. This is connected with the fact that both Einstein and Planck were strict determinists. For if there were undetermined domains of events, it would be impossible completely to capture (know) objects in their relations by means of observation (measurement).

This realist interpretation of the content of systems of scientific propositions was not essentially altered by the conventionalist account of the empirical concept or aim of knowledge. According to moderate conventionalism there must be amongst the propositions of the descriptive system some that are analytic (stipulations, conventions), but these have merely the character of grammatical rules to be fixed in each case to allow unambiguous application of the concepts used in description.

Extreme conventionalism does indeed go further and assigns analytic conventionalist status to all laws of nature, but this conception too, as we shall presently show, is compatible with the view that the system of scientific laws (analytically construed, on the conventionalist view) is co-ordinated with objects persisting through time, although their elementary order is now supposed to be fixed by the simplest forms of concepts and laws.[31]

The conventionalist view that laws of nature are stipulations (conventions) rests mainly on a method that is certainly very often applied in physical enquiry, namely not to declare a law as false if it disagrees with observed facts but to remove the disagreement by setting up auxiliary hypotheses. Poincaré, the founder of conventionalism, though rejecting the extreme position, has supported a conventionalist account of general laws of nature by emphasizing that in describing a domain of phenomena we are free which of several possible and mutually incompatible laws we wish to choose as valid. If this or that law thus declared valid disagrees with observed fact, we can always 'explain' the discrepancy by suitable auxiliary hypotheses. In choosing which laws of nature conceived as conventions are to be set up we let ourselves be guided by expediency, aiming for example at the simplest possible total system of descriptive statements.

We need not here enlarge on the epistemological difficulties that must arise for an extreme conventionalist account of procedures of physical enquiry or of the systems of descriptive statements.[32] What interests us is that this orientation is compatible with the above-mentioned view that physical knowledge aims at exactly and unambiguously marking real permanent objects in their properties and relations. Whether in order to achieve this goal one opts for the empiricist or the conventionalist method does not affect the nature of the object.

Moderate conventionalism merely points out that in order unambiguously to define any metrical quantity we must make certain stipulations, and further, that we require agreements regarding the measuring instruments (for example under what conditions we wish to say of measured values of lengths that they were obtained by means of 'absolutely rigid' rulers) if we are to attain to intersubjectively valid results of measurement. Some predications, that uncritical empiricism had construed as predicates assigned to objects, are here recognized as stipulations needed to define and apply the concepts. That there are permanent objects whose properties and relations are to be cognized is also presupposed here.

In extreme conventionalism a much greater number still of propositions is accounted analytic. Of course this orientation must admit that the system of conventions, which are now the basic laws of nature, is, to begin with, an empty schema and not a complete description of events involving objects. Only with the addition of auxiliary hypotheses that are to explain the deviations of observed values from the schema of natural laws conceived as conventions (particularly in decisive cases this is achieved by assuming 'disturbing' forces) do we obtain a complete description of what actually happens. In its most general form, this orientation leaves it open which laws are to be set up by convention. It might be the laws of proximate action of classical physics but perhaps also some differential probability laws. This shows that the extreme conventionalist account makes no prior assumptions as to the 'objects of knowledge', it merely tries to determine the method that will lead us to a system of descriptive statements. Therefore the system of statements of classical physics, as well as those of relativistic or quantum physics, may equally well be construed as systems of conventions.[33] The view that there are permanent objects with certain qualities and related in certain ways can therefore readily serve as a basis for choosing corresponding conventions to be thereafter counted as laws of nature.

That the propositions of physics concern permanent objects is a view that has suffered a change at the hands of the positivist interpretation of the content of empirical propositions, namely that physical propositions always express only empirically testable relations in the sequence of observed data. What we have imprecisely expressed by the term 'objects persisting through time' thus reveals itself as a certain regularity in the order of appearance of observed data under prevailing conditions of observation. So long as this regularity is such that we can 'localise' the connection between observed data, that is, recognize them as the same complex data through different spatio-temporal domains, there is no objection to the term 'object persisting through time'.

However, there are experimental conditions (of observation) under which the data of observation appear in distributions in which it seems impossible to recognize regularly connected complexes of data. To go on speaking of 'permanent objects' here, as is possible on a realist and on a conventionalist view alike, would be most inappropriate and would force us perhaps to set up improbable and untestable hypotheses. That is why positivists wish to speak of observed data, and of the laws according to which they appear under various conditions of observation, as the only objects of knowledge.

However, even this account, which is widespread amongst modern scientists, makes prior assumptions about the relations between observational data and descriptive concepts and propositions, which do not seem to be fulfilled given the results of logical analysis of certain fundamental concepts of relativistic and quantum physics. If physical propositions are construed as concerning the order of observed data, these latter with their qualities here count as the objects of knowledge. However, this account would have exact and unambiguous meaning only if what we call 'data of observation' can be individually marked and thus asserted to have empirical existence; a condition fulfilled only if we could characterize the data and their connections with the help of exact metrical quantities and their differential changes alone.

Yet this is impossible just when precision does matter. What we call 'empirical states' is marked not only by metrical quantities but also by assignment of fictitious values (for example probability values) all characterizing values being set into mutual relations through functions. The observational data themselves, let us say those to which we assign the values read off from measuring instruments, thus in principle appear merely as an incomplete part of 'objects' (states, phenomena) about which the propositions of physics say something. The 'objects' themselves can be characterized only by a system of values of which some must be taken as possible values of observation and others as fictitious values that are in principle not ascertainable by measurement. By means of suitably defined functions (many-place predicates) these values are combined into a system.

The real existence of any states marked by values of measurement thus always involves the relation of these states to fictitious values given by a corresponding function. Values of observation or measurement thus always constitute only a part of the objects about which physics pronounces or whose relations it describes. The question what is denoted by fictitious value domains may be answered in a certain sense if we reflect which empirical circumstances compel us to choose and apply fictitious values. As we know already, these circumstances are the necessary appearance of uncertainty ranges in connection with greatest and least possible constants involved in measuring procedures. Fictitious values or value domains are 'assigned' to processes that lie inside the uncertainty ranges and therefore cannot be captured by any measurement, though they must nevertheless be assumed. Of course in general it is not a one-one assignment as in marking by values of measurement. Rather, assigning fictitious values means a provisional statement that nevertheless denotes phenomena (states). Proof of this will be seen in the testable predictions obtained with the help of fictitious value assignments. To use a less precise mode of expression, we may say metaphorically that fictitious values denote phenomena occuring inside uncertainty ranges.

Of course we must not misunderstand this, because it is inadmissible to speak of what happens in uncertainty domains independently of observable phenomena. The attempt to describe observable states and their relations compels us to assume events in uncertainty domains, and only in that context of applying measuring procedures does it make sense to speak of 'events in uncertainty ranges' or 'assignment of fictitious values' to such fictitious events. What is epistemologically even more important, empirical existence of 'observable' states and phenomena, to which value ranges of metrical quantities are assigned for sufficient characterization, cannot be sensibly asserted by itself but only in necessary epistemological connection with what happens in uncertainty domains.

Accordingly the 'object' about which physics speaks, and whose empirical existence it asserts, consists of the 'observable' part of the states (so far as the latter can be marked by values of measurement) as well as of the part lying in the uncertainty domain, which can be marked only by introducing fictitious values and their being set in relation to observational values. The indissoluble connection between the two parts in the empirical physical object of knowledge manifests itself especially in quantum physics, where in sufficiently characterizing phenomena (states) one generally assigns a probability value to every possible value of measurement. The probability is the value of a function that defines a relation between observational and fictitious values, where the latter 'mark' states or changes of state that lie in uncertainty domains and are in principle unmeasurable. Thus an 'observable' ('measurable') state cannot be marked without setting observational and fictitious values into relations (for example by probability functions) which, as it were, characterize the part of phenomena that occurs in the uncertainty domain.

A complete characterization of the object of knowledge is thus given by a function (many-place predicate) whose argument values belong to a system partly of possible values of measurement partly of fictitious values. If at the beginning of this essay we said that scientific knowledge is of predicates, we can now characterize more closely what epistemological kind of predicates empirical physical enquiry aims at cognizing: it is many-place predicates in the form of functions (with arbitrarily many arguments) describing empirical-fictitious states and their changes. If we call such functions in general ψ-functions, then a state is sufficiently characterized by values of measurement and the values ('probability values') assigned to them by ψ, whose argument values are a system of possible measured and purely fictitious ones. Those values of ψ that count as probability values express the very relation between observational and fictitious values that entitles us to denote states marked by the two kinds of value as empirically-fictitious.

If then empirical knowledge aims at describing empirical-fictitious states and their changes with the help of suitable many-place predicates, we can call these descriptive linguistic forms themselves 'empirically-fictitious' predicates: cognizing these may in the sense explained be described as the goal of empirical enquiry.

If we take into account further special conditions of present day physical enquiry, the empirical-fictitious predicates that quantum physics always provisionally sets up and formally adapts to newly obtained results present themselves as 'probability functions'. If we like we might thus state in a somewhat simplified form that the aim of physical enquiry consists in discovering suitable probability functions. However, this is too special a characterization that takes account only of expressions particularly important for current physical enquiry, without doing justice to the general epistemological significance of the steps and conceptual elements required to constitute the relevant expressions. Indeed, our own example for explaining the empirical-fictitious predicates used in describing physical objects of knowledge has mainly been that of physical probability functions. Still, we must not overlook that while probability functions play a commanding role in quantum physics, they are after all only a special form of empirical-fictitious predicates. For epistemology it is of interest what general conditions require such predicates for the purpose of description: in some cases they might well lack the character of probability functions.

We have repeatedly emphasized that in relativistic physics the phenomena there appearing in uncertainty domains can be marked only by introducing fictitious values. There too the real existence of phenomena marked by fictitious values (for example fictitious time values) is only ever significantly assertable in the light of the functional relation of these values to observational ones. Conversely, there too, to characterize states sufficiently always requires not only values of measurement but also their connection by stipulated relations to corresponding fictitious values.

The stipulation, for example, that the fastest empirical signals used in measurement are to have the same constant speed on all paths and irrespective of the translational speeds of the reference frames, first makes it possible to assign fictitious time values to phenomena occurring in relativistic uncertainty domains; and secondly brings fictitious and observational values into quite definite functional relation. In the formulae that do this (for example the Lorentz transformation) we can enter values of either kind and the values of the function thus obtained are likewise in part possible values of measurement and in part purely fictitious values (which mark phenomena in uncertainty domains and are in principle unmeasurable). This difference between values of measurement and fictitious values does not show so obviously in relativistic description as in quantum physical characterization by values of measurement and probability. For according to the relativistic theory of measurement, the phenomena that lie in uncertainty domains are indeed marked by fictitious quantities, but these last have the same external form as the metrical quantities and are treated by means of the same operational rules in further derivations. Moreover they count as differentiable, and arbitrarily small changes in them are assumed possible. Thus in relativistic description fictitious quantities marking phenomena occurring in uncertainty domains are called 'metrical quantities' too, so that in the usual terminology the basic epistemological difference between observational values (values of measurement proper) and fictitious values (marking 'fictitious states', that is, phenomena in uncertainty domains) appears to be ignored.

This identity of form must not deceive us as to the epistemological difference between the two kinds of quantity. Where relativistic methods or formulae are used to determine characterizing values and their relations, the sufficient characterization of states is necessarily given by the two kinds of values and further always expresses a relation between the two partial states marked severally by each kind. An example of such relations between values of measurement and fictitious values is to be seen in relativistic changes in length and time, which are in principle unascertainable by measurement.

If we here speak of partial states of phenomena, this must not be misconstrued as meaning that 'states' thus marked have no reality and, as opposed to phenomena marked by values of measurement, are merely auxiliary concepts used to obtain a simpler mode of expression. This would be a fundamental error. The position is rather this, that we must regard phenomena occurring in uncertainty domains (which compel us to introduce fictitious values) as just as real as processes marked directly by values of measurement. More precisely, where fictitious values must be additionally adduced for sufficient characterization, one can speak of the reality of the partial states (defined by measured and fictitious values respectively) only in their mutual relation in the total state. The partial phenomena marked by measured values have empirical reality only in regular complementary connection with phenomena occurring in uncertainty domains, which latter provoke the introduction of complementary fictitious values and lead to the setting up of functions (fictitious predicates) in the total domain of both kinds of values.

In view of the content of the forms of concepts and laws in modern physics, we think that the knowledge aimed at by physical enquiry is not adequately characterized by the older empiricist realist and positivist concepts of knowledge. For this we need a new concept of knowledge. The decisive features of the findings expressed in the laws of recent physics, as emerging from our epistemological analysis, suggest that we might here speak of 'empirical-fictitious knowledge'. This new concept takes into account the decisive feature of the object of knowledge described by modern physical laws. It consists in this: we can speak of the reality of the object described only if we indicate both kinds of value (measured and fictitious) and their regular functional relations.

31. THE MATRIX FIELD AND THE PROBABILITY WAVE FIELD

By logically analysing what quantities and expressions modern physics uses to mark phenomena we have come to recognize forms of concept whose empirical application always characterizes states only partially by values of measurement and in addition requires that these values be set in relation with fictitious ones. As to empirical content, this corresponds to the fact that to partial states marked by values of measurement there always belong other partial states in uncertainty domains. If, in the case of quantum physical phenomena, we seek to characterize states sufficiently, the theory of measurement concerned makes it impossible to measure arbitrarily small changes in them. Hence the uncertainty domains in which we must indeed assume phenomena (changes in state) but can no longer describe them by continuous metrical quantities. In view of this we sometimes say in recent physics that processes in between the discrete 'states' (which states alone we can ascertain by measurement) defy all observation (measurement) and therefore must remain unknown to us. To the discrete 'states' correspond jumps in the characterizing values of state, and to these jumps spatio-temporal domains that count as uncertain with regard to the changes of state occuring in them. However, this way of putting it is incomplete and not very precise, referring only to characterization by metrical quantities and their changes, whereas we saw that for complete characterization physics is here compelled to introduce fictitious values.

One way of adducing fictitious values for describing phenomena is to range the possible values of measurement into a system extended by fictitious values with suitably chosen functions fixing the ordering relations of the system – above all the connections between the two kinds of value. In quantum physics the values of these functions are called 'probability values' and the functions themselves we have called 'probability functions'. The way these last are defined shows that wherever we carry out measurements in order to characterize states, we can assign such functions to the metrical quantities. If we do this in describing states throughout a spatial domain (field), we obtain the mode of description that we previously called a 'probability field'.

The most common way of defining probability functions for metrical quantities in theoretical physics is to assign a matrix to a characterizing quantity. Matrices are value systems, so that we might even choose systems with infinitely many values. Matrices obey special operational rules whose application leads to new matrices. In the value system of a matrix some values count as possible values of measurement while others are purely fictitious. By entering suitable values or applying suitable operations, matrices yield individual values counting as the probabilities assigned to the possible individual values of measurement. Thus we may regard this matrix procedure as a special way of defining probability functions.

We can imagine a probability field in which the functions assigned to field quantities are matrices throughout. This special case we call a 'matrix field'. As in any probability field, the differential laws here too merely express the functional connection of changes in the probabilities of definite individual values of a metrical quantity for differential changes in the probabilities of individual values of another metrical quantity (or of several others). The matrix field shows with especial clarity according to what conceptual mechanism values of measurement are ranged into systems of partly fictitious values, while at the same time probability functions are defined in those systems. Moreover, the method of setting up a matrix field shows that we can continue this procedure as it were step by step. For by applying suitable operations to the matrices we can obtain individual values to which we can again assign matrices. As we know already, we thus obtain probability functions of probability functions. This formal procedure is applied in the previously mentioned 'second quantisation'.

The multiplicity of possible matrix forms available for assignment –

since it is logically possible to set up a matrix field with different forms of matrices assigned to different field quantities – and the possibility of stepwise repetition of assignment, shows what extraordinary wealth of conceptual forms is here at our disposal for empirical description. Remember, too, that these matrices are merely a special form of probability function.

Another procedure applied in empirical enquiry assigns to values of measurement the squares of wave amplitudes construed as probability values. Here the probability functions are the wave functions obtained by following special instructions as to what kind of expressions to substistute for metrical quantities. For every characterizing quantity of state (metrical quantity) we perform the specified substitution and so obtain the wave or probability function belonging to it. Here too it is formally possible to choose for every characterizing quantity a suitable expression to replace it which, in the case where we describe a field, leads to a probability wave field. Like the matrix field, this is a special form of probability field, and we need not here enlarge on what formal advantages either might have over the other. It is easy to see that with expressions for quantities occurring in a probability wave field we can again make further stepwise assignments of probability waves (perhaps of quite different form) and thus obtain probability functions of probability functions. It is worth noting that the probability functions assigned in this kind of field are, or can be, wave functions as used in physics for describing certain phenomena.

By comparing physical wave equations (say, the differential equations for the propagation of electromagnetic waves) with probability wave equations we can readily make plain what is the epistemological difference between on the one hand descriptions employing only metrical quantities and on the other those involving fictitious quantities. In the first case, each value characterizing the wave describes a fact that is in principle ascertainable by measurement. Wave length, frequency, period, amplitude, speed of propagation of the wave and attendant energy transmissions are here features of the real wave phenomenon and the quantities characterizing them are in principle ascertainable by measurement.

Not so for probability waves. To start with, these waves do not mean the wave-like spread of changes in an energy field and they do not cause a transmission of energy (action). Their speed of propagation is a merely fictitious quantity whose value is in principle unmeasurable. Even if there are certain empirically measurable phenomena that are usually denoted as interference patterns of probability waves, such data are not enough to ascribe to these waves the empirical reality of energy waves. Yet it is equally unjustified to deny them every kind of reality, since we can understand them as 'processes' giving rise to statistically ascertained corpuscle distributions. Since probability waves are, however, marked at least in part by fictitious quantities, they describe only a partial phenomenon of whose reality we can speak only in connection with the part (likewise not independently real) that is marked by values of measurement, as we know already. Where attempts to characterize states by values of measurement leads to uncertainty domains, we must introduce fictitious quantities for sufficient characterization (that is, one permitting predictions by using laws of nature). The essential epistemological feature of a description thus given by relations between measured and fictitious values is that neither partial phenomenon marked by either kind of value alone can count as a sufficiently characterized phenomenon, that is, as an actually existent state; for neither partial characterization suffices for the obtaining of predictions: this becomes possible only on the basis of the logically indissoluble connection of measured and fictitious values in the total characterization. Only the total state thus characterized can count as sufficiently grounded regarding real existence.

As to that, there is no difference between matrix and probability wave descriptions. However, the latter, as previously mentioned, rests partly on observed data characterizable by values of measurement and quite analogous to certain observed data in energy waves, namely to interference rings and bands in electromagnetic waves, which patterns appear in corpuscular rays under analogous conditions. Interference patterns are not, however, enough to allow us to construe probability waves as empirically real energy waves. As main reason why not, one usually mentions the absence of any kind of ascertainable energy transmission by the probability waves. This objection must certainly be taken into account, but for energy waves too we have the epistemological criterion that reality is assigned to states only on the basis of, and according to, sufficient characterization. If the values assigned to a 'state' allow us to derive testable predictions that are empirically confirmed, then the state is denoted as empirically real. Whether the assigned values are pure values of measurement or whether some amongst them are fictitious is irrelevant as regards the epistemological reality criterion in the general sense in which we understand it today.

If therefore one wanted to specify equations for probability waves that gave them the character of energy waves, such attempts to characterize states with the help only of quantities whose values and changes can be measured (so that no fictitious values appear in the account) can give sufficient epistemological grounding for the reality of the waves only if the procedure leads to testable and suitably confirmed predictions. Attempts of this kind have repeatedly been made in recent years[34], although so far always with the result that some of the expressions characterizing the wave have to be interpreted as probability functions, if we are to obtain testable predictions. However, probability functions are always defined in domains whose values are in part fictitious, so that the equations set up again merely describe probability waves to which we must ascribe empirical-fictitious reality.

The attempts mentioned are often interpreted as trying to eliminate probability and probability laws from empirical description and so to prove that the course of phenomena is strictly continuous and determined. Epistemologically we can call these attempts of theoretical physics attempts to eliminate the use of fictitious quantities from physical description. In this tendency such theories are linked with older attempts, now hardly undertaken, to replace the metric stipulations of relativistic physics once more by those of the classical Galilean metric.

However different the concept formations of relativistic and quantum physics or wave mechanics, these theories share the circumstance that in carrying out measurements we are led to uncertainty domains that compel us to introduce fictitious quantities and to relate them to the values of measurement. Therefore all attempts to eliminate the use of fictitious quantities from physical description (amongst them the above-mentioned wave mechanical theory that wants to replace probability waves by waves with continuously measurable states) can be understood as trying to prove that uncertainty domains do not exist (more precisely, that they do not have to be used). Since the appearance of such domains is however necessarily grounded in the kind of measuring procedure and its attendant presuppositions, theoretical attempts of the kind mentioned (whether touching conceptual formations in relativistic or in quantum physics), in contrast with description involving fictitious quantities, can lead to new results only if they indicate, along with the equations set up, new and practicable theories of measurement. It is the theory of measurement that

will decide whether measurement leads to uncertainty domains or not, and
therefore whether we must introduce fictitious quantities and relate them
to the observational values by means of suitably chosen functions (fictitious predicates). In particular, when wave mechanics sets up universal
field theories containing only laws describing continuous changes in state
(laws of proximate action) – which in principle excludes differential probability laws –, it must indicate for the quantities appearing in the equations a
theory of measurement that can be carried out in practice. From this theory
it must follow that no uncertainty domains for the characterization of the
phenomena ever occur when the measurements are carried out.

Given the present state of enquiry and the available means and methods
of measurement, it seems most unlikely that a new theory of measurement
might prove that uncertainty domains need not arise in the relevant cases.
For this epistemological reason, we see little prospect for theories that want
to eliminate the use of fictitious quantities and predicates from physical description (whether in relativistic or in quantum physics). Formally one can
of course set up mathematical theories whose equations look externally like
laws of proximate action, but without an attendant theory of measurement
on which the value and arbitrarily small changes in it of all quantities appearing in the equations are measurable in practice in all domains of phenomena (from these conditions it follows that uncertainty domains are impossible), such theories are empirically empty frameworks from which no
testable predictions of consequence can be derived. In wave mechanical
theories of this kind it has always turned out that a special device in method
was required to derive testable predictions from them and thus attain a test
of the theories themselves. This is not a matter of analytic deductive steps,
but of the interpretation as probability functions of certain expressions appearing in the equations, and as probability values of certain derived values
(as probabilities for obtaining individual values of measurement). This interpretation, however, consists always in the expressions and values concerned being related to fictitious value domains, or the possible individual
values of measurement being ranged into a domain of fictitious values. Yet
in this way the equations set up become differential probability laws as to
empirical content, even if externally they look like laws of proximate action.

The signs are that physical enquiry is developing in the direction of assigning probability functions to characterizing metrical quantities, the val-

ues of these functions indicating the probabilities that an individual value of measurement will be obtained under certain conditions. The general application of this method to all metrical quantities leads as explained to the descriptive form of the 'probability field', which is to be regarded as an epistemological extreme. The possibility of repeated stepwise assignment of probability functions to expressions of quantities, and the further possibility of choosing probability functions of various forms (for example wave functions or matrices) for the individual assignments, reveals the great variety of conceptual forms from which to choose for empirical description.

32. PROBABILITY DESCRIPTION IN BIOLOGY

The appearance of uncertainty domains through measuring procedures and conditions compels us to apply the 'method of fictitious predicates'. Under certain conditions uncertainty ranges appear in the form that through the influence of the measuring processes the 'states' to be measured suffer certain disturbances that cannot be ascertained exactly. This special case shows certain analogies with phenomena that tend to occur when we make observations by measurement on organisms. If for the purpose of observation organic structures are subjected to radiation, damage or destruction occurs precisely in those 'particle constellations' that are generally regarded as the simplest biological elements. Pursuing this analogy one marked the jump-like discontinuous and therefore not exactly predictable lesions by statistical curves ('damage curves') that made it possible to set up probability functions.[35]

Within the simplest organisms – which in the most elementary cases we tend to mark as 'living molecules' –, 'quantum biology' has succeeded in determining fairly accurately what are the smallest spatial domains (volumes) in which the impingement of energy can provoke impairing or destructive effects, and what are the smallest amounts of energy ('energy deposits') that can set off such effects. Thus for given organic constellations and given kinds of irradiation we can derive the probabilities for the individual intensities of damage. Here the 'mutual action' between quantised amounts of energy and the 'guiding centres' of the organism (that is, the smallest spatial domains for which we assume a certain distribution in the organism) lies within uncertainty domains.

The application of the method of fictitious predicates in these cases pro-

ceeds by choice and assignment of fictitious value systems to those spatio-
temporal or energy ranges in which the life impairing 'changes in state' occur. Here, 'guiding' and 'amplifier theories'[36] have been set up for explaining the phenomena given only by discontinuous values of measurement; to
the extent that these theories enable us to set up functions, they mean that
we choose probability functions by which probability values are assigned
to the possible individual values of measurement (that is, to the values that
mark possible impairing effects). The joint action of energy quanta and
spatial domains denoted as 'guiding centres', describable only by such
probability values, is the 'fictitious partial phenomenon' of the total process occurring as irradation of the organic constellation. We must be clear
as to these explanatory theories being nothing else but a choice and stipulation of probability metric systems, insofar as mathematical description is
possible. By means of them the possible values of measurement are ranged
into a system of fictitious values or are set into relation with these latter.
Accordingly within the domains concerned it is only probability values that
can be derived for the possible individual cases marked by individual values
of measurement.

A causal explanation going beyond the probability description could be
given by 'guiding', 'amplifier' or 'deposit theories' only if here too it could
be proved, in terms of suitable theories of measurement, that we can capture the measured values marking states and their changes without the
appearance of only imprecisely predictable damage to the biological organic constellations. We shall leave it an open question whether this damage, predictable only with probability, which here stands for the disturbances of state in certain quantum physical measurements of conjugate
quantities, must occur in the relevant measurements of biological objects.
Niels Bohr inclines to the view that destructive damage must occur in these
cases. We shall have to await the further results of empirical biological enquiry.

However, we can consider the possibility that imprecisely determinable
impairing effects necessarily appear when biological constellations are irradiated for the purpose of observation by measurement, so that we have here
the same epistemological conditions as in measuring conjugate quantities
of state in quantum physics. It would follow that in all quantum physical
biological measurements the disturbances of the biological constellation
provoked by measuring can be determined only by statistical laws (statisti-

cal curves). Description of the biological phenomena concerned would thus at best be possible through differential probability laws.

What is surely striking here is that quantum biology always marks radiation-caused 'disturbances' of biological states as 'damage' or 'destruction'. Purely physically, it is a matter of radiation or corpuscular energy transferred to a state denoted as a living organism, although it remains impossible, even for the simplest biological constellations, to indicate in the form of relations between physical quantities the criteria that characterize life. Still, there are certain quantitative marks for damage to or destruction of the vital functions. If, however, damage or destruction affect biological constellations through energy transfers and, for reasons stated, these phenomena can be described only by probabilities, we must not forget that energy transfers are here described only in the special form of corpuscular constellations being destroyed. However, it is not only destruction that can be brought about by such transfers, but we know cases from physics where the same kind of transfer (that is, irradiation) produces certain corpuscular arrangements, so that we have construction: both are effected by energy transfers.

If then we are certainly right to construe irradation of biological constellations as transfer of radiation or corpuscular energy to a given state or corpuscular arrangement, it is not generally true that a disturbance caused by irradation (that is, energy transfer) must always consist in damage to these constellations, that is, in life-impairing destruction of corpuscular arrangements. Given a certain state or arrangement of corpuscles, suitable irradiation may yield the construction of new constellations, perhaps even such as might have to be denoted as 'furthering' or 'building up life', rather than as impairing or destroying it. For such results, under the conditions assumed, we could theoretically determine statistical curves or set up probability laws just as for the effects of damage.

Pursuing this theoretical possibility we can contrast the two extreme cases of disturbance in corpuscular constellations by energy transfer. In the first case we irradiate a state that we denote as a biological constellation or as a living organism. As disturbing effect we assume the destruction of life (for which the descriptive laws indicate a certain probability), so that after the energy transfer the state is to be marked as 'dead' or, as we shall put it, as a non-biological constellation. We can thus describe the course of phenomena in general as follows: by irradiation of a given kind a specified biological constellation is transformed with a certain probability into a non-biological one. This transition from a biological to a non-biological state means destruction of life.

In the second case a non-biological constellation (a 'dead' state or object) is irradiated. The disturbing effect of the energy transfer is now seen in the construction of corpuscular constellations such that the state resulting from irradiation can be regarded as a biological constellation. For this course of phenomena too the descriptive laws indicate a certain probability under the conditions assumed. The general description here runs as follows: by a given irradation a specified non-biological constellation is transformed with a certain probability into a biological constellation. If we wish, we can call this transition 'construction' or 'generation of life'.

What is worth noting in this form of description is the special significance of the characterizing expressions, in particular expressions and values that characterize the transition from biological to non-biological constellations or the reverse. A complete characterization of states and changes in them is here possible in principle only if we adopt probability functions. Amongst the characterizing values there must always be ones that can be ascribed or derived only in a probable manner, so that the empirical meaning of such values can in principle be given only in connection with probability values. Amongst these last are the values marking biological constellations and the transitions involved, because of the imprecisely determinable disturbances in transitions from biological to non-biological states or the reverse. Generally speaking, therefore, under the conditions assumed biological constellations and their relations of any kind with non-biological ones can be sufficiently characterized only with the help of probability functions or quantities, that is, in such a way as to allow the derivation of testable predictions (which are then necessarily probability statements).

This rule holds particularly for expressions describing the critical changes in state previously explained. These are expressions (quantities, relations between quantities) that mark 'destruction of life' or 'construction of life'. As we know, the course of phenomena describable only with probability functions (more generally: probability expressions), partly occurs in uncertainty domains, which is why we must relate measured values to fictitious ones in order to characterize these phenomena. For this reason we also say that phenomena, states and changes in them, that we can describe only with the help of probability laws, have empirical-fictitious reality: that is, their reality can in principle be asserted only in connection with the 'partial phenomena' marked by measured and fictitious values respectively.

Applied to biological constellations and their relations to non-biological states, this means that under the prior conditions assumed above we must ascribe empirical-fictitious reality to all biological processes, and particularly to transitions from biological to non-biological constellations (destruction of life) or to processes in the reverse direction (construction of life). In these processes one part of the phenomenon always necessarily falls into uncertainty domains and can be sufficiently characterized only if we bring in fictitious value domains and correspondingly defined probability expressions (functions, quantities). In this way, biological phenomena could be described in principle only by applying the method of fictitious predicates.

NOTES

[1] Amongst those who, following American sceptics like N. Goodman and W.V.O. Quine, hold the view that it is impossible exactly to distinguish between the methods of science and those of speculative metaphysics is Wolfgang Stegmüller, *Metaphysik, Wissenschaft, Skepsis,* Frankfurt a. M. – Vienna 1954. The book gives a comprehensive discussion of writings that seek to justify a sceptical stance as to separating science and metaphysics.

[2] Many ingenious attempts to set up exact formal criteria for distinguishing between scientific and metaphysical concept formations and propositions have been undertaken by Carnap, Ayer and others. However, their critics were always able to offer examples in which the 'distinguishing criteria' proved to be inadequate.

[3] The logical peculiarity of dispositional concepts was first pointed out by Carnap, R., 'Testability and Meaning', *Philosophy of Science,* 3/4 (1936/37). In terms of his theory of reduction he there tries to give a logical empiricist justification for these concepts. Critical objections to the theory of reduction were recently raised by Pap, A., *Analytische Erkenntnistheorie,* Vienna 1955. A logical critique of dispositional concepts is given also by Valpola, V., 'Ein System der negationslosen Logik mit ausschliesslich realisierbaren Prädikaten', *Acta Philosophica Fennica,* 9 (1955) (Helsinki), p.74ff, 88ff.

[4] The steps by which, starting from 'factual observations', we attain to empirical propositions of higher form and finally to general laws of nature ('second order laws') are examined by Juhos, B., *Die Erkenntnis und ihre Leistung,* Vienna 1950.

[5] A logical analysis of the relativity principle or of the concepts 'rest', 'motion' and the 'clock paradox' I have given in my essays, 'Logische Analyse der Begriffe "Ruhe" und "Bewegung"' *Studium Generale* 10 (1957), 296–302 and 'Die Metrik als Bestandteil der empirischen Beschreibung', *Archiv für Philosophie,* 7 (1957), 209–228.

[6] Cf. v. Laue, M., *Die Relativitätstheorie,* vol. 1, Brunswick 1952, pp. 36–37.

[7] Cf. Grünbaum, A., 'The Clock Paradox in the Special Theory of Relativity', *Philosophy of Science* 21, (1954) 249–253.

[8] For what follows see Juhos, B., 'Wahrscheinlichkeitsschlüsse als syntaktische Schlussformen', *Studium Generale* **6** (1953), 206–214, this volume pp. 93–104 and 'Deduktion, Induktion und Wahrscheinlichkeit', *Methodos* **6** (1954) 259–278.

[9] The definition of 'conjunctive' and 'disjunctive' classes of propositions is given in connection with other logical syntactic problems in Carnap, R., *Formalization of Logic*, Chicago 1943. The possibility of distinguishing further kinds of such classes is not there examined.

[10] Cf. the references in Note 8.

[11] Examples how we can expediently determine probabilities in practical cases, taking into account a priori and statistical elements, are treated by Carnap, R., *Logical Foundations of Probability*, Chicago 1950; *The Continuum of Inductive Methods*, Chicago 1952, under different prior assumptions.

[12] Cf. Johnson, W.E., 'Probability', *Mind* **41** (1932), 1–16, 281–296, 408–423; and Carnap, R., *The Continuum of Inductive Methods*, Chicago 1952.

[13] In my book, *Die Erkenntnis und ihre Leistung*, Vienna 1950, I indicate ascertainability of continuous changes in the quantities characterizing phenomena as the critical presupposition for the possibility of second-order predictions. What is not mentioned there is that ascertainability must be given under the necessary presuppositions for sufficiently characterizing the states.

[14] The conditions under which probability quantities are defined and applied are examined in my essays 'Die Wahrscheinlichkeit als empirische Beschreibungsform', *Philosophia Naturalis* **4**, 2/3, pp. 297–336, and 'Das Wahrscheinlichkeitsfeld', *Archiv für Philosophie* **7**, 1/2, pp. 82–95.

[15] Arthur March speaks of 'first' and 'second' quantisation according to how often the constant h is used in the definition of characterizing quantities of state or in the setting up the formulae describing them. In what follows we shall several times return to the possibilities thus opened up. Cf. March, A., *Die physikalische Erkenntnis und ihre Grenzen*, Brunswick 1955, p. 86ff.

[16] Cf. March, A., l.c., p. 76ff.

[17] Cf. ibid., p. 86ff, 98ff. P. Jordan's method of regarding quantities of state set up with the help of h as themselves matrices of a certain kind involving h a second time, is called 'second quantisation' by March. By generalizing this method we can understand by 'quantisation' the (perhaps repeated) application of operators of a certain kind to expressions of quantities.

[18] Attempts of this kind have been made several times, as recently by L. de Broglie in his 'theory of double solution'. An account of the basic features of this theory is given in his lecture 'Une interprétation nouvelle de la mécanique ondulatoire est-elle possible?', published in the series '*Les Conférences du Palais de la Découverte*', series A, No. 201, Paris 1955, of which more later. He deals with questions of the same kind in a number of essays and comprehensively in, *Une tentative d'interprétation causale et non linéaire de la Mécanique ondulatoire: la théorie de la 'double solution'*, Paris 1955.

[19] See the writings of de Broglie given in Note 18. An epistemological critique of this theory is contained in my essays, 'Die Wahrscheinlichkeit als empirische Beschreibungsform', *Philosophia Naturalis* **4** (1957), 297–336 and 'Das Wahrscheinlichkeitsfeld', *Archiv für Philosophie* **7** (1957), 82–95.

[20] See the writings mentioned in Notes 18 and 19.

[21] Cf. L. de Broglie' lecture mentioned in Note 18, p. 14ff.

[22] That the method of 'hidden parameters' applied to quantum physical problems is not suitable for enabling us to replace probability description by a 'continuous determinist' one is proved by W. Pauli in 'Bemerkungen zum Problem der verborgenen Parameter in der Quantenmechanik und zur Theorie der Führungswelle', published in, *Louis de Boglie und die Physiker*, Hamburg 1955, p. 26ff.

[23] The expressions 'probability field' is used in a somewhat different sense by Wenzl, A., *Die philosophischen Grundlagen der modernen Naturwissenschaften*, Stuttgart 1954, Urban-Bücher No. 11. What he understands by it is merely the field of the statistically construed guiding waves of wave mechanics. However, positions in these 'fields' are marked not only by probability quantities (at least in the kinds of statistic used today) but in part also by continuous and differentiable metrical quantities. As against this, what we mean is a form of description in which field points are marked only by probability quantities.

[24] Cf. Dirac, P.A.M. *Proc. Roy. Soc. London* (A) **109** (1926), p. 642, and **110** (1926), p. 561.

[25] This interpretation is suggested by the equations of Klein-Gordon and de Broglie-Proca, which constitute an extension and completion of Dirac's theory. Cf. also March, A., l. c., p. 93ff.

[26] In discussing the method of 'second quantisation' developed by P. Jordan we follow the account given by March, A., l. c., p. 86ff.

[27] An exception to this is Dirac's theory of 'q-numbers', in which the quantity 'time' is also represented by a value distribution ('q-number'). See the writings cited in Note 11. In my conception of the 'probability field' one assumes discontinuity for all quantities of state, and therefore also for 'time'.

[28] Cf. March, A., l.c., pp. 94–95.

[29] This view is held by March, A., l.c., pp. 94–95.

[30] See Juhos, B., *Die Erkenntnis und ihre Leistung*, Vienna 1950, pp. 141ff, 147ff.

[31] This argument for his at first extreme conventionalist position is given by H. Dingler in many writings. However, in so doing he cancels the presuppositions of conventionalism. His attempt which is obscure in many points might perhaps be best denoted as an '*a priori* realism'. See Dingler, H., *Das Experiment*, Munich 1928; *Das physikalische Weltbild*, 1951; *Über die Geschichte und das Wesen des Experiments*, 1952.

[32] Perhaps the best critique of extreme conventionalism was given by Viktor Kraft in *Mathematik, Logik und Erfahrung*, Vienna 1947. He obtains amongst other things the remarkable result that in carrying through an extreme conventionalism rather more statistical laws would have to be adduced than with the empiricist method.

[33] When H. Dingler, starting from extreme conventionalism, wishes to admit the general laws of Newtonian and Maxwellian physics as the only absolutely valid conventions while rejecting as false propositions of modern physics because 'incompatible' with the classical laws, the argument for his view abandons the presuppositions of conventionalism. He argues for the absolute validity of the (classical) laws of nature, chosen by him, in terms of a somewhat unclear form of transcendental apriorism ('*a priori* realism'). See Note 31.

[34] The best-known theories on this are those of L. de Broglie. See Notes 18, 19 and 21.

[35] A detailed analysis of the application of quantum physical forms of concepts in biology is given by Dessauer, F., *Quantenbiologie*, Berlin-Göttingen-Heidelberg 1954.

[36] In all these theories quantum physical forms of concepts are applied to biological phenomena. The most far-reaching theoretical foundation of quantum biology was given by P. Jordan. Cf. his works *Die Physik und das Geheimnis des organischen Lebens*, Brunswick 1949; *Verdrängung und Komplementarität*, 1942; *Eiweissmoleküle*, Stuttgart 1947.

BIBLIOGRAPHY

(Papers included in this volume are marked with an asterisk)

I. BOOKS

Über die Grundlagen der Gewißheit des reinen Denkens, Gerold & Co., Wien, 1928.

Erkenntnisformen in Natur- und Geisteswissenschaften, Pan Verlag, Leipzig, 1940.

Die Erkenntnis und ihre Leistung, Springer Verlag, Wien, 1950.

Elemente der neuen Logik, Humboldt Verlag, Frankfurt-Wien, 1954.

Das Wertgeschehen und seine Erfassung, Verlag Anton Hain, Meisenheim am Glan, 1956.

Die erkenntnislogischen Grundlagen der klassischen Physik (together with Hubert Schleichert), Duncker & Humblot, Berlin, 1963.

Die Erkenntnislogischen Grundlagen der modernen Physik, Duncker & Humblot, Berlin, 1967.

Wahrscheinlichkeit als Erkenntnisform (together with Wolfgang Katzenberger), Duncker & Humblot, Berlin, 1970.

II. PAPERS

* Stufen der Kausalität, *Jahresber. d. philos. Ges. zu Wien* (1931/32), 1–19.

  Praktische und physikalische Kausalität, *Kant-Studien* **39**/2 (1934), 188–204.

* Kritische Bemerkungen zur Wissenschaftstheorie des Physikalismus, *Erkenntnis* **4**/6 (1934), 397–418.

* Empiricism and Physicalism, *Analysis* **2**/6 (1935), 81–92.

* Negationsformen empirischer Sätze, *Erkenntnis* **6**/1 (1936), 41–55.

  Some Modes of Speach of Empirical Science, *Analysis* **3**/5 (1936), 65–74.

  Discussion logique de certaines expressions psychologiques, *Revue de Synthese* **12**/2 (1936), 203–216.

  Über juristische und ethische Freiheit, *Archiv f. Rechts- und Sozialphilosophie* **29**/3-4 (1937), 406–431.

* Principles of Logical Empiricism, *Mind* **46**/138 (1937), 320–346.

  The Truth of Empirical Statements, *Analysis* **4**/5 (1937), 65–70.

  Der Indeterminismus als Voraussetzung der Methode der idiographischen Wissenschaften, *Archiv f. Rechts- und Sozialphilosophie* **30**/2 (1938), 238–256.

  The Empirical and the Grammatical Doubt, *Analysis* **5**/3-4 (1938), 56–59.

  Wie stellt sich die neuere Erkenntniskritik zur Philosophie Schopenhauers? in *Gedächtnisschrift für Arthur Schopenhauer zur 150. Wiederkehr seines Geburtstages,* Berlin, 1938, pp. 119–139.

  Historische Formen indeterministischer Systeme in der Ethik. Ihr logischer und psychologischer Gehalt, *Archiv f. Rechts- und Sozialphilosophie* **31**/2 (1939), 145–166.

  Empirische Sätze und logische Konstanten, *The Journal of Unified Science* **8**/5-6 (1940), 354–360.

  Geschichtsschreibung und Geschichtsgestaltung, *Archiv f. Rechts- und Sozialphilosophie* **32**/4 (1940), 429–453.

  Theorie empirischer Sätze, *Archiv für Rechts- und Sozialphilosophie* **37**/1 (1945), 59–144.

* Die erkenntnisanalytische Methode, *Zeitschrift f. philosophische Forschung* **6**/1 (1951), 42–54.

Die Anwendung der logistischen Analyse auf philosophische Probleme, *Methodos* **3**/10 (1951), 81–122.

Die 'Wahrheit' wissenschaftlicher Sätze und die Methoden ihrer Bestimmung, *Methodos* **4**/13, (1952), 19–38.

Die Voraussetzungen der 'logischen Wahrheit' in den höheren Kalkülen, *Methodos* **5**/17 (1953), 31-43.

Wahrscheinlichkeitsschlüsse als syntaktische Schlussformen, *Actes du XIe Congrès Int. de Philosophie,* Bruxelles 1953, Vol. 1, 105–108.

* Wahrscheinlichkeitsschlüsse als syntaktische Schlussformen, *Studium Generale* **6**/4 (1953), 206–214.

Die neue Logik als Voraussetzung der wissenschaftlichen Erkenntnis, *Studium Generale* **6**/10 (1953), 593–599.

Ein- und zweistellige Modalitäten, *Methodos* **6**/21–22 (1954), 69–83.

Deduktion, Induktion, Wahrscheinlichkeit, *Methodos* **6**/24 (1954), 259–278.

Erkenntnisanalytische Untersuchung physikalischer Gesetzesformen, *Actes du Congrès Int. de l'Union Int. de Phil. des Sciences,* Zürich 1954, Vol. 2, 30–38.

* Der 'positive' und der 'negative' Aussagengebrauch, *Studium Generale* **9**/2 (1956), 79–85.

Die rekursive Definition der Wahrheit, *Archiv f. Philosophie* **6** (1956), 42–59.

Über Analogieschlüsse, *Studium Generale* **9**/3 (1956), 126–129.

Mögliche Gesetzesformen der Quantenphysik, *Philosophia Naturalis* **3**/2 (1956), 211–237.

Moritz Schlick, zum 20. Todestag, *Studium Generale* **10**/2 (1957), 81–87.

Logische Analyse der Begriffe 'Ruhe' und 'Bewegung', *Studium Generale* **10**/5 (1957), 296–302.

Das 'Wahrscheinlichkeitsfeld', *Archiv f. Philosophie* **7**/1–2 (1957), 82–95.

Die 'Wahrscheinlichkeit' als physikalische Beschreibungsform, *Philosophia Naturalis* **4**/2–3 (1957), 297–336.

Die 'Metrik' als Bestandteil der physikalischen Beschreibung, *Archiv f. Philosophie* **7**/3–4 (1957), 209–228.

* Die neue Form der empirischen Erkenntnis, *Archiv f. Philosophie* **8**/3-4 (1958), 110–128.

Unbestimmtheitsbereiche als Voraussetzung der neuen Erkenntnisform, *Atti del XII° Congresso Internazionale di Filosofia,* Venezia 1958, Vol. 5, 245–251.

Die empirische Beschreibung durch eineindeutige und einmehrdeutige Relationen, *Studium Generale* **13**/5 (1960), 267–288.

Über die Definierbarkeit und empirische Anwendung der Dispositionsbegriffe, *Kant-Studien* **51**/3 (1959/1960), 272–284.

Welche begriffliche Formen stehen der empirischen Beschreibung zur Verfügung, in Ernst Topitsch (ed.) *Probleme der Wissenschaftstheorie. Festschrift für Viktor Kraft,* Springer Verlag, Wien, 1960, pp. 101–158.

* Die Methode der fiktiven Prädikate, Part 1: *Archiv f. Philosophie* **9**/1–2 (1959), 140–156; Part 2: *Ibid.* **9**/3–4 (1959), 314–347; Part 3: *Ibid.* **10**/1–2 (1960), 114–161; Part 4: *Ibid.* **10**/3–4 (1960), 228–289.

Über die 'absolute' Wahrscheinlichkeit, *Philosophia Naturalis* **6**/3 (1961), 391–410.

Die zweidimensionale Zeit, *Archiv f. Philosophie* **11**/1–2 (1961), 3–27.

Nichtmaterielle Gründe der Abwanderung einheimischer Wissenschaftler, *Österreichische Hochschulzeitung* **13**/10 (1961),.2–3.

Finite und transfinite Logik, *Der Mathematikunterricht* **8**/2 (1962), 67–84.

L'introduzione di ordini fittizii nei domini relativistici non univoci, *Rivista di Filosofia* **53**/4 (1962), 403–437.

Aktualny stan filozofii naukowej w Austrii, *Ruch Filozoficzny* **21**/4, Toruń 1962, 355–374.

Grundlagenforschung pro und contra, *Österreichische Hochschulzeitung* **14**/1 (1962), 3.

Moritz Schlick. Zum 80. Geburtstag, *Archiv für Philosophie* **12**/1–2 (1963), 123–132.

Die Kennzeichnung translatorischer Bewegungszustände, *Ratio* **6**/1 (1964), 26–44; english translation in *Ratio* **6**/1 (1964), 28–49.

Die Dualität der Erkenntnis, Paper read in Rias, Berlin, on 1964 03.11, published in printed form by Rias, Berlin.

Die logischen Ordnungsformen als Grundlage der empirischen Erkenntnis, in *The Foundations of Statements and Decisions*, Warszawa 1965, pp. 251–261.

Die zwei logischen Ordnungsformen der naturwissenschaftlichen Beschreibung, *Studium Generale* **18**/9 (1965), 582–601.

Das Prinzip der virtuellen Geschwindigkeiten, *Philosophia Naturalis* **9**/1–2 (1965), 55–113.

Die Aufgaben der Wiener erkenntnislogischen Grundlagenforschung, Jubiläumausgabe der Österreichischen Hochschulzeitung anlässlich der 600-Jahrfeier der Wiener Universität 1965, pp. 201–210.

Gibt es in Österreich eine wissenschaftliche Philosophie? in *Ist Wien eine geistige Provinz?*, Forum Verlag, Wien, 1965.

Die Dualität der Erkenntnis, *Memorias del XIII Congresso Internationale Filosofico* (Mexico 1963), Universidad Nacional Autonoma de Mexico, 1966, Vol. V, 499ff.

Zwei Bereiche der physikalischen Realität, in Paul Weingartner (ed.) *Deskription, Analytizität und Existenz*, Verlag Anton Pustet, Salzburg-München, 1966, pp. 96–111.

Die Rolle der analytischen Sätze in den Erfahrungswissenschaften, *Ibid.*, pp. 340–350.

Über die empirische Induktion, *Studium Generale* **19**/4 (1966), 259–272.

Ernst Mach und die moderne Philosophie, *Österreichische Hochschulzeitung* **18**/8 (1966), 3–4.

Moritz Schlick, in *The Encyclopedia of Philosophy*, New York, 1967, Vol. 7, pp. 319–324.

Absolutbegriffe als metaphysische Voraussetzungen empirischer Theorien und ihre Relativierung, in Paul Weingartner (ed.), *Grundfragen der Wissenschaften und ihre Wurzeln in der Metaphysik*, Universitätsverlag Anton Pustet, Salzburg-München, 1967, pp. 120–135.

Die 'intensionale' Wahrheit und die zwei Arten des Aussagengebrauchs, *Kant-Studien* **58**/2 (1967), 173–186.

Schlüsselbegriffe physikalischer Theorien, *Studium Generale* **20**/12 (1967), 785–795.

Die experimentelle Überprüfung relativistischer Zeiteffekte und die Deutung der Messergebnisse, in Alwin Diemer (ed.), *Geschichte und Zukunft. Festschrift für Anton Hain*, Anton Hain Verlag, Meisenheim/Glan, 1968, pp. 65–78.

Limit Forms of Empirical Knowledge, *Sitzungsbericht des 3. internat. Kongresses für Logik, Methodologie und Philosophie der Wissenschaften*, Amsterdam 1967.

Gespräch über das Uhrenparadoxon, (together with A.M. Moser and H. Schleichert), *Philosophia Naturalis* **10**/1 (1967), 23–41.

Die. Systemidee in der Physik, in Alwin Diemer (ed.), System und Klassifikation in Wissenschaft und Dokumentation. *Studien zur Wissenschaftstheorie*, Vol. 2, Verlag Anton Hain, Meisenheim/Glan, 1968, pp. 65–78.

The Influence of Epistemological Analysis on Scientific Research, in Imre Lakatos and Alan Musgrave (eds.) *Problems in the Philosophy of Science*, North-Holland Publishing Company, Amsterdam, 1968, pp. 266–277.

Logische und empirische Induktion, in Raymond Klibansky (ed.), *Contemporary Philosophy. A Survey*, Vol. 2, La Nuova Italia Editrice, Firenze, 1968.

Methodologie der Naturwissenschaften, *Ibid.*

Logical and Empirical Probability. A Critical Supplement on Professor Ayer's Paper 'Induction and the Calculus of Probability', *Logique et Analyse* 12/47 (1969), 277–282.

Wie gewinnen wir Naturgesetze?, *Zeitschrift für philosophische Forschung* 22/4 (1968), 534–548.

Die empirische Wahrheit und ihre Überprüfung, *Kant-Studien* 59/4 (1968), 435–447.

Drei Begriffe der Wahrscheinlichkeit, *Studium Generale* 21 (1968), 207–217.

Das tschechische Phänomen, *Conceptus* 2/4 (1968), 153–154.

Logische Analyse des Relativätsprinzips, *Philosophia Naturalis* 11/2 (1969), 207–217.

Virtuelle Geschwindigkeiten als verborgene Parameter, *Philosophia Naturalis* 11/4 (1969), 440–445.

Ernst Topitsch – Philosophie, Soziologie, Heidelberg, *Österreichische Hochschulzeitung* 21/6 (1969), 3.

Studenten, Talare und die Entstehung einer neuen Klasse, *Conceptus* 3/3-4 (1969), 141–144.

* Die methodologische Symmetrie von Verifikation und Falsifikation, *Journal for General Theory of Science* 1/2 (1970), 41–70.

Zwei Begriffe der physikalischen Realität, *Ratio* 12/1 (1970), 55–67; english translation in *Ratio* 12/1 (1970), 65–78.

* Drei Quellen der Erkenntnis, *Zeitschrift für philosophische Forschung* 26/3 (1970), 335–347.

Viktor Kraft, Philosophie, Wien, *Österreichische Hochschulzeitung* 23/13 (1970), 3.

Makrophänomene und ihre Zusammensetzung aus Mikrophänomenen, *Philosophia Naturalis* 12/4 (1970), 413–420.

Rudolf Carnap zum Gedenken 1891–1970, Paper read in the ORF on 1970 09 20, appeared in *Österreichische Hochschulzeitung* 22/19 (1970), 6–7.

* Die triadische Methode, *Studium Generale* 24 (1971), 924–945.

Geometrie und Wahrscheinlichkeit, *Zeitschrift für philosophische Forschung* 25/4 (1971), 500–510.

Formen des Positivismus, *Journal for General Philosophy of Science* 2/1 (1971), 27–62.

Triadische Erkenntnisanalyse, in Hans Lenk (ed.), *Neue Aspekte der Wissenschaftstheorie*, Friedr. Vieweg+Sohn, Braunschweig, 1971, pp. 187–194.

Ernst Mach, Physiker und Philosoph, (Kurzbiographie), in *Österreichisches Biographisches Lexikon* (ed. by the Österreichische Akademie der Wissenschaften), Wien-Köln-Graz, to appear.

III. REVIEWS

Erhard Tornier und Hans Domizlaff: 'Theorie der Versuchsvorschriften der Wahrscheinlichkeitsrechnung', W. Kohlhammer Verlag, Stuttgart 1952, in *Archiv für Philosophie* 6/3-4 (1956), 342–344.

Hubert Schleichert, 'Elemente der physikalischen Semantik', R. Oldenburg Verlag, München und Wien 1966, in *Synthese* 16/1 (1966), 107–109.

Friedrich Waismann, 'How I See Philosophy' (ed. with Introduction by R. Harré), Macmillan and Co., Ltd., London–Melbourne-Toronto; St. Martin's Press, New York, 1968, in *Synthese* 20/1 (1969), 149–153.

Paul Lorenzen, 'Methodisches Denken', Reihe Theorie 2, Suhrkamp Verlag, Frankfurt
a. M., 1968, in *Archiv für Rechts- und Sozialphilosophie* **56**/4 (1970), 575–578 and in
Journal for General Philosophy of Science **1**/2 (1970), 304–310.

Gerhard Frey, 'Einführung in die philosophischen Grundlagen der Mathematik', Schroedel-
Schöningh, Hannover-Paderborn, 1968, in *Philosophische Rundschau* **18**/1–2 (1971),
135–137.

Gottlob Frege, 'Nachgelassene Schriften' (ed. by Hans Hermes, Friedrich Kambartel,
and Friedrich Kaulbach), Felix Meiner Verlag, Hamburg, 1969, in *Synthese* **21**/3–4
(1970), 488–493; in *Philosophische Rundschau* **17**/3–4 (1970), 208–213; and in
Philosophischer Literaturanzeiger **23**/5 (1970), 265–269.

J.J. Bulloff, T. C. Holyoke and S.W. Hahn (eds.), 'Foundations of Mathematics. Sym-
posium Papers Commemorating the Sixtieth Birthday of Kurt Gödel', Springer
Verlag, Berlin, 1966, in *Philosophische Rundschau* **18**/1–2 (1971), 135–137.

Wolfgang Stegmüller, 'Theorie und Erfahrung' (= Probleme und Resultate der Wissenschaftstheorie und Analytischen Philosophie, Vol. 2), Springer Verlag, Berlin-Heidel-
berg-New York, 1970, in *Journal for General Philosophy of Science* **2**/1 (1971), 138–151,

W. v. Del-Negro, 'Konvergenzen in der Gegenwartsphilosophie und die moderne Physik',
Duncker & Humblot, Berlin 1970, in *Journal for General Philosophy of Science*
2/2 (1971), 326–331.

INDEX OF NAMES

The letter 'n' refers to a note.